T0176853

SINGULARLY PERTURBED METHODS FOR NONLINEAR ELLIPTIC PROBLEMS

This introduction to singularly perturbed methods in nonlinear elliptic partial differential equations emphasises the existence and local uniqueness of solutions exhibiting a concentration property. The authors avoid using sophisticated estimates and explain the main techniques by thoroughly investigating two relatively simple but typical non-compact elliptic problems. Each chapter then progresses to other related problems to help the reader learn more about the general theories developed from singularly perturbed methods.

Designed for PhD students and junior mathematicians intending to perform research in the area of elliptic differential equations, the text covers three main topics. The first is the compactness of the minimization sequences, or the Palais–Smale sequences, or a sequence of approximate solutions; the second is the construction of peak or bubbling solutions by using the Lyapunov–Schmidt reduction method; and the third is the local uniqueness of these solutions.

Daomin Cao is a professor at the Institute of Applied Mathematics, Chinese Academy of Sciences. His research focuses on nonlinear partial differential equations. He was awarded the first-class prize for Outstanding Young Scientists by the Chinese Academy of Sciences. He is the editor of academic mathematical journals, including *Applicable Analysis, Annales Academiac Scientiarum Fennicae Mathematica*, and *Acta Mathematicae Applicatae Sinica*.

Shuangjie Peng is a professor at the School of Mathematics and Statistics, Central China Normal University. His research focuses on nonlinear elliptic problems. He was awarded the first-class prize in Natural Sciences by the Hubei Province and the second-class prize in Natural Sciences by the Ministry of Education. He is the editor of several academic mathematical journals, including *Communications on Pure and Applied Analysis*, *Acta Mathematica Scientia*, and *Acta Mathematicae Applicatae Sinica*.

Shusen Yan is a professor at the School of Mathematics and Statistics, Central China Normal University. His main research interests are nonlinear elliptic problems.

CAMBRIDGE STUDIES IN ADVANCED MATHEMATICS

All the titles listed below can be obtained from good booksellers or from Cambridge University Press. For a complete series listing visit: www.cambridge.org/mathematics.

Already published

151 M. Viana & K. Oliveira *Foundations of Ergodic Theory*
152 V. I. Paulsen & M. Raghupathi *An Introduction to the Theory of Reproducing Kernel Hilbert Spaces*
153 R. Beals & R. Wong *Special Functions and Orthogonal Polynomials*
154 V. Jurdjevic *Optimal Control and Geometry: Integrable Systems*
155 G. Pisier *Martingales in Banach Spaces*
156 C. T. C. Wall *Differential Topology*
157 J. C. Robinson, J. L. Rodrigo & W. Sadowski *TheThree-Dimensional Navier–Stokes Equations*
158 D. Huybrechts *Lectures on K3 Surfaces*
159 H. Matsumoto & S. Taniguchi *Stochastic Analysis*
160 A. Borodin & G. Olshanski *Representations of the Infinite Symmetric Group*
161 P. Webb *Finite Group Representations for the Pure Mathematician*
162 C. J. Bishop & Y. Peres *Fractals in Probability and Analysis*
163 A. Bovier *Gaussian Processes on Trees*
164 P. Schneider *Galois Representations and*(φ, Γ)*-Modules*
165 P. Gille & T. Szamuely *Central Simple Algebras and Galois Cohomology* (*2nd Edition*)
166 D. Li & H. Queffelec *Introduction to Banach Spaces, I*
167 D. Li & H. Queffelec *Introduction to Banach Spaces, II*
168 J. Carlson, S. Müller-Stach & C. Peters *Period Mappings and Period Domains* (*2nd Edition*)
169 J. M. Landsberg *Geometry and Complexity Theory*
170 J. S. Milne *Algebraic Groups*
171 J. Gough & J. Kupsch *Quantum Fields and Processes*
172 T. Ceccherini-Silberstein, F. Scarabotti & F. Tolli *Discrete Harmonic Analysis*
173 P. Garrett *Modern Analysis of Automorphic Forms by Example, I*
174 P. Garrett *Modern Analysis of Automorphic Forms by Example, II*
175 G. Navarro *Character Theory and the McKay Conjecture*
176 P. Fleig, H. P. A. Gustafsson, A. Kleinschmidt & D. Persson *Eisenstein Series and Automorphic Representations*
177 E. Peterson *Formal Geometry and Bordism Operators*
178 A. Ogus *Lectures on Logarithmic Algebraic Geometry*
179 N. Nikolski *Hardy Spaces*
180 D.-C. Cisinski *Higher Categories and Homotopical Algebra*
181 A. Agrachev, D. Barilari & U. Boscain *A Comprehensive Introduction to Sub-Riemannian Geometry*
182 N. Nikolski *Toeplitz Matrices and Operators*
183 A. Yekutieli *Derived Categories*
184 C. Demeter *Fourier Restriction, Decoupling and Applications*
185 D. Barnes & C. Roitzheim *Foundations of Stable Homotopy Theory*
186 V. Vasyunin & A. Volberg *The Bellman Function Technique in Harmonic Analysis*
187 M. Geck & G. Malle *The Character Theory of Finite Groups of Lie Type*
188 B. Richter *Category Theory for Homotopy Theory*
189 R. Willett & Guoliang Yu *Higher Index Theory*
190 A. Bobrowski *Generators of Markov Chains*
191 D. Cao, S. Peng & S. Yan *Singularly Perturbed Methods for Nonlinear Elliptic Problems*

Singularly Perturbed Methods for Nonlinear Elliptic Problems

DAOMIN CAO
Chinese Academy of Sciences

SHUANGJIE PENG
Central China Normal University

SHUSEN YAN
Central China Normal University

CAMBRIDGE
UNIVERSITY PRESS

University Printing House, Cambridge CB2 8BS, United Kingdom

One Liberty Plaza, 20th Floor, New York, NY 10006, USA

477 Williamstown Road, Port Melbourne, VIC 3207, Australia

314–321, 3rd Floor, Plot 3, Splendor Forum, Jasola District Centre, New Delhi – 110025, India

79 Anson Road, #06–04/06, Singapore 079906

Cambridge University Press is part of the University of Cambridge.

It furthers the University's mission by disseminating knowledge in the pursuit of education, learning, and research at the highest international levels of excellence.

www.cambridge.org
Information on this title: www.cambridge.org/9781108836838

DOI: 10.1017/9781108872638

First published 2021

Printed in the United Kingdom by TJ Books Limited, Padstow Cornwall

A catalogue record for this publication is available from the British Library.

Library of Congress Cataloguing-in-Publication data
Names: Cao, Daomin, 1963- author. | Peng, Shuangjie, 1968- author. | Yan, Shusen, 1963- author.
Title: Singularly perturbed methods for nonlinear elliptic problems / Daomin Cao, Shuangjie Peng, Shusen Yan.
Description: New York : Cambridge University Press, 2020. | Series: Cambridge studies in advanced mathematics | Includes bibliographical references and index. Identifiers: LCCN 2020030221 | ISBN 9781108836838 (hardback) | ISBN 9781108872638 (ebook)
Subjects: LCSH: Differential equations, Elliptic. | Differential equations, Nonlinear.
Classification: LCC QA377 .C29 2020 | DDC 515/.3533–dc23
LC record available at https://lccn.loc.gov/2020030221

ISBN 978-1-108-83683-8 Hardback

Contents

Preface

The development of variational methods is centered around a fundamental goal – namely, to find solutions for partial differential equations with variational structure. The history of such methods dates back to the nineteenth century. Despite their long history, these methods still have a strong impact on today's research. New critical point theories, such as the mountain pass lemma of Ambrosetti and Rabinowitz and the concentration compactness principle of P. L. Lions (see also Aubin [10], Brezis and Nirenberg [24]), have led to many remarkable results that verify the existence of nontrivial solutions for superlinear elliptic problems. However, these results are not very effective for finding solutions with higher energy for non-compact elliptic problems.

Fortunately, this dilemma was resolved thanks to the celebrated works of Bahri [12], Rey [125] and W.-M. Ni-Takagi [113] in the late 1980s, which opened the door to the construction of solutions for elliptic problems. In the last three decades, many powerful techniques based on the classical Lyapunov–Schmidt reduction procedure have been developed in the study of the existence of solutions for nonlinear elliptic problems and the study of these solutions' properties. However, the discussion of these techniques has been spread out among various articles, most of which are considered too technical for PhD students or even junior working mathematicians to read. This book aims to explain the main ideas associated with these techniques in a self-contained manner by investigating two typical non-compact elliptic problems. The book consists of five chapters and an appendix that focuses on the analytical aspects of nonlinear elliptic problems.

Chapter 1 is devoted to studying the existence of least-energy solutions for some typical nonlinear elliptic problems. The emphasis here is on the Schrödinger equations in $\mathbb{R}^N$ with subcritical growth and semilinear elliptic problems involving a Sobolev critical exponent in bounded domains of $\mathbb{R}^N$.

One of the main topics of the discussion is how the compactness of the minimization sequences(or the so-called Palais-Smale sequences) can be recovered by an energy constraint. We also discuss a global compactness result, showing how a Palais–Smale sequence may lose its compactness. This chapter serves as a preliminary for the topics discussed in the subsequent chapters of this book.

The Lyapunov–Schmidt reduction method and its variants have been widely used to construct peak solutions or bubbling solutions for singularly perturbed elliptic problems. Such solutions concentrate on a finite number of points. We will discuss three main issues: the necessary condition for the location of the concentration points for the peak solutions or the bubbling solutions, the existence of such solutions, and the local uniqueness of such solutions. The first two issues will be discussed in Chapter 2, while the third one is studied in Chapter 3.

The Lyapunov–Schmidt reduction argument is an application of the implicit function theorem. The difficulty in carrying out this argument is that the corresponding linear operator for the approximate solution is not invertible on the whole space. In view of this difficulty, determining the approximate kernel for this linear operator becomes essential. To illustrate this idea, we study two typical singularly perturbed elliptic problems in Chapter 2: the nonlinear Schrödinger equations in $\mathbb{R}^N$ with subcritical growth and the Brezis–Nirenberg problem. These examples were chosen so that numerous sophisticated estimates can be avoided.

Local Pohozaev identities are used to obtain the necessary condition for the location of the peak solutions and the bubbling solutions in Chapter 2. In Chapter 3, they are used to study the local uniqueness of such solutions. The advantages of using the local Pohozaev identities to study the local uniqueness problems are twofold. Firstly, the classical degree-counting methods rely heavily on the estimates of second-order derivatives of the solutions, whereas the arguments via the local Pohozaev identities only involve the estimates of the first-order derivatives of the solutions, which simplifies the problem. Secondly, the methods via local Pohozaev identities can be adapted to study other problems, where the classical degree-counting methods do not work.

The Lyapunov–Schmidt reduction method and its variants are typically used to construct solutions for elliptic problems with small parameters. In Chapter 4, these methods are adapted to study some non-singularly perturbed elliptic problems. Again, to avoid complicated estimates that may obfuscate the main ideas, we only study the Schrödinger equations with subcritical growth. The main idea in employing a reduction argument for non-singularly perturbed elliptic problems is that we can use a large integer k as the parameter in the construction of solutions, where k is the number of the bumps or the number of bubbles in the approximate solutions. The results in Chapter 4 show that the non-compactness of some elliptic problems may give rise to the existence of infinitely many positive solutions, whose energy can be arbitrarily large. We remark that such results cannot be obtained by using the abstract critical points theories.

The singularly perturbed problems studied in Chapter 2 have concentration solutions. On the other hand, it is also important to study the non-existence of concentration solutions for other singularly perturbed elliptic problems. Results in this direction are obtained by using a local Pohozaev identity, as presented in Chapter 5. As an application of these results, the existence of infinitely many sign-changing solutions for the Brezis-Nirenberg problem is proven.

For the convenience of our readers, we collect some of the results required for this book in the Appendix. These include some basic estimates, various Pohozaev identities, some preliminary properties of Sobolev spaces, certain fundamental estimates on elliptic equations, the Kelvin transformation, the kernel of some linear operators, and the estimate for the Green's function.

The Pohozaev identity was derived in the 1960s and was used in proving the non-existence of nontrivial solutions for some nonlinear elliptic problems. In this book, we emphasise the important role that various Pohozaev identities can also play in other problems, such as in the problems of the local uniqueness of solutions and the existence of solutions.

This book is written for those who have some knowledge of Sobolev spaces, nonlinear functional analysis and various estimates for elliptic equations. It aims to introduce some typical techniques for the construction of solutions to nonlinear elliptic problems. As this is not a survey article, many important results obtained in the last three decades have not been included. Also, it is hopelessly impossible to list all the relevant papers from the last three decades in the references; thus we will only briefly discuss other related problems and provide some references for those problems at the end of the monograph.

Portions of the materials contained herein were taken from the lecture notes prepared by the authors for the 2015 summer program for partial differential equations, chaired by Prof. Zhouping Xin. We would like to take this opportunity to thank Professor Xin for his support. We would also like to thank Prof. Meng Fai Lim for his help with the English presentation of this book. Last but not least, a prior version of this book served as a reference text for the course 'Perturbed Methods in Elliptic Equations' in Central China Normal University during the autumn semester of 2017. We would like to thank the graduate students of that course for their valuable feedback on this book.

Finally, this work is supported by the National Natural Science Foundation of China under grant numbers 11629101 and 11831009.

1
Non-compact Elliptic Problems

The solvability of elliptic equations with variational structure can be studied by detecting the existence of minimizers of the corresponding constrained minimization problems or critical points of the corresponding functionals. In this chapter, we will study the possible loss of compactness of some sequences of approximating critical points for some functionals. This will lay the foundation for the construction of peak or bubbling solutions in Chapters 2 and 4, and for the discussion of the multiplicity results in Chapter 5. To show the main ideas, we will consider two typical types of elliptic problems: the Schrödinger equation in $\mathbb{R}^N$ and the semilinear elliptic problems with a critical Sobolev exponent in a bounded domain of $\mathbb{R}^N$. One of the main topics in this chapter is how one can recover the compactness of the minimization sequences, or the Palais–Smale sequences, by an energy constraint. We will also discuss a global compactness result that shows how a Palais–Smale sequence may lose its compactness.

The theories in Chapter 1 were mainly developed in the 1970s and 1980s. The readers interested in other aspects of critical point theory can refer to the books [124, 134, 151].

1.1 Minimization Problems with Compactness

Consider the following eigenvalue problem

$$\begin{cases} -\Delta u = \lambda f(x,u), & \text{in } \Omega, \\ u = 0, & \text{on } \partial\Omega, \end{cases} \tag{1.1.1}$$

for some unknown pair (λ, u), where Ω is a bounded domain in $\mathbb{R}^N$ and $f(x,t) \in C(\Omega \times \mathbb{R})$ with the property that $|f(x,t)| \leq C(1 + |t|^p)$ for some constants $C > 0$ and $p \in \big(1, 2^* - 1\big)$. Here we denote 2^* to be $\frac{2N}{N-2}$ or $+\infty$ according as $N \geq 3$ or $N = 1, 2$.

Before solving the above eigenvalue problem, let us briefly discuss the following constraint minimization problem in calculus.

Let $g(x)$ and $G(x)$ be two C^1 functions in $\mathbb{R}^N$. Suppose that $\nabla G(x) \neq 0$ for any $x \in \mathbb{R}^N$ and that the equation $G(x) = 0$ defines a compact closed hyper-surface in $\mathbb{R}^N$. Define

$$m := \inf\{g(x) : x \in \mathbb{R}^N,\ G(x) = 0\}. \tag{1.1.2}$$

Theorem 1.1.1 *The quantity m defined in problem* (1.1.2) *is achieved by some x_0. Moreover, this x_0 satisfies*

$$\nabla g(x_0) = \lambda \nabla G(x_0), \tag{1.1.3}$$

for some $\lambda \in \mathbb{R}$.

Proof. Step 1. Take a minimization sequence $\{x_n\}$ such that $G(x_n) = 0$ and

$$m \leq g(x_n) \leq m + \frac{1}{n}. \tag{1.1.4}$$

Step 2. Since the set $\{x : G(x) = 0\}$ is compact, there exists a subsequence which, by abuse of notation, is still denoted by $\{x_n\}$, and x_0 such that $x_n \to x_0$. Since $G(x_n) = 0$, we have that $G(x_0) = 0$.

Step 3. It follows from (1.1.4) and the continuity of g that

$$m \geq \lim_{n\to\infty} g(x_n) = g(x_0) \geq m.$$

This shows that m is attainable.

Step 4. It remains to prove (1.1.3). For any tangent vector $\vec{l}$ of the hyper-surface $\{x : G(x) = 0\}$ at x_0, we take a curve $x(t)$ on $\{x : G(x) = 0\}$ such that

$$\begin{cases} x'(t)|_{t=0} = \vec{l}, \\ x(0) = x_0. \end{cases}$$

Let $h(t) := g(x(t))$. Then $h(0) = g(x_0)$ and $h(t)$ achieves its minimum at $t = 0$. Therefore,

$$\langle \nabla g(x_0), \vec{l} \rangle = h'(0) = 0.$$

This implies that $\nabla g(x_0)$ is perpendicular to any tangent vector of the surface $\{x : G(x) = 0\}$. In other words, $\nabla g(x_0)$ is a normal vector of this surface $G(x) \equiv 0$ at x_0. On the other hand, for any $z \in \{x : G(x) = 0\}$, $\nabla G(z)$ is a normal vector of the surface $\{x : G(x) = 0\}$ at z. So we find that at the minimum point x_0, ∇G and ∇g are parallel, and therefore, there exists $\lambda \in \mathbb{R}$ such that

$$\nabla g(x_0) = \lambda \nabla G(x_0). \qquad \square$$

Remark 1.1.2 We note that the proof of the existence of a minimizer for the constraint minimization problem (1.1.2) only requires the following assumptions: (i). The compactness of the set $\{x : G(x) = 0\}$. (ii). $g(x)$ is lower semi-continuous, i.e., $\liminf_{x \to x_0} g(x) \geq g(x_0)$.

In the sequel, we define the space $H_0^1(\Omega)$ to be the closure of $C_0^\infty(\Omega)$ with respect to the norm $\|u\| = (\int_\Omega |\nabla u|^2)^{\frac{1}{2}}$, which is induced by the inner product $\langle u, v \rangle = \int_\Omega \nabla u \nabla v$. For some important results on the space $H_0^1(\Omega)$, the readers can refer to the appendix.

Let us now study the eigenvalue problem (1.1.1). The weak solution $u \in H_0^1(\Omega)$ of (1.1.1) satisfies the following relation:

$$\int_\Omega \nabla u \nabla \varphi = \lambda \int_\Omega f(x, u)\varphi, \quad \forall \varphi \in H_0^1(\Omega). \tag{1.1.5}$$

It is well known that the left-hand side of (1.1.5) is the derivative of the functional

$$I(u) = \frac{1}{2} \int_\Omega |\nabla u|^2, \quad u \in H_0^1(\Omega),$$

while the right-hand side is the derivative of the functional $\int_\Omega F(x, u)$, where $F(x, t) = \int_0^t f(x, s)\, ds$. Thus, a weak solution (λ, u) of (1.1.1) satisfies

$$\nabla I(u) = \lambda \nabla \int_\Omega F(x, u).$$

Hence we are led to study a problem similar to the algebraic equation (1.1.3). In view of this, it is then natural to consider the following constraint minimization problem:

$$m := \inf \left\{ I(u) : u \in H_0^1(\Omega), \int_\Omega F(x, u) = 1 \right\}. \tag{1.1.6}$$

To prove that problem (1.1.6) is achieved by some $u \in H_0^1(\Omega)$, we first take a minimization sequence $\{u_n\} \subset H_0^1(\Omega)$, satisfying

$$m \leq I(u_n) \leq m + \frac{1}{n}, \quad \int_\Omega F(x, u_n) = 1. \tag{1.1.7}$$

From the proof of Theorem 1.1.1, we see that we need to prove the following:

(i) There is a subsequence of $\{u_n\}$, still denoted by $\{u_n\}$, such that u_n converges to u in 'certain' sense.
(ii) The functional $I(u)$ is lower semi-continuous. That is, $\lim_{n\to+\infty} I(u_n) \geq I(u)$.
(iii) The functional $\int_\Omega F(x,u)$ is continuous. That is, $\lim_{n\to+\infty} \int_\Omega F(x,u_n) = \int_\Omega F(x,u)$.

Remark 1.1.3 (1) $H_0^1(\Omega)$ is generally not compact. However, we do have that $H_0^1(\Omega)$ is weakly compact (see Theorem 6.3.1). In other words, any bounded sequence $\{u_n\}$ in $H_0^1(\Omega)$ has a subsequence, which we still denote by $\{u_n\}$, such that for any $\varphi \in H_0^1(\Omega)$, one has $\langle u_n, \varphi\rangle \to \langle u, \varphi\rangle$ for some $u \in H_0^1(\Omega)$. We use the notation $u_n \rightharpoonup u$ to denote the weak convergence of u_n to u in $H_0^1(\Omega)$.

(2) $I(u)$ is weakly lower semi-continuous in $H_0^1(\Omega)$. Consequently, if $u_n \rightharpoonup u$ in $H_0^1(\Omega)$, then

$$\liminf_{n\to+\infty} I(u_n) \geq I(u).$$

(3) $H_0^1(\Omega) \hookrightarrow L^{p+1}(\Omega)$ is compact if $1 \leq p < \frac{N+2}{N-2}$ and Ω is bounded.

Theorem 1.1.4 *Suppose that $\{u : u \in H_0^1(\Omega), \int_\Omega F(x,u) = 1\}$ is not empty. Then* (1.1.6) *is achieved. Moreover, if $\nabla \int_\Omega F(x,u) \neq 0$ at u,* (1.1.5) *holds for some $\lambda \in \mathbb{R}$.*

Proof. Let $\{u_n\}$ be a minimization sequence in $H_0^1(\Omega)$ that satisfies (1.1.7). Then $\{u_n\}$ is bounded in $H_0^1(\Omega)$. By Theorems 6.3.1 and 6.3.7, there exists a subsequence of $\{u_n\}$, which one still denotes by $\{u_n\}$, and a function $u \in H_0^1(\Omega)$ such that

$$u_n \rightharpoonup u \quad \text{weakly in } H_0^1(\Omega),$$

and

$$u_n \to u \quad \text{strongly in } L^{p+1}(\Omega).$$

Recalling that $|f(x,t)| \leq C(1+|t|^p)$, we now see that there exists $\theta \in [0, 1]$,

$$\begin{aligned}
&\int_\Omega |F(x,u_n) - F(x,u)| \\
&= \left| \int_\Omega f(x, u+\theta(u_n-u))(u_n-u) \right| \\
&\leq \left(\int_\Omega |f(x, u+\theta(u_n-u))|^{\frac{p+1}{p}} \right)^{\frac{p}{p+1}} \left(\int_\Omega |u_n-u|^{p+1} \right)^{\frac{1}{p+1}}
\end{aligned}$$

$$\le C\Big(\int_\Omega (1+|u_n|^p+|u|^p)^{\frac{p+1}{p}}\Big)^{\frac{p}{p+1}}\Big(\int_\Omega |u_n-u|^{p+1}\Big)^{\frac{1}{p+1}}$$

$$\le C\Big(\int_\Omega |u_n-u|^{p+1}\Big)^{\frac{1}{p+1}} \to 0,$$

as $n \to \infty$.

On the other hand, we have

$$I(u_n) = \frac{1}{2}\int_\Omega |\nabla u|^2 - \int_\Omega \nabla(u_n-u)\nabla u + \frac{1}{2}\int_\Omega |\nabla(u-u_n)|^2$$

$$\ge \frac{1}{2}\int_\Omega |\nabla u|^2 - \int_\Omega \nabla(u_n-u)\nabla u$$

$$= I(u) + o(1),$$

where $o(1) \to 0$ as $n \to +\infty$.

Therefore, $m = \lim_{n\to\infty} I(u_n) \ge I(u) \ge m$. As a result, u achieves m.

Next, we prove (1.1.5). For any function $\varphi \in H_0^1(\Omega)$, $u+t\varphi$ may not satisfy $\int_\Omega F(x, u+t\varphi) = 1$ for $t \ne 0$. Therefore, we need to correct this perturbation suitably. From the assumption $\nabla\int_\Omega F(x,u) \ne 0$, we can find a function $\psi \in H_0^1(\Omega)$ such that

$$\int_\Omega f(x,u)\psi \ne 0.$$

Define the following function

$$H(t,s) = \int_\Omega F(x, u+t\varphi+s\psi) - 1.$$

Then one has

$$H(0,0) = 0, \qquad \frac{\partial H(0,0)}{\partial s} = \int_\Omega f(x,u)\psi \ne 0.$$

By the implicit function theorem, there is a C^1 function $s(t)$ defined for small t such that $H(t, s(t)) = 0$, $s(0) = 0$, and

$$s'(0) = -\frac{\displaystyle\int_\Omega f(x,u)\varphi}{\displaystyle\int_\Omega f(x,u)\psi}.$$

Now the function $\xi(t) = I\big(u + t\varphi + s(t)\psi\big)$ attains its minimum at $t = 0$. Hence $\xi'(0) = 0$. On the other hand, a direct calculation gives

$$\xi'(0) = \int_\Omega \nabla u \nabla \varphi - \lambda \int_\Omega f(x, u)\varphi,$$

where

$$\lambda = \frac{\displaystyle\int_\Omega \nabla u \nabla \psi}{\displaystyle\int_\Omega f(x, u)\psi}.$$

Hence the result follows. □

Remark 1.1.5 If $f(x, t) = |t|^{p-1}t$, we have $F(t) = \frac{1}{p+1}|t|^{p+1}$. In this simple case, for any $\varphi \in H_0^1(\Omega)$ and $|t|$ small, $\frac{(p+1)^{\frac{1}{p+1}}(u+t\varphi)}{|u+t\varphi|_{L^{p+1}(\Omega)}}$ satisfies the constraint.

A direct consequence of Theorem 1.1.4 is the existence of a positive solution for a nonlinear elliptic problem.

Theorem 1.1.6 *The following elliptic problem has a solution*

$$\begin{cases} -\Delta u = u^p, \ u > 0, & \text{in } \Omega, \\ u = 0, & \text{on } \partial\Omega, \end{cases} \tag{1.1.8}$$

where Ω is a bounded domain in $\mathbb{R}^N$ and $p \in (1, 2^* - 1)$.

Proof. Since $F(t) = \frac{1}{p+1}|t|^{p+1}$, we can always choose a minimization sequence, which is non-negative. It then follows from Theorem 1.1.4 that there exists a constant $\lambda \in \mathbb{R}$, $u \in H_0^1(\Omega)$ and $u \ge 0$ such that

$$-\Delta u = \lambda u^p, \quad \text{in } \Omega. \tag{1.1.9}$$

Since $\frac{1}{p+1}\int_\Omega |u|^{p+1} = 1$, we must have $u \not\equiv 0$. On the other hand, it is easy to see from (1.1.9) that

$$\lambda = \frac{\displaystyle\int_\Omega |\nabla u|^2}{\displaystyle\int_\Omega |u|^{p+1}} > 0.$$

By the maximum principle, we see that $u > 0$ in Ω. Hence $\lambda^{\frac{1}{p-1}}u$ is a solution of (1.1.8). □

1.2 Non-compact Minimization Problems: The Concentration Compactness Principle

The minimization problem discussed in the previous section is compact in the sense that the functional $\int_\Omega F(x, u)$ is continuous under the weak convergence in $H_0^1(\Omega)$. Such compactness condition is not true anymore if Ω is unbounded, or $f(x, t)$ grows at the Sobolev critical rate of $|t|^{2^*-1}$ as $|t| \to +\infty$. In this section, we will consider two typical cases, where the compactness condition is violated.

Question 1. Suppose that $V(x) \in C(\mathbb{R}^N)$ satisfies

$$0 < V_0 \le V(x) \le V_1,$$

for some constants $V_1 > V_0 > 0$. We consider the following minimization problem

$$\inf\left\{\frac{1}{2}\int_{\mathbb{R}^N}(|\nabla u|^2 + V(x)|u|^2) : u \in H^1(\mathbb{R}^N),\ \int_{\mathbb{R}^N}|u|^{p+1} = 1\right\}, \tag{1.2.1}$$

where $1 < p < \frac{N+2}{N-2}$ if $N \ge 3$, $1 < p < +\infty$ if $N = 2$, and the space $H^1(\mathbb{R}^N)$ is the closure of $C_0^\infty(\mathbb{R}^N)$ under the norm $\|u\| = (\int_{\mathbb{R}^N}(|\nabla u|^2 + |u|^2))^{\frac{1}{2}}$, which is induced by the inner product $\langle u,\ v\rangle = \int_{\mathbb{R}^N}(\nabla u \nabla v + uv)$. We point out that, in this case, the functional $\frac{1}{2}\int_{\mathbb{R}^N}(|\nabla u|^2 + V(x)|u|^2)$ is weakly lower semi-continuous in $H^1(\mathbb{R}^N)$, but the functional $\int_{\mathbb{R}^N}|u|^{p+1}$ is not continuous under the weak convergence in $H^1(\mathbb{R}^N)$.

Question 2. Suppose that Ω is a bounded domain in $\mathbb{R}^N$, $N \ge 3$. Consider

$$\inf\left\{\frac{1}{2}\int_\Omega(|\nabla u|^2 - \lambda|u|^2) : u \in H_0^1(\Omega),\ \int_\Omega|u|^{2^*} = 1\right\}, \tag{1.2.2}$$

where $\lambda \in (0, \lambda_1)$. Here, we use λ_1 to denote the first eigenvalue of the operator $-\Delta$ in Ω, subject to the homogeneous Dirichlet boundary condition. We also emphasize that, in this case, $\frac{1}{2}\int_\Omega(|\nabla u|^2 - \lambda|u|^2)$ is weakly lower semi-continuous in $H_0^1(\Omega)$, but $\int_\Omega|u|^{2^*}$ is not continuous under the weak convergence in $H_0^1(\Omega)$.

Before we study the above minimization problems, let us first discuss a simple problem in calculus for some useful ideas. Let $f(x) \in C(\mathbb{R}^N)$ be bounded from below. Consider

$$m := \inf_{x\in\mathbb{R}^N} f(x). \tag{1.2.3}$$

A solution to such problem may not be achieved. For example, $\inf_{x\in\mathbb{R}} \arctan x$ cannot be achieved.

Now we discuss the conditions which guarantee that the minimum in (1.2.3) is achieved. Choose a minimization sequence $\{x_n\} \subset \mathbb{R}^N$ such that $f(x_n) \to m$ as $n \to +\infty$. Then we have two possibilities:

(i) There exists a subsequence, which we still denote by $\{x_n\}$, such that $x_n \to x_0 \in \mathbb{R}^N$. Then $f(x)$ achieved m at x_0.

(ii) $|x_n| \to \infty$. In this case, we have

$$m = \lim_{n\to+\infty} f(x_n) \geq \liminf_{|x|\to+\infty} f(x) := m^\infty \geq m.$$

Thus, $m = m^\infty$.

The possible loss of compactness for a minimization sequence $\{x_n\}$ for (1.2.3) is when $|x_n| \to +\infty$. If this occurs, we have

$$\inf_{x\in\mathbb{R}^N} f(x) = \liminf_{|x|\to+\infty} f(x).$$

Thus, the loss of the compactness will not occur if the function $f(x)$ satisfies

$$\inf_{x\in\mathbb{R}^N} f(x) < \liminf_{|x|\to+\infty} f(x). \tag{1.2.4}$$

From the above discussion, we have the following theorem.

Theorem 1.2.1 *Suppose that $f(x)$ is bounded from below and is continuous (or lower semi-continuous). If* (1.2.4) *holds, then the minimum in* (1.2.3) *is achieved.*

Remark 1.2.2 $m^\infty := \liminf\limits_{|x|\to+\infty} f(x)$ may be regarded as the limit problem of (1.2.3).

As seen from the above discussion, to study the minimization problems (1.2.1) and (1.2.2), we need to understand the possible loss of the compactness of the minimization sequence and determine the corresponding limit problem. The compactness of the minimization sequence can then be recovered by imposing a condition on the two problems.

1.2.1 A Minimization Problem in an Unbounded Domain

In this subsection, we will discuss (1.2.1). Let

$$I(u) = \int_{\mathbb{R}^N} (|\nabla u|^2 + V(x)u^2), \quad u \in H^1(\mathbb{R}^N),$$

where $V(x) \in C(\mathbb{R}^N)$ satisfies $0 < V_0 \le V(x) \le V_1$ for some constants $V_1 > V_0 > 0$, and

$$\lim_{|x|\to+\infty} V(x) = V_\infty > 0.$$

Consider the following constraint minimization problem

$$I_a := \inf\Big\{I(u) : u \in H^1(\mathbb{R}^N), \int_{\mathbb{R}^N} |u|^{p+1} = a > 0\Big\}. \tag{1.2.5}$$

To start, we summarize some properties for I_a. Using the change of variable $u = a^{\frac{1}{p+1}} v$, we can verify that

$$I_a = a^{\frac{2}{p+1}} I_1,$$

which gives

$$I_a < I_b + I_{a-b}, \quad \forall\, b \in (0, a). \tag{1.2.6}$$

It is easy to check that for any $a > 0$, one has $I_a > 0$. If I_a is achieved by some $u \in H^1(\mathbb{R}^N)$, then $|u|$ also achieves I_a. Hence we may assume that I_a is achieved by some non-negative function $u \not\equiv 0$. Moreover, it is easy to verify that u satisfies

$$\begin{cases} -\Delta u + V(x)u = \dfrac{I_a}{a} u^p,\ u > 0,\ x \in \mathbb{R}^N, \\ u \in H^1(\mathbb{R}^N). \end{cases} \tag{1.2.7}$$

Therefore, $w = (I_a/a)^{-\frac{1}{p-1}} u$ satisfies the following nonlinear Schrödinger equation

$$\begin{cases} -\Delta w + V(x)w = w^p,\ w > 0,\ x \in \mathbb{R}^N, \\ w \in H^1(\mathbb{R}^N). \end{cases} \tag{1.2.8}$$

To study whether (1.2.5) is achieved, we choose a minimizing sequence $\{u_n\} \subset H^1(\mathbb{R}^N)$ of I_a. Then $\{u_n\}$ is bounded in $H^1(\mathbb{R}^N)$. We may assume that there exists $u \in H^1(\mathbb{R}^N)$ such that

$$u_n \rightharpoonup u \text{ weakly in } H^1(\mathbb{R}^N),$$

and

$$u_n \to u \text{ strongly in } L^q_{loc}(\mathbb{R}^N) \text{ for any } q \in [2, 2^*).$$

From the weakly lower semi-continuity of $I(u)$ in $H^1(\mathbb{R}^N)$, we conclude that

$$I(u) \le \lim_{n\to+\infty} I(u_n) = I_a,$$

while Fatou's lemma implies that

$$\int_{\mathbb{R}^N} |u|^{p+1} \le \lim_{n\to\infty} \int_{\mathbb{R}^N} |u_n|^{p+1} = a.$$

Thus, in order to prove that u achieves I_a, we just need to prove that $\int_{\mathbb{R}^N} |u|^{p+1} = a$.

Following the same idea as in the discussion of (1.2.3), we need to exclude all the possibilities that may cause $\{u_n\}$ to lose compactness in $L^{p+1}(\mathbb{R}^N)$.

Let $v_n = u_n - u$. Then, for any $R > 0$, we have

$$\int_{B_R(0)} |v_n|^{p+1} \to 0, \quad \text{as } n \to +\infty.$$

Intuitively, if v_n does not converge strongly in $L^{p+1}(\mathbb{R}^N)$, we then have the following two cases.

Case 1. The main part of u_n lies in the infinity if $u \equiv 0$.

Case 2. u_n splits into two 'separate' parts u and v_n if $u \not\equiv 0$. This case is called the dichotomy case.

In case 1, we then have

$$\begin{aligned}\int_{\mathbb{R}^N} (V(x) - V_\infty)u_n^2 =& o\Big(\int_{B_R(0)} u_n^2\Big) + \int_{\mathbb{R}^N\setminus B_R(0)} (V(x) - V_\infty)u_n^2 \\ =& o(1) + O\left(\max_{|x|\ge R} |V(x) - V_\infty|\right) \int_{\mathbb{R}^N\setminus B_R(0)} u_n^2 \to 0,\end{aligned} \tag{1.2.9}$$

as $n \to +\infty$. This in turn implies that

$$I_a = \lim_{n\to+\infty} I(u_n) = \lim_{n\to+\infty} I^\infty(u_n) \ge I_a^\infty,$$

where

$$I_a^\infty = \inf\Big\{I^\infty(u) : u \in H^1(\mathbb{R}^N),\ \int_{\mathbb{R}^N} |u|^{p+1} = a\Big\}, \tag{1.2.10}$$

and

$$I^\infty(u) = \int_{\mathbb{R}^N} \big(|\nabla u|^2 + V_\infty u^2\big).$$

Similar to the discussion of (1.2.3), we call (1.2.10) the limit problem for (1.2.5). The above discussion shows that if $I_a < I_a^\infty$, then case 1 does not occur.

To exclude the dichotomy case, we need the following Brezis–Lieb lemma (see [23]).

Lemma 1.2.3 (Brezis–Lieb Lemma [23], 1983) *Let $\{u_n\} \subset L^{p+1}(\mathbb{R}^N)$, $0 < p < \infty$. If $\{u_n\}$ is bounded in $L^{p+1}(\mathbb{R}^N)$ and $u_n \to u$ a.e. on $\mathbb{R}^N$, then*

$$\int_{\mathbb{R}^N} |u_n|^{p+1} = \int_{\mathbb{R}^N} |u|^{p+1} + \int_{\mathbb{R}^N} |u_n - u|^{p+1} + o(1),$$

where $o(1) \to 0$ as $n \to +\infty$.

Proof. Observe that by Vitali's theorem, we have

$$\begin{aligned}\int_{\mathbb{R}^N} |u_n|^{p+1} - \int_{\mathbb{R}^N} |u_n - u|^{p+1} &= -\int_{\mathbb{R}^N} \int_0^1 \frac{d}{dt} |u_n - tu|^{p+1}\, dt \\ &= (p+1) \int_{\mathbb{R}^N} \int_0^1 |u_n - tu|^{p-1} (u_n - tu) u\, dt \\ &\to (p+1) \int_{\mathbb{R}^N} \int_0^1 |u - tu|^{p-1} (u - tu) u\, dt \\ &= \int_{\mathbb{R}^N} |u|^{p+1}.\end{aligned}$$

Now suppose that $\int_{\mathbb{R}^N} |u|^{p+1} = b \in (0, a)$. Since $\int_{\mathbb{R}^N} |u_n|^{p+1} = a$, it follows from the Brezis–Lieb lemma that

$$\int_{\mathbb{R}^N} |u|^{p+1} + \int_{\mathbb{R}^N} |v_n|^{p+1} + o(1) = a.$$

Let $c_n := \int_{\mathbb{R}^N} |v_n|^{p+1}$ and $c := \lim_{n \to +\infty} c_n$. Then,

$$I_{c_n} = c_n^{\frac{2}{p+1}} I_1 \to c^{\frac{2}{p+1}} I_1 = I_c$$

and

$$a = b + c, \ c = a - b > 0.$$

On the other hand, we can check easily that

$$\begin{aligned} I(u_n) =& I(u) + I(v_n) + 2 \int_{\mathbb{R}^N} \left(\nabla v_n \nabla u + V(x) v_n u\right) \\ =& I(u) + I(v_n) + o(1) \geq I_b + I_{c_n} + o(1),\end{aligned}$$

which implies that

$$I_a \geq I_b + I_c = I_b + I_{a-b}.$$

This is a contradiction to (1.2.6). Hence, due to the strict inequality (1.2.6), the dichotomy case does not occur. Therefore, we have proved the following theorem.

Theorem 1.2.4 (P. L. Lions [103, 104], 1984) *If $I_a < I_a^\infty$, then I_a is achieved.*

The above discussion shows that if u_n converges weakly to $u \not\equiv 0$ in $H^1(\mathbb{R}^N)$, then u_n converges strongly in $L^{p+1}(\mathbb{R}^N)$. One should note that due to the strict inequality (1.2.6), the only possibility that a minimization sequence is not compact in $L^{p+1}(\mathbb{R}^N)$ is when the main part of u_n lies near infinity. By comparing I_a with its limit problem I_a^∞, we can exclude this possibility by imposing the condition $I_a < I_a^\infty$. This condition is similar to (1.2.4) in the discussion of (1.2.3).

We now consider the case that $V(x)$ is a positive constant. In this case, $V(x) \equiv V_\infty$. We will prove that I_a^∞ in (1.2.10) is achieved. For I_a^∞, we also have the following strict inequality

$$I_a^\infty < I_b^\infty + I_{a-b}^\infty, \quad \forall\, b \in (0, a), \tag{1.2.11}$$

from which we can prove that the dichotomy does not occur. Therefore, if a minimization sequence $\{u_n\}$ converges weakly to a function $u \not\equiv 0$, then $\{u_n\}$ converges strongly to u in $L^{p+1}(\mathbb{R}^N)$ and u achieves I_a^∞. Note that (1.2.10) is translation invariant which implies that $\{u_n(x + x_n)\}$ is also a minimization sequence. Hence if there exists a constant $R > 0$ and a sequence $\{x_n\} \subset \mathbb{R}^N$ such that

$$\int_{B_R(x_n)} |u_n|^{p+1} \geq \alpha > 0,$$

then the new minimization sequence $\{u_n(x + x_n)\}$ converges weakly in $H^1(\mathbb{R}^N)$ to a function $u \not\equiv 0$. Therefore, u achieves I_a^∞. So it remains to exclude the case

$$\sup_{x\in\mathbb{R}^N} \int_{B_R(x)} |u_n|^{p+1} \to 0, \quad \text{as } n \to +\infty \text{ for any fixed } R > 0. \tag{1.2.12}$$

We call this case the *vanishing* case.

To exclude the occurrence of the vanishing case, we need the following vanishing lemma (see [103, 104]).

Lemma 1.2.5 (P. L. Lions [104], 1984) *Suppose that $\{u_n\}$ is a bounded sequence in $H^1(\mathbb{R}^N)$. If* (1.2.12) *holds, then*

$$u_n \to 0 \quad \text{in } L^q(\mathbb{R}^N), \text{ for any } q \in (2, 2^*),$$

as $n \to +\infty$.

Proof. To prove this lemma, we only need to prove that there exists $s \in (2, 2^*)$ such that

$$u_n \to 0 \quad \text{in } L^s(\mathbb{R}^N).$$

This is a direct consequence of the following interpolation inequality

$$\int_{\mathbb{R}^N} |u|^p \leq \Big(\int_{\mathbb{R}^N} |u|^q\Big)^t \Big(\int_{\mathbb{R}^N} |u|^r\Big)^{1-t}, \quad \forall\, q < p < r,$$

where $t \in (0, 1)$ is determined by $p = tq + (1 - t)r$. That is, $t = \frac{r-p}{r-q}$.

Fix $r \in (p + 1, 2^*)$. For any $s \in (p + 1, r)$, we have $t = \frac{r-s}{r-p-1}$ and

$$\begin{aligned}\int_{\Omega_m} |u_n|^s &\leq \Big(\int_{\Omega_m} |u_n|^{p+1}\Big)^t \Big(\int_{\Omega_m} |u_n|^r\Big)^{1-t} \\ &\leq C\Big(\int_{\Omega_m} |u_n|^{p+1}\Big)^t \Big(\int_{\Omega_m} (|\nabla u_n|^2 + |u_n|^2)\Big)^{r(1-t)/2} \\ &= o(1)\Big(\int_{\Omega_m} (|\nabla u_n|^2 + |u_n|^2)\Big)^{r(1-t)/2},\end{aligned}$$

where Ω_m is any translation of the set $\{x : x_i \in (0, 1],\ i = 1, \cdots, N\}$. Now we choose $s \in (p + 1, r)$ such that $\frac{r(1-t)}{2} = 1$. Then we have

$$s = p + 1 + \frac{2(r - p - 1)}{r} \in (p + 1, r).$$

For such s, one has

$$\int_{\Omega_m} |u_n|^s = o(1) \int_{\Omega_m} (|\nabla u_n|^2 + |u_n|^2).$$

Covering $\mathbb{R}^N$ by countable many Ω_m, we obtain the estimate

$$\int_{\mathbb{R}^N} |u_n|^s = o(1).$$

□

Equipped with Lemma 1.2.5, we can now prove the following result.

Theorem 1.2.6 *The minimization problem* (1.2.10) *is achieved.*

Proof. Suppose that the vanishing case occurs. Then it follows from Lemma 1.2.5 that $\int_{\mathbb{R}^N} |u_n|^{p+1} \to 0$ as $n \to +\infty$. But this contradicts the fact that $\int_{\mathbb{R}^N} |u_n|^{p+1} = 1$. □

The above discussions show that a minimization sequence $\{u_n\}$ of (1.2.5) (or (1.2.10)) has one of the following three possibilities.

1. Compact up to a translation: There exists $x_n \in \mathbb{R}^N$ such that $\{u_n(y + x_n)\}$ is compact in $L^{p+1}(\mathbb{R}^N)$.
2. Dichotomy: $u_n = u + v_n$, where $u \not\equiv 0$ and $v_n \to 0$ in $L^{p+1}_{loc}(\mathbb{R}^N)$.
3. Vanishing: $\sup\limits_{x\in\mathbb{R}^N} \int_{B_R(x)} |u_n|^{p+1} \to 0$ as $n \to +\infty$ for any fixed $R > 0$.

The dichotomy case is ruled out by the strict inequality (1.2.6) (or (1.2.11)). Once the dichotomy case is ruled out, we can conclude that if $\{u_n\}$ converges to $u \not\equiv 0$ weakly in $H^1(\mathbb{R}^N)$, then $\{u_n\}$ converges to u strongly in $L^{p+1}(\mathbb{R}^N)$. This is the main idea of the concentration compactness principle. See, for example, [103, 104].

Combining Theorems 1.2.4 and 1.2.6, we can prove the following result.

Theorem 1.2.7 *If $V(x) \le V_\infty$ and $V(x) \not\equiv V_\infty$, then I_a is achieved.*

Proof. By Theorem 1.2.6, I_a^∞ is achieved by some $u \in H^1(\mathbb{R}^N)$. As a result,

$$\begin{aligned} I_a \le I(u) &= I^\infty(u) + \int_{\mathbb{R}^N} \big(V(x) - V_\infty\big)u^2 \\ &< I^\infty(u) = I_a^\infty. \end{aligned}$$

We may now apply Theorem 1.2.4 to obtain the required conclusion of the theorem. □

Before ending this subsection, let us briefly discuss some other results that can be derived from Theorem 1.2.6.

Lemma 1.2.8 *For any $V(y)$, the inequality $I_a \le I_a^\infty$ holds.*

Proof. For any $\varepsilon > 0$, there exists $\varphi \in H^1(\mathbb{R}^N)$ such that

$$I^\infty(\varphi) < I_a^\infty + \varepsilon,$$

and

$$\int_{\mathbb{R}^N} |\varphi|^{p+1} = a.$$

Take $x_n \to \infty$ and let $\varphi_n(y) := \varphi(y - x_n)$. Then $\int_{\mathbb{R}^N} |\varphi_n|^{p+1} = \int_{\mathbb{R}^N} |\varphi|^{p+1} = a$ and

$$\begin{aligned} I_a^\infty + \varepsilon > I^\infty(\varphi) &= I^\infty(\varphi_n) \\ &= \int_{\mathbb{R}^N} (|\nabla \varphi_n|^2 + V(y)\varphi_n^2) + \int_{\mathbb{R}^N} (V_\infty - V(y))\varphi_n^2 \\ &= I(\varphi_n) + o(1) \\ &\ge I_a + o(1), \end{aligned}$$

where $o(1) \to 0$ as $n \to +\infty$. This gives $I_a \le I_a^\infty$. □

Theorem 1.2.9 *If $V(y) \geq V_\infty$ and $V(y) \not\equiv V_\infty$, then I_a is not achieved.*

Proof. Assuming that I_a is achieved by u, we then have that

$$I_a = I(u) > I^\infty(u) \geq I_a^\infty,$$

contradicting Lemma 1.2.8. Therefore, I_a is not achieved. □

It follows from Theorem 1.2.9 that if $V(y) \geq V_\infty$ and $V(y) \not\equiv V_\infty$, we cannot obtain a solution for (1.2.8) via the minimization problem (1.2.5). In such situation, other methods are required to get a higher energy solutions for (1.2.8). See the brief discussion in Section 1.5.

1.2.2 A Minimization Problem with Critical Growth

In this subsection, we will discuss (1.2.2). Let

$$I(u) = \int_\Omega (|\nabla u|^2 - \lambda u^2),$$

where $\Omega \subset \mathbb{R}^N$ is bounded, $N \geq 3$, $\lambda \in (0, \lambda_1)$ and λ_1 is the first eigenvalue of $-\Delta$ in Ω subject to the homogeneous Dirichlet boundary condition.

Consider

$$I_a := \left\{ I(u) : u \in H_0^1(\Omega), \int_\Omega |u|^{2^*} = a \right\}, \tag{1.2.13}$$

where $a > 0$ is a given constant.

The minimization problem (1.2.13) shares many similarities with (1.2.5). Here, we list some of these similarities. Since $\lambda \in (0, \lambda_1)$, we have that

$$\int_\Omega (|\nabla u|^2 - \lambda u^2) \geq \left(1 - \frac{\lambda}{\lambda_1}\right) \int_\Omega |\nabla u|^2. \tag{1.2.14}$$

This implies that $I_a > 0$ for any $a > 0$. It is also easy to check that $I_a = a^{\frac{2}{2^*}} I_1 = a^{\frac{N-2}{N}} I_1$, and so we have the following strict inequality

$$I_a < I_b + I_{a-b}, \quad \forall\, b \in (0, a). \tag{1.2.15}$$

On the other hand, if (1.2.13) is achieved by some $u \in H_0^1(\Omega)$, we can assume that $u > 0$ in Ω and that u satisfies

$$\begin{cases} -\Delta u - \lambda u = \dfrac{I_a}{a} u^{2^*-1}, u > 0, \text{ in } \Omega, \\ u \in H_0^1(\Omega). \end{cases} \tag{1.2.16}$$

Then $w = (I_a/a)^{-\frac{4}{N-2}} u$ satisfies the following equation

$$\begin{cases} -\Delta w - \lambda w = w^{2^*-1}, \ w > 0, \ \text{in } \Omega, \\ w \in H_0^1(\Omega). \end{cases} \tag{1.2.17}$$

We now discuss when (1.2.13) is achieved. As a start, choose a minimization sequence $\{u_n\} \subset H_0^1(\Omega)$ of I_a. Since $\{u_n\}$ is bounded in $H_0^1(\Omega)$ by (1.2.14), we may assume that there exists $u \in H_0^1(\Omega)$ such that

$$u_n \rightharpoonup u \text{ weakly in } H_0^1(\Omega),$$

and

$$u_n \to u \text{ strongly in } L^q(\Omega), \text{ for any } q \in [2, 2^*).$$

From the weakly lower semi-continuity of $I(u)$ in $H_0^1(\Omega)$, we conclude that

$$I(u) \leq \lim_{n\to+\infty} I(u_n) = I_a,$$

while it follows from Fatou's lemma that

$$\int_\Omega |u|^{2^*} \leq \lim_{n\to\infty} \int_\Omega |u_n|^{2^*} = a.$$

Therefore, in order to prove that u achieves I_a, we just need to prove that $\int_\Omega |u|^{2^*} = a$.

Let $v_n = u_n - u$. Then $v_n \rightharpoonup 0$ weakly in $H_0^1(\Omega)$ and $v_n \to 0$ strongly in $L^q(\Omega)$ for any $q \in [2, 2^*)$. Following exactly the same argument, we can also prove the following Brezis–Lieb lemma in the critical case.

Lemma 1.2.10 (Brezis–Lieb [23], 1983) *Let $\{u_n\} \subset L^{2^*}(\Omega)$ be bounded in $L^{2^*}(\Omega)$ and $u_n \to u$ almost everywhere in Ω. Then one has*

$$\int_\Omega |u_n|^{2^*} = \int_\Omega |u|^{2^*} + \int_\Omega |v_n|^{2^*} + o(1), \tag{1.2.18}$$

where $o(1) \to 0$ as $n \to +\infty$.

It follows from Lemma 1.2.10 that if $u_n \rightharpoonup u \not\equiv 0$ weakly in $H_0^1(\Omega)$, then $u_n \to u \not\equiv 0$ strongly in $H_0^1(\Omega)$. Indeed, if $\int_\Omega |u|^{2^*} = b \in (0, a)$, then it follows from (1.2.18) that

$$\begin{aligned} I_a + o(1) &\geq I(u_n) = I(u) + I(v_n) + 2\int_\Omega (\nabla v_n \nabla u - \lambda u v_n) \\ &= I(u) + I(v_n) + o(1) \geq I_b + I_{c_n} + o(1) \\ &= I_b + I_{a-b} + o(1), \end{aligned}$$

where $c_n = \int_\Omega |v_n|^{2^*} \to c = a - b$. But this contradicts (1.2.15).

It remains to exclude the possibility that $u_n \rightharpoonup 0$ weakly in $H_0^1(\Omega)$. Similar to the discussion of (1.2.5), we need to determine the limit problem for (1.2.13) and compare I_a with its limit problem. Since $u_n \rightharpoonup 0$ weakly in $H_0^1(\Omega)$, we have

$$\int_\Omega u_n^2 \to 0, \quad \text{as } n \to +\infty.$$

Thus, the corresponding limit problem should be

$$I_a^\infty = \inf\left\{I^\infty(u) : \ u \in H_0^1(\Omega), \ \int_\Omega |u|^{2^*} = a\right\}, \tag{1.2.19}$$

and

$$I^\infty(u) = \int_\Omega |\nabla u|^2.$$

Therefore, we obtain the following result.

Theorem 1.2.11 *If $I_a < I_a^\infty$, where I_a^∞ is defined in* (1.2.19), *then I_a defined in* (1.2.13) *is achieved.*

Up to this point, the arguments for the existence of a minimizer for (1.2.13) are exactly the same as those for (1.2.5).

Now we discuss the limit problem (1.2.19). The main observation is that the value I_a^∞ is independent of the domain Ω. In fact, we have the following.

Lemma 1.2.12 *One has an equality*

$$I_a^\infty = S_a := \inf\left\{\int_{\mathbb{R}^N} |\nabla u|^2 : \int_{\mathbb{R}^N} |u|^{2^*} = a, u \in D^{1,2}(\mathbb{R}^N)\right\},$$

where $D^{1,2}(\mathbb{R}^N)$ is the completion of $C_0^\infty(\mathbb{R}^N)$ under the norm $(\int_{\mathbb{R}^N} |\nabla u|^2)^{\frac{1}{2}}$.

Proof. We use $I_a^\infty(\Omega)$ to denote the dependence of the value on Ω. By translation, we can always assume $0 \in \Omega$. Let $\delta > 0$ be sufficiently small so that $B_\delta(0) \subset \Omega$. Then

$$I_a^\infty(B_\delta(0)) \geq I_a^\infty(\Omega) \geq I_a^\infty(\mathbb{R}^N).$$

We claim that

$$I_a^\infty(B_\delta(0)) = I_a^\infty(\mathbb{R}^N), \tag{1.2.20}$$

which will yield the conclusion of the lemma.

Indeed, for any $u \in H^1(\mathbb{R}^N)$, let $v = t^{\frac{N-2}{2}} u(tx)$, where $t > 0$. We then see that

$$\int_{\mathbb{R}^N} |v|^{2^*} = \int_{\mathbb{R}^N} |u|^{2^*}, \ \int_{\mathbb{R}^N} |\nabla v|^2 = \int_{\mathbb{R}^N} |\nabla u|^2.$$

Hence, for any $u \in C_0^\infty(\mathbb{R}^N)$, we have $v(x) = R^{\frac{N-2}{2}} u(Rx) \in H_0^1(B_\delta(0))$ if $R > 0$ is large enough. Now for any $\varepsilon > 0$, we take a $u \in C_0^\infty(\mathbb{R}^N)$ such that $\int_{\mathbb{R}^N} |\nabla u|^2 < I_a^\infty(\mathbb{R}^N) + \varepsilon$ and $\int_{\mathbb{R}^N} |u|^{2^*} = a$. From

$$\int_{B_\delta(0)} |\nabla v|^2 = \int_{\mathbb{R}^N} |\nabla u|^2, \quad \int_{B_\delta(0)} |v|^{2^*} = \int_{\mathbb{R}^N} |u|^{2^*},$$

we have that $I_a^\infty(\mathbb{R}^N) + \varepsilon > \int_{B_\delta(0)} |\nabla v|^2 \geq I_a^\infty(B_\delta(0))$. This shows that (1.2.20) holds. □

Noting that $I_a = a^{\frac{N-2}{N}} I_1$ and $S_a = a^{\frac{N-2}{N}} S$, where $S = S_1$ is the best Sobolev constant of the embedding $D^{1,2}(\mathbb{R}^N) \hookrightarrow L^{2^*}(\mathbb{R}^N)$, we only need to check whether the inequality $I_1 < S$ holds or not.

For problem (1.2.5), the possible loss of the compactness is due to the fact that part of the sequence $\{u_n\}$ lies near infinity. For (1.2.13), the possible loss of the compactness is due to the concentration of the sequence $\{u_n\}$. We list here some well-known facts about the minimization problem (1.2.19).

- S is achieved by

$$\bar{U}_{x,\mu}(y) = \frac{C_0 \mu^{\frac{N-2}{2}}}{(1 + \mu^2 |y - x|^2)^{\frac{N-2}{2}}}, \quad x \in \mathbb{R}^N, \ \mu \in \mathbb{R}_+, \tag{1.2.21}$$

  where C_0 is a positive constant depending on N only. See [10, 136],
- If $\Omega \neq \mathbb{R}^N$, then $I_1^\infty(\Omega)$ is not achieved. This can be proved by using the strong maximum principle.
- $\bar{U}_{x,\mu}(x) \to +\infty$ as $\mu \to +\infty$, while for any y satisfying $|y - x| \geq \delta > 0$, one has $\bar{U}_{x,\mu}(y) \leq \frac{C_0}{\mu^{\frac{N-2}{2}} \delta^{N-2}} \to 0$ as $\mu \to +\infty$. Thus, $\bar{U}_{x,\mu}(y)$ concentrate at x as $\mu \to +\infty$. For this reason, we call $\bar{U}_{x,\mu}$ a bubble at x.

Theorem 1.2.13 *If $N \geq 4$, then $I_1 < S$. In particular, I_1 is achieved.*

Proof. We first note that although the function $\bar{U}_{x,\mu}$ achieves S, it does not lie in $H_0^1(\Omega)$ and so we cannot use this function directly. Without loss of generality, assume that $0 \in \Omega$. Observe that on the boundary $\partial\Omega$, $\bar{U}_{0,\mu}$ is very small if $\mu > 0$ is large, and so we can cut $\bar{U}_{0,\mu}$ off without changing the function too much.

Let $\delta > 0$ be small enough such that $B_{2\delta}(0) \subset \Omega$. Let $\xi \in C_0^\infty(B_{2\delta}(0))$,which satisfies $\xi = 1$ in $B_\delta(0)$, $0 \leq \xi \leq 1$ and $|\nabla \xi| \leq \frac{C}{\delta}$. Define

$$v_\mu = \frac{\xi \bar{U}_{0,\mu}}{\left(\int_\Omega |\xi \bar{U}_{0,\mu}|^{2^*} \right)^{1/2^*}} \in H_0^1(\Omega).$$

Then one has $\int_\Omega |v_\mu|^{2^*} = 1$ and so

$$I_1 \le I(v_\mu).$$

Therefore, to prove this theorem, we need to prove that for sufficiently large $\mu > 0$, the inequality

$$I(v_\mu) < S \tag{1.2.22}$$

holds.

As a start, we compute that

$$\begin{aligned}\int_\Omega |\xi \bar{U}_{0,\mu}|^{2^*} &= \int_{B_\delta(0)} \bar{U}_{0,\mu}^{2^*} + O\Big(\int_{\mathbb{R}^N\setminus B_\delta(0)} \bar{U}_{0,\mu}^{2^*}\Big)\\ &= \int_{\mathbb{R}^N} \bar{U}_{0,\mu}^{2^*} + O\Big(\int_{\mathbb{R}^N\setminus B_\delta(0)} \bar{U}_{0,\mu}^{2^*}\Big)\\ &= \int_{\mathbb{R}^N} \bar{U}_{0,\mu}^{2^*} + O\Big(\int_{\mathbb{R}^N\setminus B_\delta(0)} \Big(\frac{1}{\mu^{\frac{N-2}{2}}|y|^{N-2}}\Big)^{\frac{2N}{N-2}}\Big)\\ &= \int_{\mathbb{R}^N} \bar{U}_{0,\mu}^{2^*} + O\Big(\frac{1}{\mu^N}\Big).\end{aligned} \tag{1.2.23}$$

On the other hand, noting that $\xi = 1$ and $\nabla\xi = 0$ in $B_\delta(0)$, we can deduce that

$$\begin{aligned}\int_\Omega |\nabla(\xi \bar{U}_{0,\mu})|^2 &= \int_\Omega (\xi^2|\nabla \bar{U}_{0,\mu}|^2 + 2\xi \bar{U}_{0,\mu}\nabla\xi\nabla U_{0,\mu} + |\nabla\xi|^2\bar{U}_{0,\mu}^2)\\ &= \int_{\mathbb{R}^N} |\nabla \bar{U}_{0,\mu}|^2 + O\Big(\int_{\mathbb{R}^N\setminus B_\delta(0)} |\nabla \bar{U}_{0,\mu}|^2 + \int_{\Omega\setminus B_\delta(0)} \bar{U}_{0,\mu}^2\Big)\\ &= S + O\Big(\frac{1}{\mu^{N-2}}\Big),\end{aligned} \tag{1.2.24}$$

and

$$\begin{aligned}\int_\Omega |\xi \bar{U}_{0,\mu}|^2 &= \int_\Omega \bar{U}_{0,\mu}^2 + O\Big(\int_{\Omega\setminus B_\delta(0)} \bar{U}_{0,\mu}^2\Big)\\ &= \frac{1}{\mu^2}\int_{\Omega_\mu} \bar{U}_{0,1}^2 + O\Big(\frac{1}{\mu^{N-2}}\Big),\end{aligned}$$

where $\Omega_\mu = \{y : \mu^{-1}y \in \Omega\}$. Therefore, one has that

$$\int_\Omega |\xi \bar{U}_{0,\mu}|^2 = \begin{cases} \dfrac{A_N}{\mu^2} + O\Big(\dfrac{1}{\mu^{N-2}}\Big), & N \ge 5,\\[2ex] \dfrac{\omega_3 \ln \mu}{\mu^2} + O\Big(\dfrac{1}{\mu^2}\Big), & N = 4,\end{cases} \tag{1.2.25}$$

where $A_N = \int_{\mathbb{R}^N} \bar{U}_{0,1}^2 > 0$ for $N \ge 5$, and ω_3 is the area of the unit sphere S^3.

Combining (1.2.23), (1.2.24) and (1.2.25), we conclude that if $N \geq 5$ and μ is sufficiently large, we have

$$I(v_\mu) = \frac{S + O(\mu^{2-N}) - \lambda(A_N\mu^{-2} + O(\mu^{2-N}))}{(1 + O(\mu^{-N}))^{2/2^*}}$$
$$= S - \frac{A_N\lambda}{\mu^2} + O\Big(\frac{1}{\mu^{N-2}}\Big) < S,$$

while for $N = 4$ and sufficiently large μ, we have

$$I(v_\mu) = \frac{S + O(\mu^{-2}) - \lambda(\omega_3\mu^{-2}\ln\mu + O(\mu^{-2}))}{(1 + O(\mu^{-4}))^{1/2}}$$
$$= S - \frac{\omega_3\lambda\ln\mu}{\mu^2} + O\Big(\frac{1}{\mu^2}\Big) < S.$$

Thus (1.2.22) is proved. □

An immediate consequence of Theorem 1.2.13 is the following existence result for (1.2.17).

Theorem 1.2.14 (Brezis–Nirenberg [24],1983) *If $N \geq 4$ with $\lambda \in (0, \lambda_1)$, then the Brezis-Nirenberg problem* (1.2.17) *has a solution.*

Remark 1.2.15 If $N = 3$, the existence of a solution for (1.2.17) is much more complicated. In [24], it is proved that (1.2.17) has no positive solution $u \in H_0^1(\Omega)$ when $\lambda \in (0, \lambda^*)$ for some suitable constant $\lambda^* > 0$. Moreover, if Ω is a ball, then $\lambda^* = \frac{\lambda_1}{4}$, and (1.2.17) has a solution for $\lambda \in (\lambda^*, \lambda_1)$. $N = 3$ is called a critical dimension for (1.2.17). See [24, 73].

Our next result shows that the condition $\lambda \in (0, \lambda_1)$ is necessary for the existence of a solution for (1.2.17).

Theorem 1.2.16 *If $\lambda \geq \lambda_1$, or $\lambda \leq 0$ and Ω is star-shaped with respect to some point $x_0 \in \Omega$, then* (1.2.17) *has no positive solutions.*

Proof. Suppose that $\lambda \geq \lambda_1$. Let u be a solution of (1.2.17) and let $\varphi > 0$ be a first eigenfunction corresponding to λ_1; then,

$$\int_\Omega (u^{2^*-1}\varphi + \lambda u\varphi) = -\int_\Omega \Delta u\varphi = -\int_\Omega \Delta\varphi u = \lambda_1 \int_\Omega u\varphi.$$

From the above equality, we have

$$(\lambda - \lambda_1)\int_\Omega u\varphi + \int_\Omega u^{2^*-1}\varphi = 0,$$

which is impossible, since

$$(\lambda - \lambda_1) \int_{\Omega} u\varphi \geq 0, \quad \int_{\Omega} u^{2^*-1}\varphi > 0.$$

This gives the required contradiction, and consequently, (1.2.17) has no positive solutions if $\lambda \geq \lambda_1$.

Suppose now that $\lambda \leq 0$ and that Ω is star-shaped with respect to some point $x_0 \in \Omega$. We have the following Pohozaev identity for the solution u of (1.2.17) (see Corollary 6.2.2):

$$\frac{1}{2} \int_{\partial\Omega} \langle x - x_0, \nu \rangle \left(\frac{\partial u}{\partial \nu} \right)^2 - \lambda \int_{\Omega} u^2 = 0, \tag{1.2.26}$$

where ν is the outward unit normal of $\partial\Omega$. Since $\langle x - x_0, \nu \rangle > 0$ on $\partial\Omega$ and $\lambda \leq 0$, it follows from (1.2.26) that $\frac{\partial u}{\partial \nu} = 0$ on $\partial\Omega$. But this is a contradiction to the strong maximum principle. □

1.3 Mountain Pass Lemma

In the previous sections, we have studied the constraint minimization problems. There we see that a minimizer of such problem yields a solution for the corresponding eigenvalue problem. Moreover, in some homogenous cases, the Lagrange multiplier can be eliminated (For instances, see (1.2.7) and (1.2.8), or (1.2.16) and (1.2.17)).

However, such techniques cannot be generally used to find a nontrivial solution for the following nonlinear elliptic problem

$$\begin{cases} -\Delta u = f(x, u), & \text{in } \Omega, \\ u = 0, & \text{on } \partial\Omega. \end{cases} \tag{1.3.1}$$

Denote $F(x, t) = \int_0^t f(x, s)\, ds$. Instead of studying the following constraint minimization problem

$$\inf \left\{ I(u) : u \in H_0^1(\Omega), \int_{\Omega} F(x, u) = 1 \right\},$$

which yields a solution for the following eigenvalue problem

$$\begin{cases} -\Delta u = \lambda f(x, u), & \text{in } \Omega, \\ u = 0, & \text{on } \partial\Omega, \end{cases}$$

we need to consider the existence of a nontrivial critical point for the following functional

$$I(u) = \frac{1}{2}\int_\Omega |\nabla u|^2 - \int_\Omega F(x,u), \quad u \in H_0^1(\Omega). \tag{1.3.2}$$

A critical point u of the functional $I(u)$ satisfies

$$\int_\Omega \nabla u \nabla \varphi = \int_\Omega f(x,u)\varphi, \quad \forall\, \varphi \in H_0^1(\Omega). \tag{1.3.3}$$

So u is a weak solution of (1.3.1).

For many elliptic equations, such as (1.2.8) and (1.2.17), the corresponding functionals are neither bounded from below nor bounded from above. Therefore, we have to find a saddle point for $I(u)$. The mountain pass lemma provides one of the most powerful methods to achieve this goal.

Before stating the Mountain Pass Lemma, we introduce some terminology.

Definition 1.3.1 Let $I(u)$ be a C^1 functional defined in a Banach space H. A sequence $\{u_n\} \subset H$ is said to be a Palais-Smale sequence at level c ((P.-S.)$_c$ sequence for short) for $I(u)$ if $I(u_n) \to c$ and $I'(u_n) \to 0$ as $n \to +\infty$. If every (P.-S.)$_c$ sequence of $I(u)$ has a convergent subsequence in H, we say that $I(u)$ satisfies the Palais–Smale condition at level c, or the (P.-S.)$_c$ condition. If $I(u)$ satisfies the (P.-S.)$_c$ condition for all $c \in \mathbb{R}$, we will then say that $I(u)$ satisfies the Palais–Smale condition ((P.-S.) condition for short).

Definition 1.3.2 (Ambrosetti–Rabinowitz [9], 1973) We say that $I(u)$ satisfies the mountain pass structure if there are constants $\delta > 0$ and $c_0 > 0$, and an $e \in H$ with $\|e\| > \delta$, such that $I(0) = 0$, $I(e) \le 0$ and

$$I(u) \ge c_0, \quad \forall\, u \in H \text{ with } \|u\| = \delta.$$

Definition 1.3.3 (Ambrosetti–Rabinowitz [9], 1973) Suppose that $I(u)$ satisfies the mountain pass structure. The mountain pass value c is then defined as follows:

$$c = \inf_{\gamma \in \Gamma} \max_{t \in [0,1]} I(\gamma(t)),$$

where Γ consists of the curves $\gamma(t)$ in H connecting 0 and e, that is, curves $\gamma(t)$ in H satisfying $\gamma(0) = 0$ and $\gamma(1) = e$. It is easy to see that $c \ge c_0 > 0$.

Mountain Pass Lemma (Ambrosetti–Rabinowitz [9], 1973) *Let $I(u)$ be a C^1 functional defined in a Banach space H which satisfies the mountain pass* *structure and the (P.-S.)$_c$ condition. Then the mountain pass value c is a critical value of $I(u)$. That is, there exists a $u \in H$ such that $I'(u) = 0$ and $I(u) = c$.*

The (P.-S.)$_c$ condition is a compactness condition for the functional $I(u)$. It is used to prove that if c is not a critical value of $I(u)$, then there exist small constants $\varepsilon > 0$ and $\tau > 0$ such that

$$\|I'(u)\| \geq \tau, \quad \text{for all } u \text{ with } c - \varepsilon \leq I(u) \leq c + \varepsilon. \tag{1.3.4}$$

From (1.3.4), the mountain pass lemma can be proved by using a deformation lemma. The readers can refer to sections 1.2 and 1.3 in [151] for more details. With this observation, we can prove the following mountain pass lemma without the (P.-S.)$_c$ condition.

Mountain Pass Lemma without the (P.-S.)$_c$ Condition (Ambrosetti–Rabinowitz [9], 1973, Brezis–Nirenberg [24], 1983) *Let $I(u)$ be a C^1 functional defined in a Banach space H that satisfies the mountain pass structure. Then there exists a sequence of $\{u_n\} \subset H$ such that as $n \to +\infty$, $I'(u_n) \to 0$ and $I(u_n) \to c$.*

We now introduce the following result.

Symmetric Mountain Pass Lemma (Ambrosetti–Rabinowitz [9], 1973) *Let $I(u)$ be a C^1 functional defined in an infinitely dimensional Banach space H that satisfies $I(0) = 0$, $I(-u) = I(u)$ for any $u \in H$, and the (P.-S.) condition. Suppose that $H = H_1 \bigoplus H_2$, where H_1 is finite dimensional. Also, we assume that $I(u)$ satisfies the following conditions.*

(1) There exist $c_0 > 0$ and $\rho > 0$ such that $I(u) \geq c_0$ for any $u \in H_2$ and $\|u\| = \rho$.
(2) For any finite dimensional subspace $W \subset H$, there exists $R(W) > 0$ such that $I(u) \leq 0$ for any $u \in W$ and $\|u\| \geq R(W)$.

Then $I(u)$ has a sequence of critical values $c_k \to +\infty$.

For later applications, we give the description of this sequence of critical values $c_k \to +\infty$.

Choose a basis $\{\phi_1, \phi_2, \cdots\}$ for H_2. Let $W_k = H_1 \bigoplus span\{\phi_1, \cdots, \phi_k\}$. We can also assume $R(W_k) \leq R(W_{k+1})$, where $R(W_k)$ is the number in (2) of the assumptions for $I(u)$. Define

$$\Gamma_k = \big\{h \in C(H, H) : h(-u) = -h(u), \forall\, u \in H; h(u) = u, \forall\, u \in W_j, \\ \|u\| \geq R(W_j), j \leq k\big\}.$$

Then

$$c_k = \inf_{h\in\Gamma_k} \sup_{u\in W_k} I(h(u)) \tag{1.3.5}$$

is a critical value of $I(u)$ with $c_k \to +\infty$ as $k \to +\infty$.

1.3.1 Semilinear Elliptic Problem: The Compact Case

In this subsection, we study the existence of a non-zero solution for the following elliptic problem

$$\begin{cases} -\Delta u = f(x, u), & \text{in } \Omega, \\ u = 0, & \text{on } \partial\Omega, \end{cases} \tag{1.3.6}$$

where Ω is a bounded domain in $\mathbb{R}^N$ and $f(x, t) \in C(\bar{\Omega}, \mathbb{R})$ satisfies the following conditions:

(i) $\lim\limits_{t\to 0} \dfrac{f(x,t)}{t} = \lim\limits_{|t|\to+\infty} \dfrac{f(x,t)}{|t|^{2^*-1}} = 0$ uniformly in Ω.

(ii) There exist $p > 2$ and $T > 0$ such that

$$0 < pF(x,t) \le f(x,t)t, \quad \forall\, x \in \Omega,\ |t| \ge T,$$

where $F(x,t) = \int_0^t f(x,s)\,ds$.

The functional $I(u)$ corresponding to (1.3.6) is given by (1.3.2).

We first prove the following result which will later allow us to apply the mountain pass lemma.

Lemma 1.3.4 *Suppose that (ii) holds and that*

$$\lim_{t\to 0} \frac{f(x,t)}{t} = 0, \ \limsup_{|t|\to+\infty} \frac{f(x,t)}{|t|^{2^*-1}} < +\infty, \ \textit{uniformly in } \Omega. \tag{1.3.7}$$

Then $I(u)$ has the mountain pass structure.

Proof. It is easy to check that $I(0) = 0$. On the other hand, it follows from (1.3.7) that for any small ε, there exists a large constant C_ε such that

$$|f(x,t)| \le \varepsilon|t| + C_\varepsilon|t|^{2^*-1}, \quad \forall\, x \in \Omega, t \in \mathbb{R}.$$

This in turn implies that

$$|F(x,t)| \le \frac{\varepsilon}{2}|t|^2 + \frac{C_\varepsilon}{2^*}|t|^{2^*}, \quad \forall\, x \in \Omega, t \in \mathbb{R}. \tag{1.3.8}$$

From (1.3.8), we deduce that

$$\begin{aligned} I(u) \geq & \frac{1}{2}\|u\|^2 - \frac{\varepsilon}{2}\int_\Omega |u|^2 - \frac{C_\varepsilon}{2^*}\int_\Omega |u|^{2^*} \\ \geq & \frac{1}{2}\left(1 - C\varepsilon - CC_\varepsilon \|u\|^{2^*-1}\right)\|u\|^2 \\ \geq & c_0 > 0, \end{aligned}$$

if $\|u\| = \delta > 0$ is small enough.

On the other hand, we can rewrite (ii) as

$$t\frac{\partial F(x,t)}{\partial t} \geq pF(x,t), \quad |t| \geq T.$$

Integrating the above inequality from T to t yields

$$\ln F(x,t) \geq p \ln |t| - C, \quad |t| \geq T.$$

Consequently, for some large $M > 0$, one has

$$F(x,t) \geq e^{-C}|t|^p - M, \ \forall\ |t| > 0. \tag{1.3.9}$$

Fix $v \not\equiv 0$. Take $e = tv$ for some large $t > 0$. Then

$$I(e) \leq \frac{t^2}{2}\int_\Omega |\nabla v|^2 - e^{-C}t^p \int_\Omega |v|^p + M|\Omega| < 0.$$

□

Now we can prove the following existence result.

Theorem 1.3.5 (Ambrosetti–Rabinowitz [9], 1973) *Suppose that Ω is a bounded domain in $\mathbb{R}^N$ and that $f(x,t)$ satisfies (i) and (ii). Then* (1.3.6) *has a nontrivial solution.*

Proof. By Lemma 1.3.4, it remains to prove that $I(u)$ satisfies the $(\text{P.-S.})_c$ condition in order to apply the mountain pass lemma.

Suppose that $u_n \in H_0^1(\Omega)$,which satisfies

$$I(u_n) \to c, \quad I'(u_n) \to 0,$$

as $n \to +\infty$. Then

$$\frac{1}{2}\int_\Omega |\nabla u_n|^2 - \int_\Omega F(x,u_n) = c + o(1), \tag{1.3.10}$$

and

$$\int_\Omega \nabla u_n \nabla \varphi - \int_\Omega f(x,u_n)\varphi = o(1)\|\varphi\|, \tag{1.3.11}$$

where $o(1) \to 0$ as $n \to +\infty$.

Firstly, we claim that $\|u_n\|$ is bounded. Letting $\varphi = u_n$ in (1.3.11), we obtain

$$\int_\Omega |\nabla u_n|^2 - \int_\Omega f(x, u_n)u_n = o(1)\|u_n\|. \tag{1.3.12}$$

Using (ii), (1.3.10) and (1.3.12), we find that

$$\begin{aligned}\left(\frac{p}{2} - 1\right)\int_\Omega |\nabla u_n|^2 &= \int_\Omega \big(pF(x, u_n) - f(x, u_n)u_n\big) + pc + o(1)\|u_n\| \\ &\le \int_{|u_n|\le T} \big(pF(x, u_n) - f(x, u_n)u_n\big) + pc + o(1)\|u_n\| \\ &\le C + o(1)\|u_n\|,\end{aligned}$$

for some large constant $C > 0$. This proves our claim.

Therefore, we may assume that there exists $u \in H_0^1(\Omega)$ such that

$$u_n \rightharpoonup u \text{ weakly in } H_0^1(\Omega),$$

and

$$u_n \to u \text{ strongly in } L^q(\Omega) \text{ for any } q \in [2, 2^*).$$

Taking limit in (1.3.11), we find that

$$\int_\Omega \nabla u \nabla \varphi - \int_\Omega f(x, u)\varphi = 0. \tag{1.3.13}$$

Combining (1.3.11) and (1.3.13), we have

$$\int_\Omega (\nabla u_n - \nabla u)\nabla \varphi = \int_\Omega \big(f(x, u_n) - f(x, u)\big)\varphi + o(1)\|\varphi\|. \tag{1.3.14}$$

Taking $\varphi = u_n - u$ in (1.3.14), we obtain

$$\int_\Omega |\nabla(u_n - u)|^2 = \int_\Omega \big(f(x, u_n) - f(x, u)\big)(u_n - u) + o(1). \tag{1.3.15}$$

From (i), for any $\varepsilon > 0$, we can find a constant $C_\varepsilon > 0$ such that

$$|f(x, t)| \le \varepsilon |t|^{2^*-1} + C_\epsilon.$$

Thus, for any set $A \subset \Omega$, we have

$$\begin{aligned}\int_A |f(x, u_n)u_n| &\le \varepsilon \int_A |u_n|^{2^*} + C_\epsilon \int_A |u_n| \\ &\le \varepsilon \int_\Omega |u_n|^{2^*} + C_\epsilon \Big(\int_\Omega |u_n|^{2^*}\Big)^{1/2^*} |A|^{1-1/2^*} \\ &\le C\varepsilon + C_\varepsilon' |A|^{1-1/2^*} \to 0\end{aligned}$$

as $|A| \to 0$. By Vitali's theorem, we have that

$$\int_\Omega \big(f(x,u_n) - f(x,u)\big)(u_n - u) \to 0,$$

as $n \to +\infty$. Hence (1.3.15) implies that $u_n \to u$ strongly in $H_0^1(\Omega)$. □

Remark 1.3.6 If we let $f(x,t) = 0$ for $t \le 0$, then the mountain pass lemma will yield a non-negative nontrivial solution for (1.3.6). In fact, if we take $\varphi = u_- = \min(u,0)$ in (1.3.3), then we have

$$\int_\Omega |\nabla u_-|^2 = 0,$$

which implies that $u_- \equiv 0$.

Under an extra condition on the nonlinearity $f(x,t)$, we can prove that the mountain pass value c is the least nontrivial critical value for $I(u)$. For this reason, we call this mountain pass solution the *least energy solution*.

Theorem 1.3.7 *Suppose that $f(x,t)$ satisfies the following extra condition:*

$$\frac{f(x,t)}{t} \quad \textit{is strictly increasing for } t > 0 \textit{ and is strictly decreasing for } t < 0. \tag{1.3.16}$$

Then one has

$$c \le c_1 := \inf\{I(u) : u \in \mathcal{N}\}, \tag{1.3.17}$$

where

$$\mathcal{N} := \left\{u \in H_0^1(\Omega),\ u \not\equiv 0, \int_\Omega |\nabla u|^2 = \int_\Omega f(x,u)u\right\}$$

is called the Nehari manifold.

Moreover, if the mountain pass value c is a critical value for $I(u)$, then $c = c_1$ is the least nontrivial critical value.

Proof. Note that any nontrivial solution u of (1.3.6) belongs to $\mathcal{N}$, since

$$\int_\Omega |\nabla u|^2 = \int_\Omega f(x,u)u.$$

As a result, if c is a critical value, one then has $c \ge c_1$. Therefore, it remains to prove that $c \le c_1$.

Given any $u \not\equiv 0$, we have

$$\frac{dI(tu)}{dt} = t\int_\Omega |\nabla u|^2 - \int_\Omega f(x,tu)u = t\left(\int_\Omega |\nabla u|^2 - \int_\Omega \frac{f(x,tu)}{tu}u^2\right),$$

which, together with (1.3.16) and (1.3.9), implies that there exists a unique $t_u > 0$ such that $I(tu)$ is increasing in $(0, t_u)$ and is decreasing in $(t_u, +\infty)$. So, if $\int_\Omega |\nabla u|^2 = \int_\Omega f(x, u)u$ and $u \not\equiv 0$, then $t_u = 1$.

Define

$$c_2 = \inf_{u \not\equiv 0,\, u \in H_0^1(\Omega)} \max_{t \geq 0} I(tu).$$

We first prove the equality $c_1 = c_2$. For any $\varepsilon > 0$, we can take a $u \not\equiv 0$ with $\int_\Omega |\nabla u|^2 = \int_\Omega f(x, u)u$ and such that

$$I(u) < c_1 + \varepsilon.$$

But since one has

$$I(u) = \max_{t \geq 0} I(tu) \geq c_2,$$

the inequality $c_1 \geq c_2$ clearly follows.

On the other hand, for any $\varepsilon > 0$, we can take a $u \not\equiv 0$ such that

$$I(t_u u) = \max_{t \geq 0} I(tu) < c_2 + \varepsilon.$$

It is easy to check that

$$\int_\Omega |\nabla(t_u u)|^2 = \int_\Omega f(x, t_u u) t_u u.$$

Consequently, one has

$$I(t_u u) \geq c_1,$$

which in turn implies that $c_2 \geq c_1$.

We now prove $c \leq c_2$. Let $\varepsilon > 0$. Choose a $u \not\equiv 0$ such that

$$I(t_u u) = \max_{t \geq 0} I(tu) < c_2 + \varepsilon.$$

Denote by S a two-dimensional subspace of $H_0^1(\Omega)$ that contains u and e. By (1.3.9), and by the fact that all norms in a finite dimensional Banach space are equivalent, we see that

$$\begin{aligned} I(u) &\leq \frac{1}{2} \int_\Omega |\nabla u|^2 - C \int_\Omega |u|^p + M|\Omega| \\ &\leq \frac{1}{2} \|u\|^2 - C\|u\|^p + C, \quad \forall u \in S. \end{aligned}$$

Hence we can choose a large $R > \max(\|t_u u\|, \|e\|) > 0$ such that $I(u) < 0$ for all $u \in S$ with $\|u\| = R$.

Now we define a path in Γ connecting 0 and e as follows. If $t \in [0, \frac{1}{2}]$, let $u_t = \frac{2tRu}{\|u\|}$, which is the segment connecting 0 and $\frac{Ru}{\|u\|}$. If $t \in (\frac{3}{4}, 1]$, define $u_t = \frac{4(1-t)Re}{\|e\|} + (4t-3)e$, which is the segment connecting $\frac{Re}{\|e\|}$ and e. If $t \in (\frac{1}{2}, \frac{3}{4})$, let u_t be the arc in S, satisfying $u_{\frac{1}{2}} = \frac{Ru}{\|u\|}$, $u_{\frac{3}{4}} = \frac{Re}{\|e\|}$ and $\|u_t\| = R$. Let $\gamma(t) = u_t$, $t \in [0, 1]$. Then it is easy to check that

$$c_2 + \varepsilon > \max_{s \geq 0} I(su) = \max_{t \in [0, \frac{1}{2}]} I(u_t) = \max_{t \in [0,1]} I(\gamma(t)) \geq c,$$

which implies that $c_2 \geq c$. Hence (1.3.17) holds true. □

We end this subsection discussing a multiplicity result for the following problem

$$\begin{cases} -\Delta u = g(u) + \lambda u, & \text{in } \Omega, \\ u = 0, & \text{on } \partial\Omega, \end{cases} \tag{1.3.18}$$

where $\lambda \in \mathbb{R}$, Ω is a bounded domain in $\mathbb{R}^N$, and $g(t) \in C(\mathbb{R}, \mathbb{R})$ satisfies $g(-u) = -g(u)$ and the following conditions.

(i) $\lim_{t \to 0} \frac{g(t)}{t} = \lim_{|t| \to +\infty} \frac{g(t)}{|t|^{2^*-1}} = 0.$

(ii) There exist $p > 2$ and $T > 0$, such that

$$0 < pG(t) \leq g(t)t, \quad \forall\, |t| \geq T,$$

where $G(t) = \int_0^t g(s)$.

The functional corresponding to (1.3.18) is given by

$$I(u) = \frac{1}{2}\int_\Omega |\nabla u|^2 - \int_\Omega \Big(G(u) + \frac{\lambda}{2}u^2\Big). \tag{1.3.19}$$

Then we have the following:

Theorem 1.3.8 (Ambrosetti–Rabinowitz [9], 1973) *Problem* (1.3.18) *has infinitely many solutions* u_k *with* $I(u_k) \to +\infty$ *as* $k \to +\infty$.

Proof. We will prove that the functional $I(u)$ defined in (1.3.19) satisfies the conditions in the symmetric mountain pass lemma. Let $f(u) = g(u) + \lambda u$. By reviewing the proof of Theorem 1.3.5, we see that $I(u)$ satisfies the (P.-S.) condition, if there exists $q > 2$ such that

$$0 < qF(u) \leq uf(u), \quad \forall\, |u| \geq T, \tag{1.3.20}$$

where $F(t) = \int_0^t f(s)\,ds$ and $T > 0$ is a large constant. To prove (1.3.20), we take $q \in (2, p)$. Then

$$\begin{aligned} uf(u) - qF(u) &= ug(u) - qG(u) + \lambda\Big(1 - \frac{q}{2}\Big)u^2 \\ &\geq (p-q)G(u) + \lambda\Big(1 - \frac{q}{2}\Big)u^2, \quad |u| \geq T. \end{aligned}$$

Similarly to (1.3.9), we can deduce $G(u) \geq c_0|u|^p - M$ which gives

$$uf(u) - qF(u) \geq c_0(p-q)|u|^p + \lambda\Big(1 - \frac{q}{2}\Big)u^2 - M_1 > 0,$$

whenever $|u| \geq T_1 > 0$ for some large $T_1 > 0$. Hence, (1.3.20) holds true.

Let $\varphi_i \neq 0$ be an eigenfunction corresponding to the eigenvalue λ_i of the operator $-\Delta$ in Ω with homogeneous Dirichlet boundary condition. Take $H_2 = span\{\varphi_j,\ j \geq l\}$, where $l > 0$ is a large integer. Then for $u \in H_2$, we have

$$\int_\Omega u^2 \leq \frac{1}{\lambda_l}\|u\|^2.$$

For any $r \in (2, 2^*)$, take $s \in (r, 2^*)$ and write $r = 2t + (1-t)s$, where $t = \frac{s-r}{s-2} \in (0, 1)$. We then have

$$\|u\|_{L^r(\Omega)} \leq \|u\|^t_{L^2(\Omega)}\|u\|^{1-t}_{L^s(\Omega)} \leq C\|u\|^t_{L^2(\Omega)}\|u\|^{1-t} \leq \frac{C}{\lambda_l^{t/2}}\|u\|,$$

for $u \in H_2$. As a result,

$$\begin{aligned} I(u) &\geq \frac{1}{2}\|u\|^2 - \frac{\lambda}{2\lambda_l}\|u\|^2 - \int_\Omega |u|^p - C \\ &\geq \frac{1}{2}\|u\|^2 - \frac{\lambda}{2\lambda_l}\|u\|^2 - \frac{C'}{\lambda_l^{tp/2}}\|u\|^p - C \\ &= \Big(\frac{1}{2} - \frac{\lambda}{2\lambda_l} - \frac{C'}{\lambda_l^{tp/2}}\|u\|^{p-2} - \frac{C}{\|u\|^2}\Big)\|u\|^2 > 0, \end{aligned}$$

for any $u \in H_2$ with $\|u\| = \rho > 0$ large, whenever $l > 0$ (depending on ρ) is sufficiently large enough.

Finally, we have

$$I(u) \leq C\|u\|^2 - c_1\int_\Omega |u|^p + C, \quad u \in H_0^1(\Omega),$$

for some constants $C > 0$ and $c_1 > 0$. Let W be any finite dimensional subspace of $H_0^1(\Omega)$. Since all the norms in a finite dimensional space are equivalent, we have $\int_\Omega |u|^p \geq C'\|u\|^p$ for all $u \in W$. Thus,

$$I(u) \leq C\|u\|^2 - c_1'\|u\|^p + C < 0, \quad \text{if } \|u\| = R > 0 \text{ large.}$$

□

1.3.2 Nonlinear Schrödinger Equations

In this subsection, we will use the mountain pass lemma to study the following nonlinear Schrödinger equation

$$\begin{cases} -\Delta u + V(x)u = f(u), u > 0, \text{ in } \mathbb{R}^N, \\ u \in H^1(\mathbb{R}^N), \end{cases} \tag{1.3.21}$$

where $V(x) \in C(\mathbb{R}^N)$ satisfies $0 < V_0 \leq V(x) \leq V_1$ for some constants $V_1 > V_0 > 0$, and

$$\lim_{|x|\to+\infty} V(x) = V_\infty > 0.$$

In Section 2 of this chapter, we studied this problem for the special case $f(t) = t^p$ via a constraint minimization problem. In the general case, this method only yields a solution for the corresponding eigenvalue problem. To obtain a positive solution for (1.3.21), we assume $f(t) = 0$ for $t \leq 0$.

If $V(x) = V_\infty$, then (1.3.21) becomes

$$\begin{cases} -\Delta u + V_\infty u = f(u), u > 0, \text{ in } \mathbb{R}^N, \\ u \in H^1(\mathbb{R}^N). \end{cases} \tag{1.3.22}$$

In this section, we assume that $f(t) \in C(\mathbb{R})$ satisfies all of the following conditions.

(i) $\lim_{t\to 0} \dfrac{f(t)}{t} = \lim_{|t|\to+\infty} \dfrac{f(t)}{|t|^{2^*-1}} = 0.$

(ii) There exists $p > 2$, such that

$$0 < pF(t) \leq f(t)t, \quad \forall t > 0,$$

where $F(t) = \int_0^t f(s)\,ds$.

(iii) $\dfrac{f(t)}{t}$ is strictly increasing for $t > 0$.

The functionals corresponding to (1.3.21) and (1.3.22) are given by

$$I(u) = \frac{1}{2}\int_{\mathbb{R}^N} \left(|\nabla u|^2 + V(x)u^2\right) - \int_{\mathbb{R}^N} F(u), \quad u \in H^1(\mathbb{R}^N),$$

and

$$I^\infty(u) = \frac{1}{2}\int_{\mathbb{R}^N}\left(|\nabla u|^2 + V_\infty u^2\right) - \int_{\mathbb{R}^N} F(u), \quad u \in H^1(\mathbb{R}^N),$$

respectively. By a similar discussion to that in the previous section, we can check that if $f(t)$ satisfies (i)–(iii), $I(u)$ and $I^\infty(u)$ have the mountain pass structure. We will use $c > 0$ and $c^\infty > 0$ to denote the mountain pass value for $I(u)$ and $I^\infty(u)$, respectively. Using a similar argument to that in Theorem 1.3.7, we can prove the following result:

Theorem 1.3.9 *Suppose that $f(t)$ satisfies (i)–(iii). Then,*

$$c \leq c_1 := \inf\left\{I(u):\ u \in H^1(\mathbb{R}^N),\ u \not\equiv 0,\ \int_{\mathbb{R}^N}\left(|\nabla u|^2 + V(x)u^2\right) = \int_{\mathbb{R}^N} f(u)u\right\}$$

and

$$c^\infty \leq c_1^\infty := \inf\left\{I(u): u \in H^1(\mathbb{R}^N),\ u \not\equiv 0,\ \int_{\mathbb{R}^N}\left(|\nabla u|^2 + V_\infty u^2\right) = \int_{\mathbb{R}^N} f(u)u\right\}.$$

We are now ready to prove the following existence result.

Theorem 1.3.10 *Suppose that $f(t)$ satisfies (i)–(iii). Then the mountain pass value $c^\infty > 0$ is a critical value for $I^\infty(u)$.*

Proof. By the mountain pass lemma without the (P.-S.) condition, there exists $u_n \in H^1(\mathbb{R}^N)$, which satisfies

$$I^\infty(u_n) \to c^\infty > 0, \quad (I^\infty)'(u_n) \to 0,$$

as $n \to +\infty$. To show that the mountain pass value $c^\infty > 0$ is a critical value for $I^\infty(u)$, since $I^\infty(u)$ is translation invariant, we just need to prove that up to a translation and up to a subsequence, u_n converges strongly in $H^1(\mathbb{R}^N)$. By a similar discussion to that in Theorem 1.3.5, we can prove that u_n is bounded in $H^1(\mathbb{R}^N)$.

Step 1. We claim that the vanishing case does not occur for u_n. We will argue by contradiction. Fix $q \in (2, 2^*)$. Suppose that

$$\sup_{x\in\mathbb{R}^N}\int_{B_R(x)} |u_n|^q \to 0, \quad \text{as } n \to +\infty \text{ for any fixed } R > 0. \qquad (1.3.23)$$

It then follows from Lemma 1.2.5 that $u_n \to 0$ strongly in $L^r(\mathbb{R}^N)$ for any $r \in (2, 2^*)$. From (i), we deduce that for any $\varepsilon > 0$, there exists $C_\varepsilon > 0$ such that

$$|f(t)| \le \varepsilon(|t| + |t|^{2^*-1}) + C_\varepsilon |t|^{r-1}.$$

Hence this implies that

$$\int_{\mathbb{R}^N} f(u_n)u_n \to 0 \quad \text{and} \quad \int_{\mathbb{R}^N} F(u_n) \to 0.$$

On the other hand, we also have

$$\begin{aligned} o(1) &= \langle (I^\infty)'(u_n), u_n \rangle \\ &= \int_{\mathbb{R}^N} \left(|\nabla u_n|^2 + V_\infty u_n^2\right) - \int_{\mathbb{R}^N} f(u_n)u_n \\ &= \int_{\mathbb{R}^N} \left(|\nabla u_n|^2 + V_\infty u_n^2\right) + o(1), \end{aligned}$$

which implies that $u_n \to 0$ strongly in $H^1(\mathbb{R}^N)$ as $n \to +\infty$. This contradicts the fact that $I^\infty(u_n) \to c^\infty > 0$. Hence (1.3.23) cannot happen.

Therefore, we may assume that there exist constants $\alpha > 0$, $R > 0$ and a sequence $\{x_n\} \subset \mathbb{R}^N$ such that

$$\int_{B_R(x_n)} |u_n|^q > \alpha > 0, \quad \text{as} \quad n \to +\infty.$$

Let $\tilde{u}_n(x) = u_n(x + x_n)$. Then there exists $u \in H^1(\mathbb{R}^N)$ with $u \not\equiv 0$ such that

$$\tilde{u}_n \rightharpoonup u \text{ weakly in } H^1(\mathbb{R}^N),$$

and

$$\tilde{u}_n \to u \text{ strongly in } L^q_{loc}(\mathbb{R}^N), \text{ for any } q \in [2, 2^*).$$

Step 2. We claim that $\tilde{u}_n \to u$ strongly in $H^1(\mathbb{R}^N)$. That is, a non-zero weak limit results in a strongly convergence.

Let $u_n^{(1)} = \tilde{u}_n - u$. We first prove

$$I^\infty(u_n^{(1)}) = I^\infty(\tilde{u}_n) - I^\infty(u) + o(1), \quad (I^\infty)'(u_n^{(1)}) = o(1). \tag{1.3.24}$$

In other words, $u_n^{(1)}$ is still a (P.-S.) sequence.

Applying a similar argument to that of the proof of the Brezis–Lieb lemma, we can deduce

$$\begin{aligned}\int_{\mathbb{R}^N}\big(F(\tilde{u}_n)-F(u_n^{(1)})\big) &= \int_{\mathbb{R}^N}\int_0^1 \frac{dF(t\tilde{u}_n+(1-t)u_n^{(1)})}{dt}\,dt \\ &= \int_{\mathbb{R}^N}\int_0^1 f(t\tilde{u}_n+(1-t)u_n^{(1)})u\,dt \\ &= \int_{\mathbb{R}^N}\int_0^1 f(tu)u\,dt+o(1) \\ &= \int_{\mathbb{R}^N}\int_0^1 \frac{dF(tu)}{dt}\,dt+o(1)=\int_{\mathbb{R}^N}F(u)+o(1),\end{aligned}$$

which gives

$$I^\infty(\tilde{u}_n)=I^\infty(u_n^{(1)})+I^\infty(u)+o(1). \tag{1.3.25}$$

It follows from $I'(u_n)=o(1)$ that u satisfies

$$\int_{\mathbb{R}^N}\big(\nabla u\nabla\phi+V_\infty u\phi\big)-\int_{\mathbb{R}^N}f(u)\phi=0,\quad \forall\,\phi\in H^1(\mathbb{R}^N). \tag{1.3.26}$$

That is, u is a nontrivial solution for (1.3.22). Now, using the Vitali's theorem, we can also prove

$$\begin{aligned}o(1)\|\phi\| &= \big\langle (I^\infty)'(\tilde{u}_n),\phi\big\rangle \\ &= \big\langle (I^\infty)'(u_n^{(1)}),\phi\big\rangle+\big\langle (I^\infty)'(u),\phi\big\rangle-\int_{\mathbb{R}^N}\Big(f(\tilde{u}_n)-f(u_n^{(1)})-f(u)\Big)\phi \\ &= \big\langle (I^\infty)'(u_n^{(1)}),\phi\big\rangle-\int_{\mathbb{R}^N}\Big(\int_0^1\big(f'(t\tilde{u}_n+(1-t)u_n^{(1)}) \\ &\qquad -f'(tu)\big)u\,dt\Big)\phi \\ &= \big\langle (I^\infty)'(u_n^{(1)}),\phi\big\rangle+o(1)\|\phi\|.\end{aligned}$$

Combining these calculations, we obtain (1.3.24).

From (1.3.24),(1.3.26) and Theorem 1.3.9, we have that

$$\begin{aligned}o(1)+c^\infty &= I^\infty(u_n)=I^\infty(\tilde{u}_n)=I^\infty(u)+I^\infty(u_n^{(1)})+o(1) \\ &\geq c_1^\infty+I^\infty(u_n^{(1)})+o(1)\geq c^\infty+I^\infty(u_n^{(1)})+o(1),\end{aligned} \tag{1.3.27}$$

which gives

$$I^\infty(u_n^{(1)})\leq o(1). \tag{1.3.28}$$

On the other hand, it follows from (1.3.24) that

$$\int_{\mathbb{R}^N}\big(|\nabla u_n^{(1)}|^2+V_\infty|u_n^{(1)}|^2\big)-\int_{\mathbb{R}^N}f(u_n^{(1)})u_n^{(1)}=o(1). \tag{1.3.29}$$

Combining (1.3.28) and (1.3.29), we have

$$\left(\frac{1}{2}-\frac{1}{p}\right)\int_{\mathbb{R}^N}\left(|\nabla u_n^{(1)}|^2+V_\infty|u_n^{(1)}|^2\right) \le \int_{\mathbb{R}^N}\left(F(u_n^{(1)})-\frac{1}{p}f(u_n^{(1)})u_n^{(1)}\right) + o(1) \le o(1),$$

which in turn implies that $\tilde{u}_n^{(1)} \to 0$ strongly in $H^1(\mathbb{R}^N)$. Thus, $\tilde{u}_n \to u$ strongly in $H^1(\mathbb{R}^N)$. As a result, u is a critical point of I^∞ at level c^∞, which is what we want to show. □

We now consider (1.3.21). We will prove the following result.

Theorem 1.3.11 *Suppose that $f(t)$ satisfies (i)–(iii). Then, the mountain pass value c is a critical value for $I(u)$ if $c < c^\infty$. In particular, if $V(x) \le V_\infty$ and $V(x) \not\equiv V_\infty$, the mountain pass value c is a critical value for $I(u)$.*

Proof. It suffices to prove that $I(u)$ satisfies the (P.-S.)$_c$ condition. Suppose that $u_n \in H^1(\mathbb{R}^N)$ satisfies

$$I(u_n) \to c, \quad I'(u_n) \to 0,$$

as $n \to +\infty$. Then u_n is bounded in $H^1(\mathbb{R}^N)$. So we may assume that there exists $u \in H^1(\mathbb{R}^N)$ such that

$$u_n \rightharpoonup u \text{ weakly in } H^1(\mathbb{R}^N),$$

and

$$u_n \to u \text{ strongly in } L^q_{loc}(\mathbb{R}^N) \text{ for any } q \in [2, 2^*).$$

Let $u_n^{(1)} = u_n - u$. Then a similar argument to that of the proof of Theorem 1.3.10 and (1.2.9) proves that

$$I^\infty(u_n^{(1)}) = I(u_n^{(1)}) + o(1) = I(u_n) - I(u) + o(1), \quad (I^\infty)'(u_n^{(1)}) = o(1). \tag{1.3.30}$$

Thus, $u_n^{(1)}$ is a (P.-S.) sequence for $I^\infty(u)$.

Step 1. Suppose that $u \equiv 0$. Then (1.3.30) becomes

$$I^\infty(u_n) = I(u_n) + o(1) = c + o(1), \quad (I^\infty)'(u_n) = o(1).$$

Thus, u_n is a (P.-S.)$_c$ sequence of $I^\infty(u)$. By a similar proof to that of Theorem 1.3.10, we conclude that

$$c + o(1) \ge c^\infty + o(1).$$

See (1.3.27). But this contradicts the assumption that $c < c^\infty$.

Step 2. From Step 1, we assume $u \not\equiv 0$. Then $I'(u) = 0$ and by Theorem 1.3.9, $I(u) \geq c$. From (1.3.30), we have that

$$c + o(1) = I(u_n) = I(u) + I^{\infty}(u_n^{(1)}) + o(1) \geq c + I^{\infty}(u_n^{(1)}) + o(1),$$

which implies

$$I^{\infty}(u_n^{(1)}) \leq o(1). \tag{1.3.31}$$

Combining (1.3.31) and (1.3.30), we have $\|u_n^{(1)}\| \to 0$ as $n \to +\infty$, which shows that $u_n \to u$ strongly in $H^1(\mathbb{R}^N)$.

Now we prove that if $V(x) \leq V_\infty$ and $V(x) \not\equiv V_\infty$, then $c < c^\infty$.

Indeed, it follows from Theorem 1.3.10 that there exists $u \in H^1(\mathbb{R}^N)$ such that

$$I^{\infty}(u) = c^{\infty}, \quad (I^{\infty})'(u) = 0.$$

This in turn implies

$$c \leq \max_{t \geq 0} I(tu) = I(t_u u) < I^{\infty}(t_u u) \leq \max_{t \geq 0} I^{\infty}(tu) = c^{\infty},$$

where $t_u > 0$ is the constant such that $\max_{t \geq 0} I(tu) = I(t_u u)$. □

Remark 1.3.12 The means of proving the strong convergence for a (P.-S.)$_c$ sequence $\{u_n\}$ is very similar to that for a minimization sequence discussed in Section 2. Here, we also show that if $c > 0$ is less than the critical number c^∞, then the weak convergence of u_n to $u \not\equiv 0$ implies the strong convergence of u_n to u. The main tool to achieve this goal is the decomposition in (1.3.25) or (1.3.30). Those relations show that if $c > 0$ is below the critical number c^∞, then $I^\infty(u_n - u) = o(1)$ and $(I^\infty)'(u_n - u) = o(1)$, which in turn implies that $u_n \to u$ strongly in $H^1(\mathbb{R}^N)$.

1.3.3 Nonlinear Elliptic Problems with Critical Growth

In this subsection, we will use the mountain pass lemma to study the following nonlinear elliptic problems with critical growth

$$\begin{cases} -\Delta u = u^{2^*-1} + \lambda u^{q-1}, u > 0, \text{ in } \Omega, \\ u \in H_0^1(\Omega), \end{cases} \tag{1.3.32}$$

where Ω is a bounded domain in $\mathbb{R}^N$, $N \geq 3$, $\lambda > 0$ and $q \in (2, 2^*)$. The functional corresponding to (1.3.32) is

$$I(u) = \frac{1}{2}\int_\Omega |\nabla u|^2 - \int_\Omega \left(\frac{1}{2^*}u_+^{2^*} + \frac{\lambda}{q}u_+^q\right), \quad u \in H_0^1(\Omega), \tag{1.3.33}$$

where $u_+ = u$ for $u \geq 0$ and $u_+ = 0$ for $u < 0$.

It is easy to check that $I(u)$ has the mountain pass structure. We will also use c to denote the mountain pass value for $I(u)$.

Theorem 1.3.13 (Brezis–Nirenberg [24],1983) *Suppose that* $N \geq 4$ *and* $q \in (2, 2^*)$, *or* $N = 3$ *and* $q \in (3, 6)$. *Then the mountain pass value* c *is a critical value for* $I(u)$ *defined in* (1.3.33).

Proof. We only need to prove that $I(u)$ satisfies the $(\text{P.-S.})_c$ condition . Suppose that $u_n \in H_0^1(\Omega)$ satisfies

$$I(u_n) \to c, \quad I'(u_n) \to 0,$$

as $n \to +\infty$. Then $\{u_n\}$ is bounded in $H_0^1(\Omega)$, and so we may assume that there exists $u \in H_0^1(\Omega)$ such that

$$u_n \rightharpoonup u \text{ weakly in } H_0^1(\Omega)$$

and

$$u_n \to u \text{ strongly in } L^q(\Omega) \text{ for any } q \in [2, 2^*).$$

As in the discussion for the Schrödinger equation, we expect that if $u \not\equiv 0$, then u_n converges strongly to u in $H_0^1(\Omega)$.

Step 1. Denote

$$I^\infty(u) = \frac{1}{2}\int_\Omega |\nabla u|^2 - \frac{1}{2^*}\int_\Omega u_+^{2^*}, \quad u \in H_0^1(\Omega).$$

Let $u_n^{(1)} = u_n - u$. Then we can prove the following decomposition

$$I^\infty(u_n^{(1)}) = I(u_n) - I(u) + o(1), \quad (I^\infty)'(u_n^{(1)}) = o(1). \tag{1.3.34}$$

Hence $\{u_n^{(1)}\}$ is a (P.-S.) sequence for $I^\infty(u)$.

Step 2. Suppose that now $u \not\equiv 0$. Then

$$I'(u) = 0.$$

It follows from (1.3.34) and Theorem 1.3.7 that

$$\begin{aligned} c + o(1) &= I(u_n) = I(u) + I^\infty(u_n^{(1)}) + o(1) \\ &\geq c + I^\infty(u_n^{(1)}) + o(1), \end{aligned}$$

which gives

$$I^\infty(u_n^{(1)}) \leq o(1). \tag{1.3.35}$$

So, (1.3.34) and (1.3.35) give $\|u_n^{(1)}\| \to 0$ as $n \to +\infty$.

Step 3. By Step 2, it therefore remains to show that $u \not\equiv 0$. We prove this by contradiction. Suppose that $u \equiv 0$. Then (1.3.34) becomes

$$I^\infty(u_n) = I(u_n) + o(1) = c + o(1), \quad (I^\infty)'(u_n) = o(1). \tag{1.3.36}$$

Let $\|u_n\|^2 \to l > 0$. From $(I^\infty)'(u_n) = o(1)$, we have that

$$\|u_n\|^2 = \int_\Omega |(u_n)_+|^{2^*} + o(1) \le S^{-2^*/2}\|u_n\|^{2^*} + o(1), \tag{1.3.37}$$

where S is the best constant in the Sobolev embedding $D^{1,2}(\mathbb{R}^N) \hookrightarrow L^{2^*}(\mathbb{R}^N)$. Letting $n \to +\infty$ in (1.3.37), we obtain

$$l \le S^{-2^*/2} l^{2^*/2},$$

which gives either $l \ge S^{N/2}$ or $l = 0$. But since $l > 0$, we then have

$$c + o(1) = I(u_n) = I^\infty(u_n) + o(1) = \left(\frac{1}{2} - \frac{1}{2^*}\right)\|u_n\|^2 + o(1)$$
$$\ge \frac{1}{N} S^{N/2} + o(1).$$

This will give a desired contradiction once we can prove

$$c < \frac{1}{N} S^{N/2}. \tag{1.3.38}$$

Step 4. We now verify the inequality (1.3.38).

Let $\bar{U}_{x,\mu}(y)$ be the function defined in (1.2.21). Then

$$U_{x,\mu}(y) = S^{\frac{N-2}{4}} \bar{U}_{x,\mu}(y) \tag{1.3.39}$$

satisfies

$$-\Delta u = u^{2^*-1}, \; u \in D^{1,2}(\mathbb{R}^N).$$

It then follows that

$$\int_{\mathbb{R}^N} |\nabla U_{x,\mu}|^2 = \int_{\mathbb{R}^N} U_{x,\mu}^{2^*},$$

and

$$\frac{1}{2}\int_{\mathbb{R}^N} |\nabla U_{x,\mu}|^2 - \frac{1}{2^*}\int_{\mathbb{R}^N} U_{x,\mu}^{2^*} = \frac{1}{N} S^{N/2}.$$

Without loss of generality, we may assume $0 \in \Omega$. Choose a small enough $\delta > 0$ such that $B_{2\delta}(0) \subset \Omega$. Take $\xi \in C_0^\infty(B_{2\delta}(0))$ such that $\xi = 1$ in $B_\delta(0)$, $0 \le \xi \le 1$ and $|\nabla \xi| \le \frac{C}{\delta}$. We define

$$v_\mu = \xi U_{0,\mu} \in H_0^1(\Omega).$$

Then we have

$$c \le \max_{t\ge 0} I(tv_\mu).$$

It therefore remains to prove that for $\mu > 0$ large, the following inequality holds:

$$\max_{t\ge 0} I(tv_\mu) < \frac{1}{N}S^{N/2}.$$

By a similar calculation to that in (1.2.23), we have

$$\int_\Omega |\xi U_{0,\mu}|^{2^*} = \int_{\mathbb{R}^N} U_{0,\mu}^{2^*} + O\Big(\frac{1}{\mu^N}\Big), \tag{1.3.40}$$

while a similar calculation to that in (1.2.24) yields

$$\int_\Omega |\nabla(\xi U_{0,\mu})|^2 = \int_{\mathbb{R}^N} U_{0,\mu}^{2^*} + O\Big(\frac{1}{\mu^{N-2}}\Big). \tag{1.3.41}$$

On the other hand, supposing that $N \ge 4$ and $q \in (2, 2^*)$, or $N = 3$ and $q \in (3, 6)$, we then have

$$\begin{aligned}\int_\Omega |\xi U_{0,\mu}|^q &= \int_\Omega U_{0,\mu}^q + O\Big(\int_{\Omega\setminus B_\delta(0)} U_{0,\mu}^q\Big)\\ &= \frac{1}{\mu^{N-\frac{q(N-2)}{2}}}\int_{\Omega_\mu} U_{0,1}^q + O\Big(\frac{1}{\mu^{q(N-2)/2}}\Big)\\ &= \frac{A_{N,q}}{\mu^{N-\frac{q(N-2)}{2}}} + O\Big(\frac{1}{\mu^{q(N-2)/2}}\Big),\end{aligned} \tag{1.3.42}$$

where $\Omega_\mu = \{y : \mu^{-1}y \in \Omega\}$, $A_{N,q} = \int_{\mathbb{R}^N} U_{0,1}^q > 0$.

Let t_μ be such that $I(t_\mu v_\mu) = \max_{t\ge 0} I(tv_\mu)$. Then it is easy to check that $t_\mu \in (a_1, a_2)$ for some constants $a_2 > a_1 > 0$, and t_μ satisfies

$$t_\mu \int_\Omega |\nabla(\xi U_{0,\mu})|^2 = t_\mu^{2^*-1}\int_{\mathbb{R}^N} |\xi U_{0,\mu}|^{2^*} + \lambda t_\mu^{q-1}\int_{\mathbb{R}^N} |\xi U_{0,\mu}|^q. \tag{1.3.43}$$

Combining (1.3.40), (1.3.41), (1.3.42) and (1.3.43), one has

$$t_\mu(t_\mu^{2^*-2} - 1)\int_{\mathbb{R}^N} U_{0,\mu}^{2^*} = -\lambda t_\mu^{q-1}\int_{\mathbb{R}^N} |\xi U_{0,\mu}|^q + O\Big(\frac{1}{\mu^{N-2}} + \frac{1}{\mu^{q(N-2)/2}}\Big),$$

which implies that

$$t_\mu = 1 - \frac{\int_{\mathbb{R}^N} |U_{0,1}|^q}{(2^* - 2)\int_{\mathbb{R}^N} U_{0,1}^{2^*}} \frac{\lambda}{\mu^{N - \frac{q(N-2)}{2}}} + O\Big(\frac{1}{\mu^{N - \frac{q(N-2)}{2} + 1}} + \frac{1}{\mu^{N-2}} + \frac{1}{\mu^{q(N-2)/2}}\Big).$$

By appealing to (1.3.43) again, we see that

$$\begin{aligned} I(t_\mu v_\mu) &= \frac{1}{N}\int_{\mathbb{R}^N} |t_\mu \xi U_{0,\mu}|^{2^*} + \lambda\Big(\frac{1}{2} - \frac{1}{q}\Big)\int_{\mathbb{R}^N} |t_\mu \xi U_{0,\mu}|^q \\ &= \frac{1}{N}S^{N/2} - \frac{2^*\lambda \int_{\mathbb{R}^N} U_{0,1}^q}{N(2^*-2)\mu^{N - \frac{q(N-2)}{2}}} + \frac{\lambda\Big(\frac{1}{2} - \frac{1}{q}\Big)\int_{\mathbb{R}^N} U_{0,1}^q}{\mu^{N - \frac{q(N-2)}{2}}} \\ &\quad + O\Big(\frac{1}{\mu^{N - \frac{q(N-2)}{2} + 1}} + \frac{1}{\mu^{N-2}} + \frac{1}{\mu^{q(N-2)/2}}\Big) \\ &= \frac{1}{N}S^{N/2} - \frac{\lambda \int_{\mathbb{R}^N} U_{0,1}^q}{q\mu^{N - \frac{q(N-2)}{2}}} + O\Big(\frac{1}{\mu^{N - \frac{q(N-2)}{2} + 1}} + \frac{1}{\mu^{N-2}} \\ &\quad + \frac{1}{\mu^{q(N-2)/2}}\Big) < \frac{1}{N}S^{N/2}, \end{aligned}$$

which is what we want to show. □

1.4 Global Compactness

In the previous section, we discussed the possible convergence of the $(P.\text{-}S.)_c$ sequence and proved that non-zero weak convergence results in strong convergence if the level c is less than a critical number. From the discussions there, we know that if the level c is bigger than that critical number, the possible loss of compactness is due to the existence of nontrivial solutions for the corresponding limit problem. We call these solutions the *bumps* or the *bubbles*. These solutions carry energy that is larger than a positive constant. If we eliminate these bumps or bubbles, we still get a (P.-S.) sequence at a lower level. See the decomposition in (1.3.25), (1.3.30), or (1.3.36). Since the total energy for a $(P.\text{-}S.)_c$ sequence is bounded from above, the procedure to eliminate the bumps or bubbles must terminate after a finite number of steps.

In this section, we use two typical problems to explain this procedure.

1.4.1 Nonlinear Schrödinger Equations

In this part, we assume that $f(t) \in C(\mathbb{R})$ satisfies the following conditions.

(i) $\lim_{t\to 0} \frac{f(t)}{t} = \lim_{|t|\to+\infty} \frac{f(t)}{|t|^{2^*-1}} = 0.$

(ii) There exists $p > 2$ such that

$$0 < pF(t) \le f(t)t, \quad \forall\, t \in \mathbb{R},$$

where $F(t) = \int_0^t f(s)\,ds$.

(iii) $\frac{f(t)}{t}$ is strictly increasing for $t > 0$ and $\frac{f(t)}{t}$ is strictly decreasing for $t < 0$.

Consider the functional

$$I(u) = \frac{1}{2}\int_{\mathbb{R}^N} \left(|\nabla u|^2 + V(x)u^2\right) - \int_{\mathbb{R}^N} F(u), \quad u \in H^1(\mathbb{R}^N). \tag{1.4.1}$$

Let $\{u_n\}$ be a sequence in $H^1(\mathbb{R}^N)$ that satisfies

$$I(u_n) \to \alpha > 0, \quad I'(u_n) \to 0, \tag{1.4.2}$$

as $n \to +\infty$, where $\alpha > 0$ is a constant. Then, as discussed in the proof of Theorem 1.3.5, the sequence $\{u_n\}$ is bounded in $H^1(\mathbb{R}^N)$, and so we may assume that there exists $u \in H^1(\mathbb{R}^N)$ such that

$$u_n \rightharpoonup u \quad \text{weakly in } H^1(\mathbb{R}^N),$$

and

$$u_n \to u \quad \text{strongly in } L^q_{loc}(\mathbb{R}^N) \text{ for any } q \in [2, 2^*).$$

It is then easy to check that u satisfies

$$\begin{cases} -\Delta u + V(x)u = f(u), & x \in \mathbb{R}^N, \\ u \in H^1(\mathbb{R}^N). \end{cases} \tag{1.4.3}$$

Suppose that

$$\lim_{|x|\to\infty} V(x) = V_\infty.$$

Then the following problem is the limit problem of (1.4.3)

$$\begin{cases} -\Delta u + V_\infty u = f(u), & x \in \mathbb{R}^N, \\ u \in H^1(\mathbb{R}^N). \end{cases} \tag{1.4.4}$$

Denote the corresponding limit functional for (1.4.1) as follows

$$I^\infty(u) = \frac{1}{2}\int_{\mathbb{R}^N} \left(|\nabla u|^2 + V_\infty u^2\right) - \int_{\mathbb{R}^N} F(u), \quad u \in H^1(\mathbb{R}^N).$$

We have the following decomposition for a (P.-S.) sequence $\{u_n\} \subset H^1(\mathbb{R}^N)$.

Theorem 1.4.1 (Zhu–Cao [157], 1989) *Suppose that $\{u_n\}$ is a sequence in $H^1(\mathbb{R}^N)$ that satisfies (1.4.2). Then there exist $u \in H^1(\mathbb{R}^N)$, a non-negative integer k, $x_n^{(j)} \in \mathbb{R}^N$, and functions $v_j \in H^1(\mathbb{R}^N)$, $j = 1, \cdots, k$, such that up to a subsequence, still denoted by $\{u_n\}$,*

$$u_n(x) = u(x) + \sum_{j=1}^{k} v_j(x - x_n^{(j)}) + \omega_n,$$

$$I(u_n) = I(u) + \sum_{j=1}^{k} I^\infty(v_j) + o(1).$$

Moreover, u satisfies (1.4.3), v_j satisfies (1.4.4), $|x_n^{(j)}| \to +\infty$, $|x_n^{(j)} - x_n^{(i)}| \to \infty$ for $j \neq i$, and $\|\omega_n\| \to 0$ as $n \to +\infty$.

Proof. Suppose that $u_n \rightharpoonup u$. Fix $q \in (2, 2^*)$. Let $u_n^{(1)} = u_n - u$. Then

$$\|u_n - u\|_{L^q(B_R(0))} \to 0, \quad \text{for any } R > 0. \tag{1.4.5}$$

A similar argument to that in (1.3.30) shows that $u_n^{(1)}$ satisfies

$$I^\infty(u_n^{(1)}) = I(u_n^{(1)}) + o(1) = I(u_n) - I(u) + o(1), \quad (I^\infty)'(u_n^{(1)}) = o(1).$$

In other words, $\{u_n^{(1)}\}$ is a (P.-S.) sequence for $I^\infty(u)$.

Step 1. Suppose that

$$\sup_{x \in \mathbb{R}^N} \int_{B_R(x)} |u_n^{(1)}|^q \to 0, \quad \text{as } n \to +\infty \text{ for any fixed } R > 0. \tag{1.4.6}$$

Then the argument in Step 1 of the proof of Theorem 1.3.10 shows that $u_n^{(1)} \to 0$ strongly in $H^1(\mathbb{R}^N)$ as $n \to +\infty$. In the event of such, we will take $k = 0$.

Step 2. Suppose now that (1.4.6) does not hold. We may then assume that there exist $c_0 > 0$, $R > 0$ and $x_n^{(1)} \in \mathbb{R}^N$ with $|x_n^{(1)}| \to +\infty$ such that

$$\int_{B_R(x_n^{(1)})} |u_n^{(1)}|^q \geq c_0, \quad \text{as} \quad n \to +\infty.$$

Set $\tilde{u}_n^{(1)}(x) = u_n^{(1)}(x + x_n^{(1)})$. Then there exists $v_1 \in H^1(\mathbb{R}^N)$ such that $v_1 \not\equiv 0$ with

$$\tilde{u}_n^{(1)} \rightharpoonup v_1 \quad \text{weakly in } H^1(\mathbb{R}^N),$$

and

$$\tilde{u}_n^{(1)} \to v_1 \quad \text{strongly in } L^q_{loc}(\mathbb{R}^N) \text{ for any } q \in [2, 2^*). \tag{1.4.7}$$

The argument in Step 2 of the proof of Theorem 1.3.10 shows that $u_n^{(2)} := \tilde{u}_n^{(1)} - v_1$ is still a (P.-S.) sequence for $I^\infty(u)$. Consequently, one has the estimates

$$I^\infty(u_n^{(2)}) = I^\infty(u_n^{(1)}) - I^\infty(v_1) + o(1) \quad \text{and} \quad (I^\infty)'(u_n^{(2)}) = o(1).$$

Moreover, v_1 satisfies (1.4.4). Therefore, it follows that

$$I^\infty(v_1) \geq c^\infty > 0,$$

where c^∞ is the mountain pass value corresponding to (1.4.4).

From (1.4.7) and (1.4.5), we have that

$$\|u_n - u - v_1(x - x_n^{(1)})\|_{L^q(B_R(x_n^{(i)}))} \to 0,$$

for any $R > 0$ and $i = 0, 1$, where $x_n^{(0)} = 0$.

Step 3. Continue this procedure till we reach $I^\infty(u_n^{(k+1)}) \leq o(1)$. From

$$I^\infty(u_n^{(k+1)}) \leq o(1) \quad \text{and} \quad (I^\infty)'(u_n^{(k+1)}) = o(1),$$

we can deduce that $\|u_n^{(k+1)}\| \to 0$ as $n \to +\infty$.

Note that for each $l = 1, \cdots, k$, one has

$$\left\| u_n - u - \sum_{t=1}^{l} v_t(x - x_n^{(t)}) \right\|_{L^q(B_R(x_n^{(i)}))} \to 0$$

for any $R > 0$ and $i = 0, 1, \cdots, l$. This shows that for $u_n^{(l+1)}(x) = u_n - u - \sum_{t=1}^{l} v_t(x - x_n^{(t)})$, if

$$\int_{B_R(x_n^{(l+1)})} |u_n^{(l+1)}|^q \geq c_0, \quad \text{as} \quad n \to +\infty,$$

then $|x^{(l+1)} - x_n^{(i)}| > R_1$ for any large $R_1 > 0$. This proves $|x_n^{(i)} - x_n^{(j)}| \to \infty$, whenever $i \neq j$. □

1.4.2 Nonlinear Elliptic Problems with Critical Growth

In this subsection, we will discuss the (P.-S.) sequence for the following functional

$$I(u) = \frac{1}{2}\int_\Omega (|\nabla u|^2 - \lambda u^2) - \frac{1}{2^*}\int_\Omega |u|^{2^*}, \quad u \in H_0^1(\Omega),$$

where $0 < \lambda < \lambda_1$, and Ω is a bounded domain in $\mathbb{R}^N$ with $N \geq 3$.

The limit functional corresponding to $I(u)$ is given by

$$I^\infty(u) = \frac{1}{2}\int_\Omega |\nabla u|^2 - \frac{1}{2^*}\int_\Omega |u|^{2^*}, \quad u \in H_0^1(\Omega).$$

Suppose that $\{u_n\}$ is a sequence in $H_0^1(\Omega)$ satisfying

$$I(u_n) \to \alpha > 0, \quad I'(u_n) \to 0, \quad \text{as} \quad n \to +\infty.$$

Then we may assume that there exists $u \in H_0^1(\Omega)$ such that

$$u_n \rightharpoonup u \text{ weakly in } H_0^1(\Omega),$$

and

$$u_n \to u \text{ strongly in } L^q(\Omega), \quad \text{for any } q \in [2, 2^*).$$

A similar argument to that in Step 1 of the proof of Theorem 1.3.13 shows that for $u_n^{(1)} = u_n - u$, we have

$$I^\infty(u_n^{(1)}) = I(u_n) - I(u) + o(1), \quad (I^\infty)'(u_n^{(1)}) = o(1).$$

Thus, $\{u_n^{(1)}\}$ is a (P.-S.) sequence for $I^\infty(u)$.

It is also easy to show that u is a weak solution of

$$-\Delta u = |u|^{2^*-2}u + \lambda u, \quad u \in H_0^1(\Omega). \tag{1.4.8}$$

To obtain the decomposition for the sequence $\{u_n\}$, we only need to discuss the (P.-S.) sequence $\{u_n^{(1)}\}$ for $I^\infty(u)$ such that

$$u_n^{(1)} \rightharpoonup 0 \quad \text{weakly in } H_0^1(\Omega).$$

If $u_n^{(1)}$ does not converge strongly in $H_0^1(\Omega)$, $u_n^{(1)}$ will concentrate at some points as shown by the following concentration lemma.

Lemma 1.4.2 (P. L. Lions [105, 106], 1985) *Let $\{w_n\}$ be bounded in $H_0^1(\Omega)$ and suppose that $w_n \rightharpoonup w$ weakly in $H_0^1(\Omega)$. Then*

$$|w_n|^{2^*} \rightharpoonup \nu = |w|^{2^*} + \sum_{j\in J} \nu_j \delta_{P_j},$$

$$|\nabla w_n|^2 \rightharpoonup \mu \geq |\nabla w|^2 + \sum_{j\in J} \mu_j \delta_{P_j},$$

and

$$\mu_j \geq S\nu_j^{\frac{2}{2^*}}, \tag{1.4.9}$$

where the indexing set J is at most countable, δ_{P_j} is the Dirac function at $P_j \in \overline{\Omega}$ and $P_i \neq P_j$ for $i \neq j$.

Lemma 1.4.2 applies to any bounded sequence in $H_0^1(\Omega)$. We now apply this lemma to the (P.-S.) sequence $\{u_n^{(1)}\}$ for $I^\infty(u)$. In fact, since $\{u_n^{(1)}\} \subset H_0^1(\Omega)$ satisfies $(I^\infty)'(u_n^{(1)}) \to 0$ as $n \to +\infty$, we can show that the set J is finite.

Proposition 1.4.3 *The set J is finite and $\nu_j \geq S^{\frac{N}{2}}$.*

Proof. We have

$$\int_\Omega \nabla u_n^{(1)} \nabla \varphi = \int_\Omega |u_n^{(1)}|^{2^*-2} u_n^{(1)} \varphi + o_n(1)||\varphi||, \quad \forall\, \varphi \in H_0^1(\Omega). \quad (1.4.10)$$

Firstly, we take a small $\sigma > 0$. For any point P_j, let $\xi \in C_0^\infty(B_{2\sigma}(P_j))$ be such that $\xi(x) = 1$ if $x \in B_\sigma(P_j)$, $0 \leq \xi \leq 1$. Inserting $\varphi = \xi u_n^{(1)} \in H_0^1(\Omega)$ into (1.4.10), we see

$$\int_\Omega \nabla u_n^{(1)} \nabla(\xi u_n^{(1)}) = \int_\Omega \xi |u_n^{(1)}|^{2^*} + o_n(1)||\xi u_n^{(1)}|| \to \int_\Omega \xi\, d\nu, \quad (1.4.11)$$

as $n \to +\infty$.

On the other hand, from $u_n^{(1)} \to 0$ strongly in $L^2(\Omega)$, we find

$$\int_\Omega \nabla u_n^{(1)} \nabla(\xi u_n^{(1)}) = \int_\Omega \xi |\nabla u_n^{(1)}|^2 + \int_\Omega u_n^{(1)} \nabla u_n^{(1)} \nabla \xi \to \int_\Omega \xi d\mu \geq \mu_j. \quad (1.4.12)$$

Combining (1.4.11) and (1.4.12) yields

$$\int_\Omega \xi\, d\nu \geq \mu_j. \quad (1.4.13)$$

Letting $\sigma \to 0$ in (1.4.13), we obtain

$$\nu_j \geq \mu_j,$$

which, together with (1.4.9), gives

$$\nu_j \geq \mu_j \geq S \nu_j^{\frac{N-2}{N}}.$$

This in turn implies that $\nu_j \geq S^{\frac{N}{2}}$. Since $\int_\Omega |u_n^{(1)}|^{2^*} < +\infty$, the set J must therefore be finite. □

By Lemma 1.4.2, if $\{u_n^{(1)}\}$ does not converge strongly in $H_0^1(\Omega)$, there is a bubble near P_j. One should take note here that there may be several bubbles near the same points P_j. The questions we need to answer now are the following:

(1) How does one find a bubble?
(2) What does a bubble look like?

Intuitively, a typical bubble should be close to U_{x_n,μ_n}, which is defined in (1.3.39) for some $x_n \to P_j$ and $\mu_n \to +\infty$. Take a small constant $\tau > 0$. If $\int_{B_{r_n}(P_j)} U_{x_n,\mu_n}^{2^*} = \tau$, then x_n must be close to P_j since $\tau > 0$. Moreover, r_n must be small since $\tau > 0$ is small. Hence we may proceed as follows to determine the location of the bubble.

Suppose that

$$\int_{\Omega} |u_n^{(1)}|^{2^*} \ge C_0 > 0.$$

We set $u_n^{(1)}(x) = 0$ if $x \notin \bar{\Omega}$. For $r > 0$, we define

$$Q_n(r) := \sup_{x \in \bar{\Omega}} \int_{B_r(x)} |u_n^{(1)}|^{2^*}.$$

Then $Q_n(0) = 0$ and there exists a large constant $r > 0$ such that

$$\int_{B_r(x)} |u_n^{(1)}|^{2^*} = \int_{\Omega} |u_n^{(1)}|^{2^*} \ge C_0, \quad \forall\, x \in \Omega.$$

Take $\tau \in (0, S^{N/2})$. Since $Q_n(r)$ is continuous and is increasing with respect to r, there exists $r_n^{(1)} > 0$ such that $Q_n(r_n^{(1)}) = \tau$. We then have that

$$\int_{B_{r_n^{(1)}}(x)} |u_n^{(1)}|^{2^*} \le Q_n(r_n^{(1)}) = \tau, \quad \forall\, x \in \Omega, \tag{1.4.14}$$

and that there exists $x_n^{(1)} \in \bar{\Omega}$ with

$$\int_{B_{r_n^{(1)}}(x_n^{(1)})} |u_n^{(1)}|^{2^*} = \tau < S^{\frac{N}{2}}. \tag{1.4.15}$$

We will prove that $x_n^{(1)}$ is the location of a bubble in the sense that if we blow up $u_n^{(1)}$ at $x_n^{(1)}$, the limit of the blow-up sequence is non-zero.

We first prove that $r_n^{(1)} \to 0$ as $n \to +\infty$. Assume that $r_n^{(1)} \ge r_0 > 0$. Then, in view of Proposition 1.4.3, one has

$$\int_{B_{r_n^{(1)}}(P_j)} |u_n^{(1)}|^{2^*} \ge \int_{B_{\frac{1}{2}r_0}(P_j)} |u_n^{(1)}|^{2^*} \ge \nu_j + o(1) \ge S^{\frac{N}{2}} + o(1),$$

which contradicts (1.4.14). Therefore, $r_n^{(1)} \to 0$.

To simplify notation, we omit the superscript in $u_n^{(1)}$, $r_n^{(1)}$ and $x_n^{(1)}$. Define the blow-up sequence by

$$\tilde{u}_n(y) = r_n^{\frac{N-2}{2}} u_n(r_n y + x_n). \tag{1.4.16}$$

Then we have

$$\int_{\Omega_n} \nabla \tilde{u}_n \nabla \phi = \int_{\Omega_n} |\tilde{u}_n|^{2^*-2} \tilde{u}_n \phi + o_n(1)\|\phi\|, \quad \forall\, \phi \in H_0^1(\Omega_n), \tag{1.4.17}$$

where $\Omega_n = \{y \in \mathbb{R}^N : r_n y + x_n \in \Omega\}$,

$$\int_{B_1(0)} |\tilde{u}_n|^{2^*} = \int_{B_{r_n}(x_n)} |u_n|^{2^*} = \tau,$$

and

$$\int_{B_1(x_0)} |\tilde{u}_n|^{2^*} \le \tau, \quad \text{for any } x_0 \in \Omega_n,$$

in view of (1.4.14) and (1.4.15).

On the other hand, one has

$$\int_{\Omega_n} |\nabla \tilde{u}_n|^2 = \int_{\Omega} |\nabla u_n|^2 \le C.$$

Hence we may assume that there exists $v_1 \in D^{1,2}(\mathbb{R}^N)$ such that

$$\tilde{u}_n \to v_1, \quad \text{weakly in } D^{1,2}(\mathbb{R}^N),$$

and

$$\tilde{u}_n \to v_1, \quad \text{strongly in } L^2_{loc}(\mathbb{R}^N).$$

To show that x_n is really a concentration point, we need to prove $v_1 \not\equiv 0$.

Suppose to the contrary that $v_1 \equiv 0$. For any $x_0 \in B_1(0)$, let $\xi \in C_0^\infty(B_1(x_0))$ with $\xi \equiv 1$ for $x \in B_{\frac{1}{2}}(x_0)$ and $0 \le \xi \le 1$. Then we have

$$\int_{B_1(x_0)} \nabla \tilde{u}_n \nabla (\xi^2 \tilde{u}_n) = \int_{B_1(x_0)} \xi^2 |\tilde{u}_n|^{2^*} + o_n(1),$$

which implies that

$$\int_{B_1(x_0)} (\xi^2 |\nabla \tilde{u}_n|^2 + \tilde{u}_n \nabla \tilde{u}_n \nabla \xi^2) = \int_{B_1(x_0)} \xi^2 |\tilde{u}_n|^{2^*} + o_n(1).$$

Since $\tilde{u}_n \to v_1 \equiv 0$ strongly in $L^2(B_1(x_0))$, we obtain

$$\int_{B_1(x_0)} \tilde{u}_n \nabla \tilde{u}_n \nabla \xi^2 = o_n(1), \tag{1.4.18}$$

and

$$\int_{B_1(x_0)} \xi^2 |\nabla \tilde{u}_n|^2 = \int_{B_1(x_0)} |\nabla(\xi \tilde{u}_n)|^2 + o(1). \tag{1.4.19}$$

In conclusion, we have proved that

$$S\Big(\int_{B_1(x_0)} |\xi\tilde{u}_n|^{2^*}\Big)^{2/2^*} \le \int_{B_1(x_0)} |\nabla(\xi\tilde{u}_n)|^2 = \int_{B_1(x_0)} \xi^2|\tilde{u}_n|^{2^*} + o_n(1). \tag{1.4.20}$$

On the other hand, from (1.4.14), we see that

$$\begin{aligned}\int_{B_1(x_0)} \xi^2|\tilde{u}_n|^{2^*} &= \int_{B_1(x_0)} (\xi\tilde{u}_n)^2|\tilde{u}_n|^{2^*-2} \\ &\le \Big(\int_{B_1(x_0)} |\xi\tilde{u}_n|^{2^*}\Big)^{\frac{2}{2^*}} \Big(\int_{B_1(x_0)} |\tilde{u}_n|^{2^*}\Big)^{1-\frac{2}{2^*}} \\ &\le \tau^{1-\frac{2}{2^*}}\Big(\int_{B_1(x_0)} |\xi\tilde{u}_n|^{2^*}\Big)^{\frac{2}{2^*}},\end{aligned} \tag{1.4.21}$$

which, together with (1.4.20), gives $\int_{B_1(x_0)} |\xi\tilde{u}_n|^{2^*} = o_n(1)$ for any $x_0 \in B_1(0)$. This is a contradiction to $\int_{B_1(0)} |\tilde{u}_n|^{2^*} = \tau > 0$. Hence $v_1 \not\equiv 0$. Therefore, we have found a bubble $r_n^{-\frac{N-2}{2}} v_1\big(r_n^{-1}(y - x_n)\big)$.

For the limit of the domain Ω_n as $n \to +\infty$, we distinguish two cases:

(1) $dist(x_n, \partial\Omega)r_n^{-1} \to +\infty$ as $n \to +\infty$. In this case, $\Omega_n \to \mathbb{R}^N$.
(2) $dist(x_n, \partial\Omega)r_n^{-1} \to c_0 \ge 0$ as $n \to +\infty$. In this case, after suitable translation and rotation, $\Omega_n \to \mathbb{R}^N_+$.

We prove that case (2) does not occur. In fact, it is easy to check from (1.4.17) that v_1 satisfies

$$-\Delta v_1 = |v_1|^{2^*-2}v_1, \quad v_1 \in D^{1,2}(\mathbb{R}^N_+).$$

But it follows from the Pohozaev identity (Corollary 6.2.2) that $v_1 \equiv 0$. Thus, case (2) does not occur.

In conclusion, we have proved that the bubble is determined by the following limit problem:

$$-\Delta v_1 = |v_1|^{2^*-2}v_1, \quad v_1 \in D^{1,2}(\mathbb{R}^N). \tag{1.4.22}$$

Note that the bubble $r_n^{-\frac{N-2}{2}} v_1\big(r_n^{-1}(y - x_n)\big)$ does not belong to $H_0^1(\Omega)$. Technically, we need to modify it so that we can eliminate this bubble from the original sequence u_n to obtain a new (P.-S.) sequence for $I^\infty(u)$ in $H_0^1(\Omega)$ at lower level. For this purpose, we take a large $R_n > 0$ such that $R_n \to +\infty$ as $n \to +\infty$, $B_{2R_n}(0) \subset \Omega_n$ and

$$\int_{\Omega_n} |\nabla(v_1 - \xi_n v_1)|^2 \to 0,$$

as $n \to +\infty$, where $\xi_n \in C_0^\infty(B_{2R_n}(0))$, $\xi_n = 1$ in $B_{R_n}(0)$ and $0 \le \xi_n \le 1$. Now we modify the bubble as follows:

$$r_n^{-\frac{N-2}{2}} \xi\big(r_n^{-1}(y - x_n)\big) v_1\big(r_n^{-1}(y - x_n)\big) \in H_0^1(\Omega). \tag{1.4.23}$$

We eliminate the bubble defined in (1.4.23) from u_n:

$$u_n^{(2)} = u_n - r_n^{-\frac{N-2}{2}} \xi\big(r_n^{-1}(y - x_n)\big) v_1\big(r_n^{-1}(y - x_n)\big).$$

Then it is easy to check

$$\begin{aligned} I^\infty(u_n) &= I^\infty\Big(r_n^{-\frac{N-2}{2}} \xi\big(r_n^{-1}(y - x_n)\big) v_1\big(r_n^{-1}(y - x_n)\big)\Big) + I^\infty(u_n^{(2)}) + o_n(1) \\ &= J(v_1) + I^\infty(u_n^{(2)}) + o_n(1), \end{aligned}$$

and

$$(I^\infty)'(u_n^{(2)}) = o_n(1),$$

where

$$J(u) = \frac{1}{2} \int_{\mathbb{R}^N} |\nabla u|^2 - \frac{1}{2^*} \int_{\mathbb{R}^N} |u|^{2^*}.$$

Thus, we obtain a new (P.-S.) sequence for $I^\infty(u)$ at level $I^\infty(u_n) - J(v_1)$. Since $J(v) \ge \frac{1}{N} S^{N/2}$ for any nontrivial solution of (1.4.22), we conclude that this procedure must terminate after a finite number of times.

From the above analysis, we can obtain the following global compactness result:

Theorem 1.4.4 (M. Struwe [135], 1984) *Assume that $\{u_n\}$ is a (P.-S.) sequence of $I(u)$; then there exist $u \in H_0^1(\Omega)$ satisfying (1.4.8), a non-negative integer k, $x_n^{(j)} \in \Omega$, $r_n^{(j)} \to 0$, and nontrivial solution v_j of (1.4.22), $j = 1, \cdots, k$, such that*

$$dist(x_n^{(j)}, \partial\Omega)(r_n^{(j)})^{-1} \to +\infty,$$

$$I(u_n) = I(u) + \sum_{j=1}^{k} J(v_j) + o_n(1),$$

and

$$u_n - u - \sum_{j=1}^{k} (r_n^{(j)})^{-\frac{N-2}{2}} v_j\big((r_n^{(j)})^{-1}(y - x_n^{(j)})\big) \to 0 \text{ in } D^{1,2}(\mathbb{R}^N).$$

Remark 1.4.5 Similar to the proof of Theorem 1.4.1, the above discussions also lead to

$$\frac{r_n^{(j)}}{r_n^{(i)}} + \frac{r_n^{(i)}}{r_n^{(j)}} + \frac{|x_n^{(i)} - x_n^{(j)}|^2}{r_n^{(j)} r_n^{(i)}} \to +\infty, \quad \forall\, j \neq i,$$

as $n \to +\infty$.

In fact, by similar discussions to those in (1.4.18)–(1.4.21), we can prove that for the first bubble, we have

$$\int_{B_{Rr_n^{(1)}}(x_n^{(1)})} |u_n - (r_n^{(1)})^{-\frac{N-2}{2}} \xi\big((r_n^{(1)})^{-1}(y - x_n^{(1)})\big) v_1\big((r_n^{(1)})^{-1}(y - x_n^{(1)})\big)|^{2^*}$$

$$= o_n(1),$$

for any $R > 0$. This implies that for the second bubble,

$$(r_n^{(2)})^{-\frac{N-2}{2}} \xi\big((r_n^{(2)})^{-1}(y - x_n^{(2)})\big) v_2\big((r_n^{(2)})^{-1}(y - x_n^{(2)})\big),$$

if $0 < a_0 \leq \frac{r_n^{(1)}}{r_n^{(2)}} \leq a_1 < +\infty$, then we must have $r_n^{(1)} |x_n^{(1)} - x_n^{(2)}| \to +\infty$ as $n \to +\infty$.

Remark 1.4.6 Since v_j is a nontrivial solution of (1.4.22), it is easy to check that $J(v_j) \geq \frac{1}{N} S^{\frac{N}{2}}$. From (1.4.4), we conclude that $I(u)$ satisfies the (P.-S.)$_c$ condition if $c < \frac{1}{N} S^{\frac{N}{2}}$.

Before we end this section, let us introduce a profile decomposition result for sequence of functions in $H_0^1(\Omega)$ that improves the result in Lemma 1.4.2.

Lemma 1.4.7 *Suppose that $\{w_n\}$ is bounded in $H_0^1(\Omega)$ and that $w_n \rightharpoonup w$ weakly in $H_0^1(\Omega)$. Then there exist at most countable $x_n^{(j)} \in \bar{\Omega}$, $r_n^{(j)} \to 0$, $v_j \not\equiv 0$, $j = 1, 2, \cdots$, such that*

$$\frac{r_n^{(j)}}{r_n^{(i)}} + \frac{r_n^{(i)}}{r_n^{(j)}} + \frac{|x_n^{(i)} - x_n^{(j)}|^2}{r_n^{(j)} r_n^{(i)}} \to +\infty, \quad \forall\, j \neq i,$$

$$(r_n^{(j)})^{\frac{N-2}{2}} w_n\big(r_n^{(j)} y + x_n^{(j)}\big) \rightharpoonup v_j(y), \quad \text{in } D^{1,2}(\mathbb{R}^N), \text{ as } n \to \infty,$$

$$\|w_n\|^2 = \|w\|^2 + \sum_{j=1}^{\infty} \|v_j\|^2_{D^{1,2}(\mathbb{R}^N)} + \|r_n\|^2_{D^{1,2}(\mathbb{R}^N)} + o_n(1),$$

where $r_n = w_n - w - \sum_{j=1}^{\infty} (r_n^{(j)})^{-\frac{N-2}{2}} v_j\big((r_n^{(j)})^{-1}(y - x_n^{(j)})\big)$ *satisfies*

$$\|r_n\|_{L^{2^*}(\mathbb{R}^N)} \to 0, \quad \text{as } n \to +\infty.$$

For further discussion of Lemma 1.4.7, refer to [137, 138] and the references therein.

Note that if we also know that $\|v_j\|^2_{D^{1,2}(\mathbb{R}^N)} \geq a > 0$ for some constant a independent of n, then the number of the bubbles must be finite and $\|r_n\|_{D^{1,2}(\mathbb{R}^N)} \to 0$ as $n \to +\infty$. This is the typical case if $\{w_n\}$ is a (P.-S.)$_c$ sequence (or a sequence of approximate solutions) of some critical elliptic problems (see Theorem 1.4.4). Let us point out that if the nonlinearities in the problems are not so simple, it may be difficult to determine the limit problem for v_j as we did in the proof of Theorem 1.4.4. Finally, we will make use of Lemma 1.4.7 in the study of the asymptotic behavior of the approximate solutions for (1.4.8) in Chapter 5.

1.5 Further Results and Comments

Consider the nonlinear Schrödinger equation

$$\begin{cases} -\Delta u + V(x)u = |u|^{p-1}u, & \text{in } \mathbb{R}^N, \\ u \in H^1(\mathbb{R}^N), \end{cases} \tag{1.5.1}$$

where $p \in (1, 2^* - 1)$, and $V(x) \in C(\mathbb{R}^N)$ satisfies $0 < V_0 \leq V(x) \leq V_1$ for some constants $V_1 > V_0 > 0$ with

$$\lim_{|x| \to +\infty} V(x) = V_\infty > 0.$$

The limit problem for (1.5.1) is given by

$$\begin{cases} -\Delta u + V_\infty u = |u|^{p-1}u, & \text{in } \mathbb{R}^N, \\ u \in H^1(\mathbb{R}^N). \end{cases} \tag{1.5.2}$$

The functionals corresponding to (1.5.1) and (1.5.2) are given by

$$I(u) = \frac{1}{2}\int_{\mathbb{R}^N} \left(|\nabla u|^2 + V(x)u^2\right) - \frac{1}{p+1}\int_{\mathbb{R}^N} |u|^{p+1}, \quad u \in H^1(\mathbb{R}^N),$$

and

$$I^\infty(u) = \frac{1}{2}\int_{\mathbb{R}^N} \left(|\nabla u|^2 + V_\infty u^2\right) - \frac{1}{p+1}\int_{\mathbb{R}^N} |u|^{p+1}, \quad u \in H^1(\mathbb{R}^N),$$

respectively.

Similar to Theorem 1.2.9, we can prove easily that if $V(y) \geq V_\infty$ and $V(y) \not\equiv V_\infty$, then $c = c^\infty$, where c and c^∞ are the mountain pass values of the functionals $I(u)$ and $I^\infty(u)$, respectively. This shows that c is not a critical value of $I(u)$, and that any nontrivial solution u for (1.5.1) must satisfy

$I(u) > c^\infty$. Therefore, we can use Theorem 1.4.1, together with the uniqueness of a positive solution for (1.5.2) (see [92]), to prove that $I(u)$ satisfies the (P.-S.)$_\alpha$ condition whenever $\alpha \in (c^\infty, 2c^\infty)$.

The following result is due to Cao [28].

Theorem 1.5.1 *There exists a constant $\delta > 0$, such that if $V(x)$ satisfies $V_\infty \leq V(x) \leq V_\infty + \delta$, then (1.5.1) has a positive solution u, which satisfies $I(u) \in (c^\infty, 2c^\infty)$.*

Interested readers can also refer to [14] for a similar result.

There are other non-compact elliptic problems that share similar properties as (1.5.1). We will now discuss some of these.

Elliptic problem with critical growth in bounded domains with a small hole

Consider

$$\begin{cases} -\Delta u = u^{2^*-1}, \ u > 0, & \text{in } \Omega, \\ u \in H_0^1(\Omega), \end{cases} \tag{1.5.3}$$

where Ω is a bounded domain in $\mathbb{R}^N$. The functional corresponding to (1.5.3) is

$$I(u) = \frac{1}{2}\int_\Omega |\nabla u|^2 - \frac{1}{2^*}\int_\Omega u_+^{2^*}, \quad u \in H_0^1(\Omega).$$

Let c be the mountain pass value for $I(u)$. It is easy to prove that $c = \frac{1}{N}S^{\frac{N}{2}}$, where S is the best Sobolev constant of the embedding from $D^{1,2}(\mathbb{R}^N)$ to $L^{2^*}(\mathbb{R}^N)$, and that c is not a critical value of $I(u)$. By Theorem 1.4.4, we can prove that $I(u)$ satisfies the (P.-S.)$_\alpha$ condition for $\alpha \in (\frac{1}{N}S^{\frac{N}{2}}, \frac{2}{N}S^{\frac{N}{2}})$.

The following result is proved by Coron [48]:

Theorem 1.5.2 *Suppose that there are constants $R_2 > R_1 > 0$ such that the following two statements hold.*

- $B_{R_2}(0) \setminus B_{R_1}(0) \subset \Omega$.
- $B_{B_1}(0)$ *is not a subset of* Ω.

Then, if $\frac{R_2}{R_1}$ is sufficiently large, (1.5.3) *has a solution u, which satisfies $I(u) \in \left(\frac{1}{N}S^{\frac{N}{2}}, \frac{2}{N}S^{\frac{N}{2}}\right)$.*

The condition on Ω implies that Ω has a small hole. More generally, Bahri and Coron [13] prove that if Ω has nontrivial homology, (1.5.3) still has a solution.

Elliptic problem in exterior domains

We now discuss the following problem

$$\begin{cases} -\Delta u + u = u^p, \ u > 0, & \text{in } \mathbb{R}^N \setminus \Omega, \\ u \in H_0^1(\mathbb{R}^N \setminus \Omega), \end{cases} \tag{1.5.4}$$

where $p \in (1, 2^* - 1)$, and Ω is a bounded domain in $\mathbb{R}^N$. The functional corresponding to (1.5.4) is

$$I(u) = \frac{1}{2} \int_{\mathbb{R}^N \setminus \Omega} \left(|\nabla u|^2 + u^2\right) - \frac{1}{p+1} \int_{\mathbb{R}^N \setminus \Omega} u_+^{p+1}, \quad u \in H_0^1(\mathbb{R}^N \setminus \Omega).$$

Let

$$I^\infty(u) = \frac{1}{2} \int_{\mathbb{R}^N} \left(|\nabla u|^2 + u^2\right) - \frac{1}{p+1} \int_{\mathbb{R}^N} u_+^{p+1}, \quad u \in H_0^1(\mathbb{R}^N).$$

Then $c = c^\infty$, where c and c^∞ are the mountain pass values for $I(u)$ and $I^\infty(u)$, respectively, and c is not a critical value of $I(u)$.

In [20], Benci and Cerami proved the following existence result for (1.5.4).

Theorem 1.5.3 *Suppose that $\Omega \subset B_\delta(0)$. Then, if $\delta > 0$ is sufficiently small,* (1.5.4) *has a solution u, which satisfies $I(u) \in (c^\infty, 2c^\infty)$.*

Elliptic problem in $\mathbb{R}^N$ with critical growth

The last problem we discuss is the following:

$$\begin{cases} -\Delta u + a(x)u = u^{2^*-1}, \ u > 0, & \text{in } \mathbb{R}^N, \\ u \in D^{1,2}(\mathbb{R}^N), \end{cases} \tag{1.5.5}$$

where $a \geq 0$ and $a \in L^{\frac{N}{2}}(\mathbb{R}^N)$. The functional corresponding to (1.5.5) is

$$I(u) = \frac{1}{2} \int_{\mathbb{R}^N} \left(|\nabla u|^2 + a(x)u^2\right) - \frac{1}{2^*} \int_{\mathbb{R}^N} u_+^{2^*}, \quad u \in D^{1,2}(\mathbb{R}^N).$$

It is easy to prove that the mountain pass value c of $I(u)$ satisfies $c = \frac{1}{N} S^{\frac{N}{2}}$, and c is not a critical value of $I(u)$.

In [21], Benci and Cerami prove the following existence result:

Theorem 1.5.4 *There exists a constant $\delta > 0$, such that if $\|a\|_{L^{\frac{N}{2}}(\mathbb{R}^N)} \leq \delta$, then* (1.5.5) *has a solution u, which satisfies $I(u) \in (\frac{1}{N} S^{\frac{N}{2}}, \frac{2}{N} S^{\frac{N}{2}})$.*

2
Perturbation Methods

In Chapter 1, by either studying the corresponding minimization problem or by using the mountain pass lemma, we prove the existence of a solution for the following elliptic problem involving critical Sobolev exponent:

$$\begin{cases} -\Delta u = u^{2^*-1} + \lambda u, \quad u > 0, \quad \text{in } \Omega, \\ u \in H_0^1(\Omega), \end{cases} \tag{2.0.1}$$

where Ω is a bounded domain in $\mathbb{R}^N$, $N \geq 4$ and $\lambda \in (0, \lambda_1)$. In this chapter, we will study the asymptotic behavior of the solutions for (2.0.1) as $\lambda \to 0$, and the effect of the domain on the multiplicity of the solutions.

From Chapter 1, for a star-shaped domain Ω, (2.0.1) has no solution if $\lambda = 0$. From this, we conclude that if u_λ is any solution of (2.0.1) with $\|u_\lambda\| \leq C$, then $u_\lambda \rightharpoonup 0$ weakly in $H_0^1(\Omega)$ as $\lambda \to 0$. Hence it follows that if $\|u_\lambda\| \geq c_0 > 0$, then u_λ cannot be uniformly bounded in Ω. Consequently, u_λ must blow up at some points as $\lambda \to 0$.

Let

$$\mu_\lambda^{\frac{N-2}{2}} := \max_{x \in \Omega} u_\lambda(x) \to +\infty,$$

and let $x_\lambda \in \Omega$ be the maximum point of u_λ. In other words, we have $u_\lambda(x_\lambda) = \max_{x\in\Omega} u_\lambda(x)$. Intuitively, from the zero boundary condition, we expect $x_\lambda \to x_0 \in \Omega$. We define the blow-up sequence (see (1.4.16) in Chapter 1) as

$$\tilde{u}_\lambda(y) = \mu_\lambda^{-\frac{N-2}{2}} u_\lambda\big(\mu_\lambda^{-1} y + x_\lambda\big).$$

Then one has $\max_{y\in\Omega_\lambda} \tilde{u}_\lambda(y) = 1$, where $\Omega_\lambda = \big\{y \in \mathbb{R}^N : \mu_\lambda^{-1} y + x_\lambda \in \Omega\big\}$, and $\tilde{u}_\lambda$ satisfies

$$\begin{cases} -\Delta \tilde{u}_\lambda = \tilde{u}_\lambda^{2^*-1} + \lambda \mu_\lambda^{-2} \tilde{u}_\lambda, \quad \tilde{u}_\lambda > 0, \quad y \in \Omega_\lambda, \\ \tilde{u}_\lambda \in H_0^1(\Omega_\lambda). \end{cases}$$

Since $x_0 \in \Omega$, $\|\tilde{u}_\lambda\| \le C$ and $\tilde{u}_\lambda(0) = \max_{y\in\Omega_\lambda} \tilde{u}_\lambda(y) = 1$, we may apply a combination of the L^p estimate (Theorem 6.4.5), the Sobolev embedding theorem (Theorem 6.3.6) and the Schauder estimate (Theorem 6.4.4) to conclude that as $\lambda \to 0$, $\tilde{u}_\lambda$ converges in $C^2_{loc}(\mathbb{R}^N)$ to u, which satisfies

$$\begin{cases} -\Delta u = u^{2^*-1}, \quad u > 0, \quad \text{in } \mathbb{R}^N, \\ u(0) = \max\limits_{y\in\mathbb{R}^N} u(y) = 1. \end{cases}$$

Thus, $u = U_{0,a}$ for some $a > 0$ satisfying $U_{0,a}(0) = 1$, where $U_{x,\mu}$ is defined as in (1.3.39). As a result, we find

$$u_\lambda(x) \approx U_{x_\lambda, a_0\mu_\lambda}(x), \quad \text{for } x \text{ near } x_\lambda. \tag{2.0.2}$$

We will investigate the following problems.

- Necessary condition for x_0. We want to find the condition that determines the location of x_0.
- Existence. We want to study the existence of solutions that satisfy $u_\lambda(x) \approx U_{x_\lambda, a_0\mu_\lambda}(x)$ in Ω for some $x_\lambda \to x_0$, with x_0 satisfying the above necessary condition. Here we also like to study the effect of the domain on the number of solutions for (2.0.1) if $\lambda > 0$ is small.
- Local uniqueness. We want to prove that for each x_0 satisfying the necessary condition above, there exists exactly one solution u_λ concentrating at x_0.

The first two problems will be studied in this chapter, and the third one will be discussed in Chapter 3.

Another problem we will study is the following nonlinear Schrödinger equations:

$$\begin{cases} -\varepsilon^2\Delta u + V(x)u = u^p, \quad u > 0, \quad \text{in } \mathbb{R}^N, \\ u \in H^1(\mathbb{R}^N), \end{cases} \tag{2.0.3}$$

where $\varepsilon > 0$ is a small parameter, $1 < p < 2^* - 1$, and $V(x)$ is a C^2 function satisfying $V_1 \ge V(x) \ge V_0 > 0$ in $\mathbb{R}^N$.

Let us point out that (2.0.3) may not have solutions at all. In fact, upon multiplying $\frac{\partial u}{\partial x_j}$ to (2.0.3) and integrating, we obtain

$$\int_{\mathbb{R}^N} \frac{\partial V(x)}{\partial x_j} u^2 = 0.$$

Thus, if $\frac{\partial V}{\partial x_j}$ has fixed sign and is not identically zero, then $u \equiv 0$.

We now suppose that (2.0.3) has a solution u_ε. Let $x_\varepsilon \in \mathbb{R}^N$ be a maximum point of u_ε. Then one has $\Delta u_\varepsilon(x_\varepsilon) \le 0$. As a result,

$$V(x_\varepsilon)u_\varepsilon(x_\varepsilon) \le u_\varepsilon^p(x_\varepsilon),$$

which implies

$$u_\varepsilon(x_\varepsilon) \ge \big(V(x_\varepsilon)\big)^{\frac{1}{p-1}} \ge V_0^{\frac{1}{p-1}}.$$

We define

$$\tilde{u}_\varepsilon(y) = u_\varepsilon\big(\varepsilon y + x_\varepsilon\big).$$

Then $\max_{y\in\mathbb{R}^N} \tilde{u}_\varepsilon(y) \ge V_0^{\frac{1}{p-1}}$ and $\tilde{u}_\varepsilon$ satisfies

$$\begin{cases} -\Delta\tilde{u}_\varepsilon + V\big(\varepsilon y + x_\varepsilon\big)\tilde{u}_\varepsilon = \tilde{u}_\varepsilon^p, \quad \tilde{u}_\varepsilon > 0, \text{ in } \mathbb{R}^N, \\ \tilde{u}_\varepsilon \in H^1(\mathbb{R}^N). \end{cases}$$

Suppose that $\tilde{u}_\varepsilon$ is bounded in $H^1(\mathbb{R}^N)$, which in turn is equivalent to saying that

$$\int_{\mathbb{R}^N} \big(\varepsilon^2|\nabla u_\varepsilon|^2 + u_\varepsilon^2\big) \le C\varepsilon^N,$$

for some constant $C > 0$. Assume that $x_\varepsilon \to x_0 \in \mathbb{R}^N$ as $\varepsilon \to 0$. Since $\tilde{u}_\varepsilon$ is bounded in $H^1(\mathbb{R}^N)$, we may apply the Moser iteration (Lemma 6.4.6) to conclude that $\tilde{u}_\varepsilon$ is bounded in $L^\infty(\mathbb{R}^N)$. Thus, as $\varepsilon \to 0$, $\tilde{u}_\varepsilon$ converges to u in $C^2(B_R(0))$ for any $R > 0$, and u satisfies

$$\begin{cases} -\Delta u + V(x_0)u = u^p, \quad u > 0, \text{ in } \mathbb{R}^N, \\ u(0) = \max_{y\in\mathbb{R}^N} u(y) \ge (V_0)^{\frac{1}{p-1}}, \quad u \in H^1(\mathbb{R}^N). \end{cases} \tag{2.0.4}$$

Let w be the solution of

$$\begin{cases} -\Delta w + w = w^p, \quad w > 0, \text{ in } \mathbb{R}^N, \\ w(0) = \max\limits_{y\in\mathbb{R}^N} w(y), \quad w \in H^1(\mathbb{R}^N). \end{cases} \tag{2.0.5}$$

Then the solution u of (2.0.4) is given by

$$u(y) = \big(V(x_0)\big)^{\frac{1}{p-1}} w\big(\sqrt{V(x_0)}\,y\big).$$

Therefore, we see that

$$u_\varepsilon(x) \approx \big(V(x_0)\big)^{\frac{1}{p-1}} w\Big(\frac{\sqrt{V(x_0)}(x - x_\varepsilon)}{\varepsilon}\Big), \quad \text{for } x \text{ near } x_\varepsilon. \tag{2.0.6}$$

Similar to (2.0.1), we will investigate the following problems for (2.0.3):

- Necessary condition for x_0. We want find the condition that determines the location of x_0 for solution satisfying (2.0.6).
- Existence. We want to study the effect of the function $V(x)$ on the number of solutions for (2.0.3) if $\varepsilon > 0$ is small.
- Local uniqueness. We want to prove the uniqueness of solutions u_ε concentrating at x_0, which satisfies the above necessary condition.

We will also leave the third problem for Chapter 3.

We remark that the solution w of (2.0.5) is radially symmetric (see [76]), and

$$\lim_{|y|\to\infty} w(y)e^{|y|}|y|^{(N-1)/2} = C > 0, \qquad \lim_{|y|\to\infty} \frac{w}{w'} = -1.$$

Moreover, the solution w is non-degenerate in the sense that the kernel of the linear operator $-\Delta\omega+\omega-pw^{p-1}\omega$ in $H^1(\mathbb{R}^N)$ is spanned by $\{\frac{\partial w}{\partial y_1}, \cdots, \frac{\partial w}{\partial y_N}\}$ (Theorem 6.6.3).

2.1 Locating the Concentration Points

The main tool to determine the location of the concentration points is the following local Pohozaev identities (Theorem 6.2.1) for solutions of the following elliptic problem:

$$-\Delta u = f(y, u) \quad \text{in } D,$$

where D is a domain in $\mathbb{R}^N$. Here, we do not impose any boundary condition for u on ∂D.

Let $S \subset\subset D$. Then u satisfies

$$-\int_{\partial S} \frac{\partial u}{\partial \nu}\frac{\partial u}{\partial y_i} + \frac{1}{2}\int_{\partial S} \nu_i |\nabla u|^2 = \int_{\partial S} \nu_i F(y, u) - \int_S F_{y_i}(y, u), \tag{2.1.1}$$

where $F(y, t) = \int_0^t f(y, s)ds$ and ν is the unit outward normal of ∂S.

2.1.1 Nonlinear Schrödinger Equations

It follows from (2.1.1) that any solution u_ε of (2.0.3) satisfies

$$\begin{aligned} &-\varepsilon^2 \int_{\partial S} \frac{\partial u_\varepsilon}{\partial \nu}\frac{\partial u_\varepsilon}{\partial y_i} + \frac{1}{2}\varepsilon^2 \int_{\partial S} \nu_i |\nabla u_\varepsilon|^2 \\ =&\frac{1}{p+1}\int_{\partial S} u_\varepsilon^{p+1}\nu_i - \frac{1}{2}\int_{\partial S} V(y)u_\varepsilon^2 \nu_i + \int_S \frac{\partial V}{\partial y_i} u_\varepsilon^2. \end{aligned} \tag{2.1.2}$$

In this subsection, we will consider a solution u_ε of (2.0.3) satisfying (2.0.6). More precisely, our solution u_ε is assumed to satisfy the following conditions.

(i) u_ε has k local maximum points $x_{\varepsilon,j} \in \mathbb{R}^N$ such that $x_{\varepsilon,j} \to x_j \in \mathbb{R}^N$, as $\varepsilon \to 0$, and $x_i \neq x_j$ for $i \neq j$.

(ii) For any given small $\tau > 0$, there exists a large constant $R > 0$, such that

$$|u_\varepsilon(x)| \leq \tau, \quad \forall x \in \mathbb{R}^N \setminus \cup_{j=1}^k B_{R\varepsilon}(x_{\varepsilon,j}).$$

(iii) There exists $C > 0$ such that

$$\|u_\varepsilon\|_\varepsilon^2 := \int_{\mathbb{R}^N} \left(\varepsilon^2 |\nabla u_\varepsilon|^2 + V(y) u_\varepsilon^2\right) \leq C\varepsilon^N.$$

Here, let us make some remarks on the three provided conditions. Set

$$\tilde{u}_\varepsilon(y) = u_\varepsilon(\varepsilon y + x_{\varepsilon,j}). \tag{2.1.3}$$

Then it follows from condition (iii) that

$$\int_{\mathbb{R}^N} \left(|\nabla \tilde{u}_\varepsilon|^2 + \tilde{u}_\varepsilon^2\right) \leq C.$$

Using the Moser iteration argument (Lemma 6.4.6), we see that

$$0 < \tilde{u}_\varepsilon \leq C, \tag{2.1.4}$$

for some constant C independent of ε. By the L^r estimates in the theory of elliptic equations (Theorem 6.4.4), we can find that for any given $r > 2$, there exists a constant $C > 0$ such that $\|\tilde{u}_\varepsilon\|_{W^{2,r}(B_2(x_0))} \leq C$ for any $x_0 \in \mathbb{R}^N$. By the Sobolev embedding theory (Theorem 6.3.6), we see that $\|\tilde{u}_\varepsilon\|_{C^{1,\theta}(B_2(x_0))} \leq C$ for $0 < \theta < 1$. It follows from the Schaulder estimates (Theorem 6.4.4) that $\|\tilde{u}_\varepsilon\|_{C^{2,\theta}(B_2(x_0))} \leq C$. As a result, $\tilde{u}_\varepsilon$ converges in C^2 to U in $B_R(0)$ for any $R > 0$. Since $x_{\varepsilon,j}$ is a local maximum point of u_ε, one has $\tilde{u}_\varepsilon(0) = u_\varepsilon(x_{\varepsilon,j}) \geq V_0^{\frac{1}{p-1}} > 0$. Thus, we have $U > 0$.

Concerning the location of the local maximum points $x_{\varepsilon,j}$ of a solution u_ε of (2.0.3) satisfying (i)–(iii), we will prove the following:

Theorem 2.1.1 *Let $x_{\varepsilon,j}$ be the local maximum points of a solution u_ε of* (2.0.3) *satisfying (i)–(iii). Then $\nabla V(x_j) = 0$, $j = 1, \cdots, k$.*

Let $\delta > 0$ be a small constant such that $\delta < \frac{1}{2}\min_{i\neq j}|x_i - x_j|$. To prove Theorem 2.1.1, we take $S = B_\delta(x_{\varepsilon,j})$ in (2.1.2) to obtain

$$\begin{aligned} & -\varepsilon^2 \int_{\partial B_\delta(x_{\varepsilon,j})} \frac{\partial u_\varepsilon}{\partial \nu}\frac{\partial u_\varepsilon}{\partial y_i} + \frac{1}{2}\varepsilon^2 \int_{\partial B_\delta(x_{\varepsilon,j})} \nu_i |\nabla u_\varepsilon|^2 \\ &= \frac{1}{p+1}\int_{\partial B_\delta(x_{\varepsilon,j})} u_\varepsilon^{p+1}\nu_i - \frac{1}{2}\int_{\partial B_\delta(x_{\varepsilon,j})} V(y)u_\varepsilon^2 \nu_i + \int_{B_\delta(x_{\varepsilon,j})} \frac{\partial V}{\partial y_i} u_\varepsilon^2. \end{aligned} \tag{2.1.5}$$

We will prove

$$\int_{B_\delta(x_{\varepsilon,j})} \frac{\partial V}{\partial y_i} u_\varepsilon^2 = A\varepsilon^N \Big(\frac{\partial V(y)}{\partial y_i}\Big|_{y=x_{\varepsilon,j}} + o(1)\Big),$$

for some constant $A > 0$, and that all the other terms in (2.1.5) are of the order $O(e^{-\frac{\theta}{\varepsilon}})$ for some suitably small $\theta > 0$. For this purpose, we need to estimate u_ε on $\partial B_\delta(x_{\varepsilon,j})$, as well as in $B_\delta(x_{\varepsilon,j})$.

Lemma 2.1.2 *Suppose that u_ε is the solution of (2.0.3) satisfying (i)–(iii). Then for any $\alpha \in \big(0, \min_{x\in\mathbb{R}^N} V(x)\big)$, there is a constant $C > 0$ such that*

$$u_\varepsilon(x) \le C \sum_{j=1}^{k} e^{-\frac{\sqrt{\alpha}|x - x_{\varepsilon,j}|}{\varepsilon}}, \quad \forall\, x \in \mathbb{R}^N. \tag{2.1.6}$$

Moreover, we have

$$|\nabla u_\varepsilon(x)| \le C e^{-\frac{\sqrt{\alpha}\delta}{4\varepsilon}}, \quad \forall x \in \partial B_\delta(x_{\varepsilon,j}),\ j = 1, \cdots, k. \tag{2.1.7}$$

Proof. Set $V_m = \min_{x\in\mathbb{R}^N} V(x)$. Write

$$-\varepsilon^2 \Delta u_\varepsilon + (V(x) - u_\varepsilon^{p-1})u_\varepsilon = 0.$$

By (ii), for any $\alpha \in \big(0, V_m\big)$, there exists $R > 0$ such that

$$V(x) - u_\varepsilon^{p-1} \ge \alpha, \quad x \in \mathbb{R}^N \setminus \cup_{j=1}^k B_{R\varepsilon}(x_{\varepsilon,j}).$$

Therefore, we have

$$-\varepsilon^2 \Delta u_\varepsilon + \alpha u_\varepsilon \le 0 \quad \text{in } \mathbb{R}^N \setminus \cup_{j=1}^k B_{R\varepsilon}(x_{\varepsilon,j}). \tag{2.1.8}$$

Let

$$L_\varepsilon v = -\varepsilon^2 \Delta v + \alpha v. \tag{2.1.9}$$

Then it follows that

$$L_\varepsilon e^{-\frac{\sqrt{\alpha} r}{\varepsilon}} = -\varepsilon^2 \Big(\frac{\alpha}{\varepsilon^2} - \frac{(N-1)}{r}\frac{\sqrt{\alpha}}{\varepsilon} \Big) e^{-\frac{\sqrt{\alpha} r}{\varepsilon}} + \alpha e^{-\frac{\sqrt{\alpha} r}{\varepsilon}} > 0, \tag{2.1.10}$$

where $r = |x - x_{\varepsilon,j}|$. By (2.1.4), one has $u_\varepsilon \le M$ for some $M > 0$. Let

$$w_\varepsilon(x) = Me^{\sqrt{\alpha}R}\sum_{j=1}^{k} e^{-\frac{\sqrt{\alpha}|x-x_{\varepsilon,j}|}{\varepsilon}} - u_\varepsilon(x).$$

Then it follows from (2.1.8) and (2.1.10) that

$$L_\varepsilon w_\varepsilon \ge 0, \quad \text{in } \mathbb{R}^N \setminus \cup_{j=1}^k B_{\varepsilon R}(x_{\varepsilon,j}).$$

But on $\partial B_{\varepsilon R}(x_{\varepsilon,j})$, one has

$$w_\varepsilon \ge M - \max_{x\in\mathbb{R}^N} u_\varepsilon(x) \ge 0.$$

Hence, by the comparison theorem (Theorem 6.4.3), we obtain

$$w_\varepsilon \ge 0, \quad \text{in } \mathbb{R}^N \setminus \cup_{j=1}^k B_{\varepsilon R}(x_{\varepsilon,j}),$$

while in $B_{\varepsilon R}(x_{\varepsilon,j})$, we have the estimate

$$Me^{\sqrt{\alpha}R}\sum_{j=1}^{k} e^{-\frac{\sqrt{\alpha}|x-x_{\varepsilon,j}|}{\varepsilon}} \ge M \ge u_\varepsilon(x).$$

This completes the proof of (2.1.6).

On the other hand, u_ε satisfies

$$-\Delta u_\varepsilon = \frac{1}{\varepsilon^2}(u_\varepsilon^p - V(x)u_\varepsilon).$$

It follows from an application of L^q estimate (Theorem 6.4.5) that whenever $z \in \partial B_\delta(x_{\varepsilon,j})$, one has

$$\begin{aligned}\|u_\varepsilon\|_{W^{2,q}(B_{\frac{1}{4}\delta}(z))} &\le \frac{C}{\varepsilon^2}\|(u_\varepsilon^p - V(x)u_\varepsilon)\|_{L^q(B_{\frac{1}{2}\delta}(z))} + C\|u_\varepsilon\|_{L^q(B_{\frac{1}{2}\delta}(z))}\\ &\le \frac{C}{\varepsilon^2}e^{-\frac{\sqrt{\alpha}\delta}{2\varepsilon}} \le Ce^{-\frac{\sqrt{\alpha}\delta}{4\varepsilon}}.\end{aligned} \tag{2.1.11}$$

Taking $q > N$, (2.1.7) then follows from (2.1.11) and the Sobolev embedding theorem (Theorem 6.3.6). □

Proof of Theorem 2.1.1. It follows from (2.1.7) that (2.1.5) is equivalent to

$$\int_{B_\delta(x_{\varepsilon,j})} \frac{\partial V}{\partial y_i}u_\varepsilon^2 = O\big(e^{-\frac{\sqrt{\alpha}\delta}{4\varepsilon}}\big). \tag{2.1.12}$$

On the other hand, by (2.1.6) and the fact that $\tilde{u}_\varepsilon \to U$ in $C^2_{loc}(\mathbb{R}^N)$, we have

$$\begin{aligned}\int_{B_\delta(x_{\varepsilon,j})} \frac{\partial V}{\partial y_i} u_\varepsilon^2 &= \varepsilon^N \int_{B_{\frac{\delta}{\varepsilon}}(0)} \frac{\partial V(\varepsilon y + x_{\varepsilon,j})}{\partial y_i} \tilde{u}_\varepsilon^2 \\ &= \varepsilon^N \Big(\frac{\partial V(y)}{\partial y_i}\Big|_{y=x_{\varepsilon,j}} \int_{\mathbb{R}^N} U^2 + o(1) \Big).\end{aligned} \tag{2.1.13}$$

Combining (2.1.12) and (2.1.13), we obtain

$$\nabla V(x_{\varepsilon,j}) = o(1),$$

which in turn yields the required conclusion of the theorem. □

2.1.2 Nonlinear Elliptic Problems with Critical Growth

Recall that any solution u_λ of (2.0.1) satisfies

$$-\int_{\partial S} \frac{\partial u_\lambda}{\partial \nu} \frac{\partial u_\lambda}{\partial y_i} + \frac{1}{2} \int_{\partial S} \nu_i |\nabla u_\lambda|^2 = \frac{1}{2^*} \int_{\partial S} u_\lambda^{2^*} \nu_i + \frac{\lambda}{2} \int_{\partial S} u_\lambda^2 \nu_i. \tag{2.1.14}$$

In this subsection, we will consider a solution u_λ of (2.0.1) satisfying (2.0.2). For simplicity, we only consider the case that u_λ has one bubble. More precisely, the solution u_λ is assumed to satisfy the following conditions.

(i) $\|u_\lambda\|_{L^\infty(\Omega)} \to +\infty$ as $\lambda \to 0$.
(ii) There exists a constant $C > 0$, independent of λ, such that

$$|u_\lambda(x)| \le C U_{x_\lambda, \mu_\lambda}(x), \quad \forall x \in \Omega,$$

where $\mu_\lambda^{\frac{N-2}{2}} = \max_{x\in\Omega} u_\lambda(x)$ and $x_\lambda \in \Omega$ is the maximum point of u_λ. That is to say, $u_\lambda(x_\lambda) = \max_{x\in\Omega} u_\lambda(x)$.

It follows from (ii) that

$$\|u_\lambda\|^2 \le C \int_\Omega |U_{x_\lambda,\mu_\lambda}|^{2^*} \le C.$$

Let $G(x, P)$ be the solution of

$$\begin{cases} -\Delta u = \delta_P, & \text{in } \Omega, \\ u = 0, & \text{on } \partial\Omega, \end{cases} \tag{2.1.15}$$

where δ_P is the Dirac measure at $P \in \Omega$. In other words, $G(x, P)$ is the Green's function for $-\Delta$ in Ω with homogeneous Dirichlet boundary condition of the form

$$G(x, P), = \frac{1}{(N-2)\omega_{N-1}} \frac{1}{|x-P|^{N-2}} - H(x, P),$$

where ω_{N-1} is the area of the the unit sphere $\mathbb{S}^{N-1}$ in $\mathbb{R}^N$. Here $H(x,P)$ is the regular part of the Green's function $G(x,P)$, satisfying

$$\begin{cases} \Delta H(x,P) = 0, & \text{in } \Omega, \\ H(x,P) = \dfrac{1}{(N-2)\omega_{N-1}} \dfrac{1}{|x-P|^{N-2}}, & \text{on } \partial\Omega. \end{cases}$$

We define the Robin function $\varphi(x)$ by the formula $\varphi(x) = H(x,x)$.

Concerning the location of the maximum point x_λ of a solution u_λ of (2.0.1) satisfying (i)–(ii), we will prove the following theorem.

Theorem 2.1.3 *Let x_λ be the local maximum point of a solution u_λ of* (2.0.1) *satisfying (i)–(ii). Suppose that $x_\lambda \to x_0 \in \overline{\Omega}$ as $\lambda \to 0$. Then $x_0 \in \Omega$ and $\nabla\varphi(x_0) = 0$.*

Similar to the discussion in the last subsection, we will use the local Pohozaev identities (2.1.14) to prove Theorem 2.1.3. For this purpose, we need to estimate u_λ away from the maximum point x_λ. To begin, we prove the following:

Lemma 2.1.4 *Suppose that u_λ is the solution of* (2.0.1) *satisfying (i)–(ii). Then as $\lambda \to 0$, one has*

$$\mu_\lambda d(x_\lambda, \partial\Omega) \to +\infty. \tag{2.1.16}$$

Proof. Let

$$\tilde{u}_\lambda(y) = \mu_\lambda^{-\frac{N-2}{2}} u_\lambda\big(\mu_\lambda^{-1} y + x_\lambda\big).$$

Then $\tilde{u}_\lambda(0) = \max_{y\in\Omega_\lambda} \tilde{u}_\lambda(y) = 1$, where $\Omega_\lambda = \big\{y : \mu_\lambda^{-1} y + x_\lambda \in \Omega\big\}$, and $\tilde{u}_\lambda$ satisfies

$$\begin{cases} -\Delta\tilde{u}_\lambda = \tilde{u}_\lambda^{2^*-1} + \lambda\mu_\lambda^{-2}\tilde{u}_\lambda, \tilde{u}_\lambda > 0, & \text{in } \Omega_\lambda, \\ \tilde{u}_\lambda \in H_0^1(\Omega_\lambda). \end{cases}$$

Suppose that $\mu_\lambda d(x_\lambda, \partial\Omega) \to a < +\infty$ as $\lambda \to 0$. By the L^p estimate (Theorem 6.4.5), the Sobolev embedding theorem (Theorem 6.3.6) and the Schauder estimate (Theorem 6.4.4), we have that as $\lambda \to 0$, after translation and rotation, $\tilde{u}_\lambda$ converges in $C^2_{loc}(\mathbb{R}^N_+)$ to u, which satisfies

$$\begin{cases} -\Delta u = u^{2^*-1}, \quad u > 0, \quad \text{in } \mathbb{R}^N_+ := \{y \in \mathbb{R}^N : y_N > 0\}, \\ u(0) = \max\limits_{y\in\mathbb{R}^N_+} u(y) = 1, \quad u \in H_0^1(\mathbb{R}^N_+). \end{cases} \tag{2.1.17}$$

But it follows from the Pohozaev identity (Corollary 6.2.2) that (2.1.17) only has the trivial solution $u \equiv 0$. This is a contradiction to $u(0) = 1$. Hence we have proved (2.1.16). □

Next, we estimate $u_\lambda(x)$ for $x \in \Omega \setminus B_{R\mu_\lambda^{-1}}(x_\lambda)$. Intuitively, since $|u_\lambda|$ is small in $\Omega \setminus B_{R\mu_\lambda^{-1}}(x_\lambda)$, the super-linear term $u_\lambda^{2^*-1}$ is negligible. Thus, $-\Delta u_\lambda \approx 0$ in $\Omega \setminus B_{R\mu_\lambda^{-1}}(x_\lambda)$. On the other hand, $\int_\Omega u_\lambda^{2^*-1} = \mu_\lambda^{-\frac{N-2}{2}} \big(\int_{\mathbb{R}^N} U_{0,a}^{2^*-1} + o(1)\big)$. Hence, we expect

$$-\Delta u_\lambda \approx \Big(\mu_\lambda^{-\frac{N-2}{2}} \int_{\mathbb{R}^N} U_{0,a}^{2^*-1}\Big)\delta_{x_\lambda} \quad \text{in } \Omega.$$

This, in turn, yields $u_\lambda(x) \approx \mu_\lambda^{-\frac{N-2}{2}} \int_{\mathbb{R}^N} U_{0,a}^{2^*-1} G(x, x_\lambda)$. In fact, we have the following result:

Lemma 2.1.5 *Suppose that u_λ is a solution of* (2.0.1) *satisfying (i)–(ii). Then for $x \in \Omega \setminus B_{R\mu_\lambda^{-1}}(x_\lambda)$, we have*

$$u_\lambda(x) = A_\lambda \mu_\lambda^{-\frac{N-2}{2}} G(x, x_\lambda) + O\Big(\frac{\lambda}{\mu_\lambda^{\frac{N-2}{2}} d^{N-2}} + \frac{1}{\mu_\lambda^{\frac{N+2}{2}} d^N} + \frac{1}{\mu_\lambda^{\frac{N}{2}} d^{N-1}}\Big), \tag{2.1.18}$$

and

$$\nabla u_\lambda(x) = A_\lambda \mu_\lambda^{-\frac{N-2}{2}} \nabla_x G(x, x_\lambda) + O\Big(\frac{\lambda}{\mu_\lambda^{\frac{N-2}{2}} d^{N-1}} + \frac{1}{\mu_\lambda^{\frac{N+2}{2}} d^{N+1}} + \frac{1}{\mu_\lambda^{\frac{N}{2}} d^N}\Big), \tag{2.1.19}$$

where $A_\lambda = \int_{B_{\frac{1}{2}d\mu_\lambda}(0)} \tilde{u}_\lambda^{2^*-1}(y)\,dy$, $d = |x_\lambda - x|$.

Proof. We have

$$u_\lambda(x) = \int_\Omega G(x, y)\Big(u_\lambda^{2^*-1}(y) + \lambda u_\lambda(y)\Big)\,dy.$$

Set $d = |x_\lambda - x|$. Using (ii), we calculate that

$$\begin{aligned}
&\lambda \int_\Omega G(x, y) u_\lambda(y)\,dy \le C\lambda \int_\Omega G(x, y) U_{x_\lambda,\mu_\lambda}(y)\,dy \\
&\le C\lambda \int_\Omega \frac{1}{|y - x|^{N-2}} \frac{1}{\mu_\lambda^{\frac{N-2}{2}} |y - x_\lambda|^{N-2}}\,dy
\end{aligned}$$

$$
\begin{aligned}
&= C\lambda \int_{B_{\frac{1}{2}d}(x_\lambda)} \frac{1}{|y-x|^{N-2}} \frac{1}{\mu_\lambda^{\frac{N-2}{2}} |y-x_\lambda|^{N-2}}\, dy \\
&\quad + C\lambda \int_{\Omega \backslash B_{\frac{1}{2}d}(x_\lambda)} \frac{1}{|y-x|^{N-2}} \frac{1}{\mu_\lambda^{\frac{N-2}{2}} |y-x_\lambda|^{N-2}}\, dy \\
&\le \frac{C\lambda}{\mu_\lambda^{\frac{N-2}{2}} d^{N-2}} \Big(\int_{B_{\frac{1}{2}d}(x_\lambda)} \frac{1}{|y-x_\lambda|^{N-2}}\, dy + \int_{\Omega \backslash B_{\frac{1}{2}d}(x_\lambda)} \frac{1}{|y-x|^{N-2}}\, dy \Big) \\
&\le \frac{C\lambda}{\mu_\lambda^{\frac{N-2}{2}} d^{N-2}}.
\end{aligned}
$$

On the other hand,

$$
\begin{aligned}
&\int_{\Omega \backslash B_{\frac{1}{2}d}(x_\lambda)} G(x,y) u_\lambda^{2^*-1}(y)\, dy \le C \int_{\Omega \backslash B_{\frac{1}{2}d}(x_\lambda)} G(x,y) U_{x_\lambda,\mu_\lambda}^{2^*-1}(y)\, dy \\
&\le \frac{C}{\mu_\lambda^{\frac{N+2}{2}}} \int_{\Omega \backslash B_{\frac{1}{2}d}(x_\lambda)} \frac{1}{|y-x|^{N-2}} \frac{1}{|y-x_\lambda|^{N+2}}\, dy.
\end{aligned}
$$

But we also have

$$
\begin{aligned}
&\frac{C}{\mu_\lambda^{\frac{N+2}{2}}} \int_{(\Omega \backslash B_{\frac{1}{2}d}(x_\lambda)) \backslash B_{2d}(x)} \frac{1}{|y-x|^{N-2}} \frac{1}{|y-x_\lambda|^{N+2}}\, dy \\
&\le \frac{C}{\mu_\lambda^{\frac{N+2}{2}} d^{N-2}} \int_{\Omega \backslash B_{\frac{1}{2}d}(x_\lambda)} \frac{1}{|y-x_\lambda|^{N+2}}\, dy \le \frac{C}{\mu_\lambda^{\frac{N+2}{2}} d^N}
\end{aligned}
$$

and

$$
\begin{aligned}
&\frac{C}{\mu_\lambda^{\frac{N+2}{2}}} \int_{(\Omega \backslash B_{\frac{1}{2}d}(x_\lambda)) \cap B_{2d}(x)} \frac{1}{|y-x|^{N-2}} \frac{1}{|y-x_\lambda|^{N+2}}\, dy \\
&\le \frac{C}{\mu_\lambda^{\frac{N+2}{2}} d^{N+2}} \int_{B_{2d}(x)} \frac{1}{|y-x|^{N-2}}\, dy \le \frac{C}{\mu_\lambda^{\frac{N+2}{2}} d^N}.
\end{aligned}
$$

Hence we obtain the inequality

$$
\int_{\Omega \backslash B_{\frac{1}{2}d}(x_\lambda)} G(x,y) u_\lambda^{2^*-1}(y)\, dy \le \frac{C}{\mu_\lambda^{\frac{N+2}{2}} d^N}.
$$

Moreover,

$$
\begin{aligned}
&\int_{B_{\frac{1}{2}d}(x_\lambda)} G(x,y) u_\lambda^{2^*-1}(y)\, dy \\
&= \frac{1}{\mu_\lambda^{\frac{N-2}{2}}} \int_{B_{\frac{1}{2}d\mu_\lambda}(0)} G(x, \mu_\lambda^{-1} y + x_\lambda) \tilde{u}_\lambda^{2^*-1}(y)\, dy
\end{aligned}
$$

$$
\begin{aligned}
&= \frac{G(x,x_\lambda)}{\mu_\lambda^{\frac{N-2}{2}}} \int_{B_{\frac{1}{2}d\mu_\lambda}(0)} \tilde{u}_\lambda^{2^*-1} \\
&+ \frac{1}{\mu_\lambda^{\frac{N-2}{2}}} \int_{B_{\frac{1}{2}d\mu_\lambda}(0)} \Big(G(x,\mu_\lambda^{-1}y+x_\lambda) - G(x,x_\lambda)\Big)\tilde{u}_\lambda^{2^*-1}(y)\,dy \\
&= \frac{G(x,x_\lambda)}{\mu_\lambda^{\frac{N-2}{2}}} \int_{B_{\frac{1}{2}d\mu_\lambda}(0)} \tilde{u}_\lambda^{2^*-1} + \frac{1}{\mu_\lambda^{\frac{N-2}{2}}} O\Big(\int_{B_{\frac{1}{2}d\mu_\lambda}(0)} \frac{|y|}{\mu_\lambda d^{N-1}} U_{0,1}^{2^*-1}\Big) \\
&= \frac{G(x,x_\lambda)}{\mu_\lambda^{\frac{N-2}{2}}} \int_{B_{\frac{1}{2}d\mu_\lambda}(0)} \tilde{u}_\lambda^{2^*-1} + O\Big(\frac{1}{\mu_\lambda^{\frac{N}{2}} d^{N-1}}\Big).
\end{aligned}
$$

Therefore, we have proved (2.1.18).

Similarly, from

$$
\nabla u_\lambda(x) = \int_\Omega \nabla_x G(x,y)\Big(u_\lambda^{2^*-1}(y) + \lambda u_\lambda(y)\Big)\,dy,
$$

we can prove (2.1.19). □

Proof of Theorem 2.1.3. Assume that as $\lambda \to 0$, $x_\lambda \to x_0 \in \bar{\Omega}$.

Suppose that $x_0 \in \Omega$. Then we can take $\tau > 0$ small enough so that $B_\tau(x_0) \subset \Omega$. Since

$$
\int_{B_{\frac{1}{2}\tau\mu_\lambda}(0)} \tilde{u}_\lambda^{2^*-1} = A + o(1),
$$

for some constant $A > 0$, it follows from Lemma 2.1.5 that (2.1.14) is equivalent to

$$
\int_{\partial B_\tau(x_\lambda)} \frac{\partial G(y,x_\lambda)}{\partial \nu}\frac{\partial G(y,x_\lambda)}{\partial y_i} - \frac{1}{2}\int_{\partial B_\tau(x_\lambda)} \nu_i |\nabla G(y,x_\lambda)|^2 = O\Big(\lambda + \frac{1}{\mu_\lambda}\Big).
$$

By Proposition 6.2.3, this is the same as saying that

$$
\frac{\partial H(y,x_\lambda)}{\partial y_i}\Big|_{y=x_\lambda} = O\Big(\lambda + \frac{1}{\mu_\lambda}\Big), \quad i = 1, \cdots, N. \tag{2.1.20}
$$

Hence we have $\nabla\varphi(x_0) = 0$.

Suppose, on the contrary, that $x_0 \in \partial\Omega$. Taking $d_\lambda = \frac{1}{2}d(x_\lambda, \partial\Omega)$, it then follows from Lemma 2.1.4 that $d_\lambda\mu_\lambda \to +\infty$ as $\lambda \to 0$. Therefore, (2.1.18) and (2.1.19) hold on $\partial B_{d_\lambda}(x_\lambda)$. Now let $S = B_{d_\lambda}(x_\lambda)$ in (2.1.14). Then (2.1.14) is equivalent to

$$
\begin{aligned}
\int_{\partial B_{d_\lambda}(x_\lambda)} \frac{\partial G(y,x_\lambda)}{\partial \nu}\frac{\partial G(y,x_\lambda)}{\partial y_i} &- \frac{1}{2}\int_{\partial B_{d_\lambda}(x_\lambda)} \nu_i |\nabla G(y,x_\lambda)|^2 \\
&= O\Big(\frac{\lambda}{d_\lambda^{N-1}} + \frac{1}{\mu_\lambda d_\lambda^N}\Big).
\end{aligned}
$$

This, together with Proposition 6.2.3, gives

$$\frac{\partial H(y,x_\lambda)}{\partial y_i}\Big|_{y=x_\lambda} = O\Big(\frac{\lambda}{d_\lambda^{N-1}} + \frac{1}{\mu_\lambda d_\lambda^N}\Big), \quad i=1,\cdots,N.$$

But this is impossible in view of Proposition 6.7.1. Thus, we have completed the proof of the theorem. □

2.2 Peak Solutions for Nonlinear Schrödinger Equations

In Section 2.1, we obtain a necessary condition for the concentration points of the peak solutions for the nonlinear Schrödinger equations and the bubbling solutions for an elliptic equation involving critical Sobolev exponent. We now consider the converse of such problem: the existence of such solutions concentring at the points satisfying the necessary condition.

To present the idea more clearly, let us consider the nonlinear Schrödinger equation (2.0.3). Let x_0 be a critical point of $V(x)$. Now, we want to find a solution of the form

$$u_\varepsilon(x) = U_{\varepsilon,x_\varepsilon} + \omega_\varepsilon, \tag{2.2.1}$$

where $U_{\varepsilon,x_\varepsilon}$ is the solution of

$$\begin{cases} -\varepsilon^2\Delta u + V(x_\varepsilon)u = u^p, \quad u>0, \quad \text{in } \mathbb{R}^N, \\ u\in H^1(\mathbb{R}^N), \;\; u(x_\varepsilon) = \max\limits_{x\in\mathbb{R}^N} u(x), \end{cases} \tag{2.2.2}$$

and as $\varepsilon\to 0$, $x_\varepsilon\to x_0$, ω_ε is small (comparing with $U_{\varepsilon,x_\varepsilon}$) in the sense that $\|\omega_\varepsilon\|_\varepsilon^2 = o(\varepsilon^N)$, where

$$\|\omega_\varepsilon\|_\varepsilon := \Big(\int_{\mathbb{R}^N}\big(\varepsilon^2|\nabla\omega_\varepsilon|^2 + V(x)\omega_\varepsilon^2\big)\Big)^{\frac{1}{2}}$$

is induced by the inner product

$$\langle u,v\rangle_\varepsilon := \int_{\mathbb{R}^N}\big(\varepsilon^2\nabla u\nabla v + V(x)uv\big), \quad \forall\, u,\, v\in H^1(\mathbb{R}^N).$$

By using the scaling on both y and u, we find easily that

$$U_{\varepsilon,x_\varepsilon}(y) = \big(V(x_\varepsilon)\big)^{\frac{1}{p-1}} w\left(\frac{\sqrt{V(x_\varepsilon)}}{\varepsilon}(y-x_\varepsilon)\right),$$

where w is the solution of (2.0.5).

It is easy to verify that ω_ε satisfies the following equation:

$$\begin{cases} L_\varepsilon\omega_\varepsilon = l_\varepsilon + R_\varepsilon(\omega_\varepsilon), & \text{in } \mathbb{R}^N, \\ \omega_\varepsilon \in H^1(\mathbb{R}^N), \end{cases} \tag{2.2.3}$$

where L_ε is a bounded linear operator in $H^1(\mathbb{R}^N)$ given by

$$\langle L_\varepsilon\omega, \psi\rangle_\varepsilon = \int_{\mathbb{R}^N} \left(\varepsilon^2\nabla\omega\nabla\psi + V(x)\omega\psi - pU_{\varepsilon,x_\varepsilon}^{p-1}\omega\psi\right), \quad \forall\, \psi \in H^1(\mathbb{R}^N),$$

$l_\varepsilon \in H^1(\mathbb{R}^N)$ satisfying

$$\langle l_\varepsilon, \psi\rangle_\varepsilon = \int_{\mathbb{R}^N} \big(V(x_\varepsilon) - V(x)\big)U_{\varepsilon,x_\varepsilon}\psi, \quad \forall\, \psi \in H^1(\mathbb{R}^N),$$

and R_ε is a map from $H^1(\mathbb{R}^N)$ to $H^1(\mathbb{R}^N)$ that is defined by

$$\langle R_\varepsilon(\omega), \psi\rangle_\varepsilon = \int_{\mathbb{R}^N} \left((U_{\varepsilon,x_\varepsilon} + \omega)_+^p - U_{\varepsilon,x_\varepsilon}^p - pU_{\varepsilon,x_\varepsilon}^{p-1}\omega\right)\psi, \quad \forall\, \psi \in H^1(\mathbb{R}^N).$$

We can check that

$$\|l_\varepsilon\|_\varepsilon^2 = o\big(\varepsilon^N\big).$$

(See Lemma 2.2.4.)

On the other hand, we can easily verify that

$$R_\varepsilon(\omega) = o(|\omega|), \quad \text{as } \omega \to 0.$$

Since the left-hand side of (2.2.3) is linear in ω, (2.2.3) is a perturbation of the following problem for small ω

$$\begin{cases} L_\varepsilon\omega_\varepsilon = l_\varepsilon, & \text{in } \mathbb{R}^N, \\ \omega_\varepsilon \in H^1(\mathbb{R}^N). \end{cases} \tag{2.2.4}$$

Suppose that L_ε is a bijective map in some space. Then (2.2.4) has a solution $\omega_\varepsilon = L_\varepsilon^{-1}l_\varepsilon$. Moreover, if $\|L_\varepsilon^{-1}\|_\varepsilon \le C$ for some $C > 0$ independent of ε, or equivalently,

$$\|L_\varepsilon\omega\|_\varepsilon \ge \frac{1}{C}\|\omega\|_\varepsilon, \tag{2.2.5}$$

we then can use the contraction mapping theorem in the following small ball

$$\left\{\omega \in H^1(\mathbb{R}^N) : \|\omega\|_\varepsilon \le \varepsilon^{\frac{N+\tau}{2}},\ 0 < \tau < 1\right\}$$

in $H^1(\mathbb{R}^N)$ to prove that (2.2.3) has a solution near 0.

But (2.2.5) should not be true. Otherwise, (2.0.3) would have solutions concentrating at any given point $x_0 \in \mathbb{R}^N$, which is impossible in view of Theorem 2.1.1.

To present the main idea on how to construct a solution for (2.0.3) with the form (2.2.1), let us revisit the implicit function theorem.

2.2.1 Implicit Function Theorem

Suppose that $F_j(x, y)$, $j = 1, \cdots, m$, $x \in \mathbb{R}^n$, $y \in \mathbb{R}^m$, is C^1 and satisfies $F_j(0, 0) = 0$. We want to solve the following equation

$$F_j(x, y_1, \cdots, y_m) = 0, \quad x \in \mathbb{R}^n, \quad j = 1, 2, \cdots, m, \tag{2.2.6}$$

to get $y_j = y_j(x)$, with $x \in B_\delta(0)$ for some small $\delta > 0$, such that $F_j(x, y_1(x), \cdots, y_m(x)) = 0$, and $y_j(x) \to 0$ as $|x| \to 0$.

For this purpose, we linearize $F_j(x, y)$ at $y = 0$ to obtain

$$F_j(x, y) = F_j(x, 0) + \langle \nabla_y F_j(x, 0), y \rangle + G_j(x, y),$$

where

$$G_j(x, y) = O(|y|^2), \quad \nabla_y G_j(x, y) = O(|y|), \quad j = 1, \cdots, m. \tag{2.2.7}$$

This shows that $G_j(x, y)$ is super-linear at $y = 0$. Let

$$A = \left(\frac{\partial F_j(x, 0)}{\partial y_i} \right)_{m \times m}.$$

Then (2.2.6) is equivalent to

$$Ay = -F(x, 0) - G(x, y), \tag{2.2.8}$$

where

$$F(x, 0) = (F_1(x, 0), \cdots, F_m(x, 0))^T,$$

and

$$G(x, y) = (G_1(x, y), \cdots, G_m(x, y))^T.$$

Note that (2.2.8) is similar to (2.2.3). In view of (2.2.7), we can regard (2.2.8) as a perturbation of

$$Ay = -F(x, 0),$$

at $(x, y) = (0, 0)$.

Case (i). Suppose that A^{-1} exists. Then we can rewrite (2.2.8) as

$$y = By := -A^{-1} F(x, 0) - A^{-1} G(x, y). \tag{2.2.9}$$

Note that $F(x,0) \to 0$ as $|x| \to 0$. It then follows from this that for any given small enough $\delta > 0$, we have $|A^{-1}F(x,0)| \le \frac{1}{2}\delta$ if $|x| \le \tau$ for some small $\tau > 0$. Consequently, one has

$$|By| \le \frac{1}{2}\delta + C|y|^2 < \delta, \quad \forall\, y \in B_\delta(0).$$

Moreover, for $y,\ z \in B_\delta(0)$, it follows from (2.2.7) that

$$|By - Bz| = |A^{-1}\big(G(x,y) - G(x,z)\big)| \le \frac{1}{2}|y - z|.$$

Thus, B is a contraction map from $B_\delta(0)$ to $B_\delta(0)$. By the contraction mapping theorem, (2.2.9) has a solution in $B_\delta(0)$ and

$$|y| \le C|F(x,0)| + C|y|^2,$$

which gives $|y| \le C|F(x,0)|$.

Remark 2.2.1 The above argument can be used to prove that the following problem

$$\begin{cases} -\Delta u = f(x) + |u|^{p-1}u, & \text{in } \Omega, \\ u = 0, & \text{on } \partial\Omega, \end{cases}$$

has a solution if $\|f\|_{L^\infty(\Omega)}$ is small, where $p > 1$ and $\Omega \subset \mathbb{R}^N$ is bounded.

Indeed, if we use $(-\Delta)^{-1}g$ to denote the solution u of

$$\begin{cases} -\Delta u = g(x), & \text{in } \Omega, \\ u = 0, & \text{on } \partial\Omega, \end{cases}$$

we can rewrite the problem as

$$u = (-\Delta)^{-1}f + (-\Delta)^{-1}|u|^{p-1}u.$$

The existence of a solution in $\{u : \|u\|_{L^\infty(\Omega)} \le \delta\}$ is now a direct consequence of the contraction mapping theorem.

Case (ii). A is not invertible. Here we only discuss the case that A is symmetric. Denote

$$\begin{aligned} Ker\ A &= \{y \in \mathbb{R}^m\ : Ay = 0\}, \\ Im\ A &= \{b \in \mathbb{R}^m\ : \exists\ y \in \mathbb{R}^m \text{ s.t. } Ay = b\}. \end{aligned}$$

Then it is straightforward to check that $Ker\ A \perp Im\ A$ and $\mathbb{R}^m = Ker A \oplus Im A$. Thus, $Ay = b$ has a solution if and only if $b \perp Ker\ A$. Moreover, if $b \perp Ker\ A$, then $Ay = b$ has a unique solution $y \in (Ker A)^\perp$.

For any $b \in \mathbb{R}^m$, $Ay = b$ may not have solution. So we can consider the following problem

$$Ay = Pb \in (KerA)^{\perp},$$

where P is the projection of $\mathbb{R}^m$ to $(KerA)^{\perp}$. To define this operator P, we suppose

$$KerA = \left\{ \sum_{i=1}^{k} a_i e_i : \ a_i \in \mathbb{R} \right\} = Span\{e_1, ..., e_k\},$$

where $\{e_1, \cdots, e_k\}$ forms a basis for $Ker\ A$. Then for any $b \in \mathbb{R}^m$, we set

$$Pb = b - \sum_{i=1}^{k} a_i e_i,$$

where a_i is chosen such that $\langle Pb, e_j \rangle = 0$ for $j = 1, \cdots, k$. It then follows that

$$\sum_{i=1}^{k} \langle e_i, e_j \rangle a_i = \langle b, e_j \rangle, \quad j = 1, \cdots, k.$$

Thus, for any $b \in \mathbb{R}^m$, we can always find a unique $y \in (KerA)^{\perp}$, and some $a_i \in \mathbb{R}^1$, such that $Ay = b - \sum_{i=1}^{k} a_i e_i$.

For (2.2.9), if A is not invertible, we consider

$$Ay = PF(x, 0) + PG(x, y), \quad y \in (KerA)^{\perp}. \tag{2.2.10}$$

Then we can use the contraction mapping theorem to prove that whenever $|x| \le \delta$, (2.2.10) has a solution in $B_{L\delta}(0)$ for some large $L > 0$. That is to say, there exists $y \in (Ker\ A)^{\perp}$, satisfying

$$Ay = F(x, 0) + G(x, y) - \sum_{i=1}^{k} a_i e_i,$$

for some $a_i \in \mathbb{R}$.

In application, the exact kernel of A is not easy to find. Here, what we actually need is that in $(KerA)^{\perp}$, $A = PA$ is invertible. So, for any subspace $K_\varepsilon = span\{e_{\varepsilon,1}, \cdots, e_{\varepsilon,k}\}$, if $P_\varepsilon A$ is invertible in $(K_\varepsilon)^{\perp}$, where P_ε is the projection of $\mathbb{R}^m$ to $(K_\varepsilon)^{\perp}$, then we can carry out the same argument to prove that there exists $y_\varepsilon \in (K_\varepsilon)^{\perp}$, such that

$$Ay_\varepsilon = F(x, 0) + G(x, y) - \sum_{i=1}^{k} a_{\varepsilon,i} e_{\varepsilon,i},$$

for some $a_{\varepsilon,i} \in \mathbb{R}$. Usually, if K_ε is close to $Ker A$ in certain sense – that is, K_ε is the approximate kernel of A – we can expect that $P_\varepsilon A$ is invertible in $(K_\varepsilon)^\perp$. Such y_ε is called a solution of (2.2.8) modulo an approximate kernel K_ε of A.

2.2.2 Existence of Peak Solutions for Nonlinear Schrödinger Equations

In this subsection, we consider the converse of Theorem 2.1.1. That is, we will construct peak solutions for the following nonlinear Schrödinger equations:

$$\begin{cases} -\varepsilon^2 \Delta u + V(y)u = u_+^p, & \text{in } \mathbb{R}^N, \\ u \in H^1(\mathbb{R}^N), \end{cases} \tag{2.2.11}$$

where $\varepsilon > 0$ is a small parameter, $V(y)$ satisfies $0 < V_0 \le V(y) \le V_1$ and $u_+ = u$ if $u \ge 0$, $u_+ = 0$ if $u < 0$. Note that in view of the maximum principle, any nontrivial solution of (2.2.11) is positive.

We use $U_{\varepsilon,x_\varepsilon}$ to denote the solution of (2.2.2). Then

$$U_{\varepsilon,x_\varepsilon}(y) = \big(V(x_\varepsilon)\big)^{\frac{1}{p-1}} w\left(\frac{\sqrt{V(x_\varepsilon)}}{\varepsilon}(y - x_\varepsilon)\right),$$

where w is the solution of

$$\begin{cases} -\Delta w + w = w^p, \quad w > 0, \quad w \in H^1(\mathbb{R}^N), \\ w(0) = \max\limits_{y \in \mathbb{R}^N} w(y). \end{cases} \tag{2.2.12}$$

Let p_j, $j = 1, \cdots, k$, be the critical points of $V(y)$. We want to construct a solution u_ε of the form

$$u_\varepsilon = \sum_{j=1}^{k} U_{\varepsilon,x_{\varepsilon,j}} + \omega_\varepsilon,$$

where as $\varepsilon \to 0$, $x_{\varepsilon,j} \to p_j$, and $\|\omega_\varepsilon\|_\varepsilon^2 = o(\varepsilon^N)$. Then, ω_ε satisfies the following equation

$$\begin{cases} L_\varepsilon \omega_\varepsilon = l_\varepsilon + R_\varepsilon(\omega_\varepsilon), \quad x \in \mathbb{R}^N, \\ \omega_\varepsilon \in H^1(\mathbb{R}^N), \end{cases} \tag{2.2.13}$$

where L_ε is a bounded linear operator in $H^1(\mathbb{R}^N)$, defined by

$$\langle L_\varepsilon\omega, \psi\rangle_\varepsilon = \int_{\mathbb{R}^N}\Big(\varepsilon^2\nabla\omega\nabla\psi + V(y)\omega\psi - p\Big(\sum_{j=1}^k U_{\varepsilon,x_{\varepsilon,j}}\Big)^{p-1}\omega\psi\Big),$$

$$\forall\, \psi \in H^1(\mathbb{R}^N), \tag{2.2.14}$$

$l_\varepsilon \in H^1(\mathbb{R}^N)$ satisfying

$$\langle l_\varepsilon, \psi\rangle_\varepsilon = \int_{\mathbb{R}^N}\Big(\sum_{j=1}^k (V(x_{\varepsilon,j}) - V(y))U_{\varepsilon,x_{\varepsilon,j}} + \Big(\sum_{j=1}^k U_{\varepsilon,x_{\varepsilon,j}}\Big)^p - \sum_{j=1}^k U^p_{\varepsilon,x_{\varepsilon,j}}\Big)\psi,$$

$$\forall\, \psi \in H^1(\mathbb{R}^N),$$

and $R_\varepsilon(\omega) \in H^1(\mathbb{R}^N)$ satisfying

$$\langle R_\varepsilon(\omega), \psi\rangle_\varepsilon = \int_{\mathbb{R}^N}\Big(\Big(\sum_{j=1}^k U_{\varepsilon,x_{\varepsilon,j}} + \omega\Big)_+^p - \Big(\sum_{j=1}^k U_{\varepsilon,x_{\varepsilon,j}}\Big)^p - p\Big(\sum_{j=1}^k U_{\varepsilon,x_{\varepsilon,j}}\Big)^{p-1}\omega\Big)\psi, \quad \forall\, \psi \in H^1(\mathbb{R}^N).$$

Note that $R_\varepsilon(\omega)$ is super-linear at $\omega = 0$.

To construct a multi-peak solution for (2.2.11), we need the following facts about the solution w of (2.2.12) (see [76], for example).

(i) w is radially symmetric, $w'(r) < 0$ and

$$\lim_{|y|\to\infty} w(y)e^{|y|}|y|^{(N-1)/2} = C > 0, \qquad \lim_{|y|\to\infty}\frac{w}{w'} = -1.$$

(ii) The kernel of $-\Delta\omega + \omega - pw^{p-1}\omega$, $\omega \in H^1(\mathbb{R}^N)$, is given by

$$span\Big\{\frac{\partial w}{\partial y_i},\ i = 1, \cdots, N\Big\}. \tag{2.2.15}$$

See Theorem 6.6.3 in the Appendix. In view of (ii), we call the solution w non-degenerate.

Problem (2.2.13) is not always solvable. The best we can do is to solve it modulo an approximate kernel of L_ε. By the definition of L_ε, we can regard the following linear operator as its 'limit' operator at p_m,

$$L_m\omega = -\Delta\omega + V(p_m)\omega - pU_m^{p-1}\omega,$$

where $U_m(y) = \big(V(p_m)\big)^{\frac{1}{p-1}} w\big(\sqrt{V(p_m)}y\big)$. In view of (2.2.15), we expect the approximate kernel K_ε of L_ε is given by

$$K_\varepsilon = span\Big\{\frac{\partial U_{\varepsilon,x_{\varepsilon,j}}}{\partial y_i}, \quad j = 1, \cdots, k;\ i = 1, \cdots, N\Big\}.$$

We will define

$$E_\varepsilon = K_\varepsilon^\perp = \Big\{\omega : \omega \in H^1(\mathbb{R}^N),\ \Big\langle \omega, \frac{\partial U_{\varepsilon,x_j}}{\partial y_i}\Big\rangle_\varepsilon = 0, \quad j = 1, \cdots, k;$$
$$i = 1, \cdots, N\Big\},$$

where $\langle u, v\rangle_\varepsilon = \int_{\mathbb{R}^N} \big(\varepsilon^2 \nabla u \nabla v + V(y)uv\big)\, dy$.

The procedure to construct a k-peak solution for (2.0.3) consists of two parts.

(i) Finite dimensional reduction. We solve (2.2.13) modulo an approximate kernel K_ε of L_ε. That is, for any given x_j, we prove the existence of $\omega_\varepsilon \in E_\varepsilon$, such that

$$L_\varepsilon \omega_\varepsilon = l_\varepsilon + R_\varepsilon(\omega_\varepsilon) + \sum_{j=1}^{k}\sum_{i=1}^{N} a_{\varepsilon,i,j} \frac{\partial U_{\varepsilon,x_j}}{\partial y_i}, \tag{2.2.16}$$

for some constants $a_{\varepsilon,i,j}$. Note that $a_{\varepsilon,i,j}$ depends on x_j, $j = 1, \cdots, k$.

(ii) Solve the finite dimensional problem. We need to choose x_j, $j=1, \cdots, k$, suitably such that all the constants $a_{\varepsilon,i,j}$ in (2.2.16) are zero.

To carry out the reduction argument, we define the projection Q_ε from $H^1(\mathbb{R}^N)$ to E_ε as follows:

$$Q_\varepsilon u = u - \sum_{j=1}^{k}\sum_{i=1}^{N} b_{\varepsilon,i,j} \frac{\partial U_{\varepsilon,x_j}}{\partial y_i}, \tag{2.2.17}$$

where $b_{\varepsilon,i,j}$ is chosen in such a way that $\big\langle Q_\varepsilon u, \frac{\partial U_{\varepsilon,x_j}}{\partial y_i}\big\rangle_\varepsilon = 0$, $i = 1, \cdots, N$, $j = 1, \cdots, k$. Therefore, every $b_{\varepsilon,i,j}$ is determined by the following equations

$$\sum_{j=1}^{k}\sum_{i=1}^{N} \Big\langle \frac{\partial U_{\varepsilon,x_j}}{\partial y_i}, \frac{\partial U_{\varepsilon,x_m}}{\partial y_\ell}\Big\rangle_\varepsilon b_{\varepsilon,i,j} = \Big\langle u, \frac{\partial U_{\varepsilon,x_m}}{\partial y_\ell}\Big\rangle_\varepsilon, \quad m = 1, \cdots, k,$$
$$\ell = 1, \cdots, N. \tag{2.2.18}$$

We now prove that (2.2.18) is solvable. Firstly, we introduce the following lemma.

Lemma 2.2.2 *For any $\alpha > 0$, $\beta > 0$, and $\ell \neq j$, there exists a constant $\tau > 0$ such that*

$$\int_{\mathbb{R}^N} U^\alpha_{\varepsilon,x_{\varepsilon,\ell}} U^\beta_{\varepsilon,x_{\varepsilon,j}} \leq C\varepsilon^N e^{-\tau|x_{\varepsilon,\ell}-x_{\varepsilon,j}|/\varepsilon}.$$

Proof. Suppose that $\alpha \geq \beta$. Note that $U_{\varepsilon,x_{\varepsilon,\ell}}(x) \leq Ce^{-\frac{\sqrt{V_0}}{\varepsilon}|x-x_{\varepsilon,\ell}|}$. Now, applying Lemma 6.1.3, we obtain

$$\begin{aligned}\int_{\mathbb{R}^N} U^{\alpha}_{\varepsilon,x_{\varepsilon,\ell}} U^{\beta}_{\varepsilon,x_{\varepsilon,j}} &\leq C\int_{\mathbb{R}^N} e^{-\frac{\alpha\sqrt{V_0}}{\varepsilon}|x-x_{\varepsilon,\ell}|} e^{-\frac{\beta\sqrt{V_0}}{\varepsilon}|x-x_{\varepsilon,j}|}\\ &= C\,\varepsilon^N \int_{\mathbb{R}^N} e^{-\alpha\sqrt{V_0}|x|} e^{-\beta\sqrt{V_0}|x-\frac{x_{\varepsilon,j}-x_{\varepsilon,\ell}}{\varepsilon}|}\\ &\leq \begin{cases} C\varepsilon^N e^{-\beta\sqrt{V_0}|\frac{x_{\varepsilon,j}-x_{\varepsilon,\ell}}{\varepsilon}|}, & \text{if } \alpha > \beta,\\ C\varepsilon^N e^{-(\beta-\theta)\sqrt{V_0}|\frac{x_{\varepsilon,j}-x_{\varepsilon,\ell}}{\varepsilon}|}, & \text{if } \alpha = \beta.\end{cases}\end{aligned}$$

Here $\theta > 0$ is any small constant. □

Since

$$-\varepsilon^2\Delta\frac{\partial U_{\varepsilon,x_j}}{\partial y_i} + V(x_j)\frac{\partial U_{\varepsilon,x_j}}{\partial y_i} = pU^{p-1}_{\varepsilon,x_j}\frac{\partial U_{\varepsilon,x_j}}{\partial y_i},$$

one has

$$\left\langle\frac{\partial U_{\varepsilon,x_j}}{\partial y_i}, \psi\right\rangle_\varepsilon = p\int_{\mathbb{R}^N} U^{p-1}_{\varepsilon,x_j}\frac{\partial U_{\varepsilon,x_j}}{\partial y_i}\psi + \int_{\mathbb{R}^N}\Big(V(y) - V(x_j)\Big)\frac{\partial U_{\varepsilon,x_j}}{\partial y_i}\psi. \tag{2.2.19}$$

Using (2.2.19) and the symmetry of U_{ε,x_j}, we find that

$$\left\langle\frac{\partial U_{\varepsilon,x_j}}{\partial y_h}, \frac{\partial U_{\varepsilon,x_j}}{\partial y_i}\right\rangle_\varepsilon = \delta_{hi}\varepsilon^{N-2}\Big(c_j + o(1)\Big), \tag{2.2.20}$$

where $\delta_{hi} = 0$ if $h \neq i$, and $\delta_{ii} = 1$, $c_j > 0$ is a constant. Moreover, it follows from Lemma 2.2.2 that

$$\left\langle\frac{\partial U_{\varepsilon,x_m}}{\partial y_h}, \frac{\partial U_{\varepsilon,x_j}}{\partial y_i}\right\rangle_\varepsilon = O\big(e^{-\frac{1}{2}\sqrt{V_0}|x_m-x_j|/\varepsilon}\big)\varepsilon^{N-2}, \quad j \neq m. \tag{2.2.21}$$

Hence, (2.2.18) is solvable and we have the following estimate:

$$|b_{\varepsilon,i,j}| \leq C\varepsilon^{2-N}\sum_{j,i}\left|\left\langle u, \frac{\partial U_{\varepsilon,x_j}}{\partial y_i}\right\rangle_\varepsilon\right| \leq C\varepsilon^{-\frac{N}{2}+1}\|u\|_\varepsilon. \tag{2.2.22}$$

From (2.2.22), we then have

$$\|Q_\varepsilon u\|_\varepsilon \leq C\|u\|_\varepsilon. \tag{2.2.23}$$

The following proposition plays an essential role in carrying out the reduction argument.

Proposition 2.2.3 *There exist ε_0, $\theta_0 > 0$ and $\rho > 0$, which are independent of x_j, $j = 1, \cdots, k$, such that for any $\varepsilon \in (0, \varepsilon_0]$ and $x_j \in B_{\theta_0}(p_j)$, $Q_\varepsilon L_\varepsilon$ is bijective in E_ε. Moreover, one has*

$$\|Q_\varepsilon L_\varepsilon \omega\|_\varepsilon \geq \rho \|\omega\|_\varepsilon, \quad \forall \omega \in E_\varepsilon. \tag{2.2.24}$$

Proof. To prove (2.2.24), we argue by contradiction. Suppose there exists $\varepsilon_n \to 0$, $x_{\varepsilon_n,j} \to p_j$, $\omega_n \in E_{\varepsilon_n}$ such that

$$\|Q_{\varepsilon_n} L_{\varepsilon_n} \omega_n\|_{\varepsilon_n} \leq \frac{1}{n} \|\omega_n\|_{\varepsilon_n}. \tag{2.2.25}$$

We may assume $\|\omega_n\|^2_{\varepsilon_n} = \varepsilon_n^N$. Using (2.2.25), we know that for any $\phi \in E_{\varepsilon_n}$, one has

$$\begin{aligned}
&\int_{\mathbb{R}^N} \Big(\varepsilon_n^2 \nabla \omega_n \nabla \phi + V(y)\omega_n \phi - p\Big(\sum_{j=1}^k U_{\varepsilon_n, x_{\varepsilon_n,j}}\Big)^{p-1} \omega_n \phi\Big) \\
&= \big\langle L_{\varepsilon_n} \omega_n, \phi \big\rangle_{\varepsilon_n} = \langle Q_{\varepsilon_n} L_{\varepsilon_n} \omega_n, \phi \rangle_{\varepsilon_n} \\
&= o(1) \|\omega_n\|_{\varepsilon_n} \|\phi\|_{\varepsilon_n} = o(\varepsilon_n^{\frac{N}{2}}) \|\phi\|_{\varepsilon_n}.
\end{aligned} \tag{2.2.26}$$

Taking $\phi = \omega_n$ in (2.2.26), we see that

$$\int_{\mathbb{R}^N} \Big(\varepsilon_n^2 |\nabla \omega_n|^2 + V(x)\omega_n^2 - p\Big(\sum_{j=1}^k U_{\varepsilon_n, x_{\varepsilon_n,j}}\Big)^{p-1} \omega_n^2\Big) = o(\varepsilon_n^N). \tag{2.2.27}$$

On the other hand, we can take a large $R > 0$ so that

$$p\Big(\sum_{j=1}^k U_{\varepsilon_n, x_{\varepsilon_n,j}}\Big)^{p-1} \leq \frac{1}{2} V(x), \quad \text{in } \mathbb{R}^N \setminus \cup_{j=1}^k B_{\varepsilon_n R}(x_{\varepsilon_n,j}).$$

Thus,

$$\begin{aligned}
&\int_{\mathbb{R}^N} \Big(\varepsilon_n^2 |\nabla \omega_n|^2 + V(y)\omega_n^2 - p\Big(\sum_{j=1}^k U_{\varepsilon_n, x_{\varepsilon_n,j}}\Big)^{p-1} \omega_n^2\Big) \geq \|\omega_n\|^2_{\varepsilon_n} \\
&- p \int_{\cup_{j=1}^k B_{\varepsilon_n R}(x_{\varepsilon_n,j})} \Big(\sum_{j=1}^k U_{\varepsilon_n, x_{\varepsilon_n,j}}\Big)^{p-1} \omega_n^2 - \frac{1}{2} \int_{\mathbb{R}^N \setminus \cup_{j=1}^k B_{\varepsilon_n R}(x_{\varepsilon_n,j})} V(y)\omega_n^2 \\
&\geq \frac{1}{2} \varepsilon_n^N - p \int_{\cup_{j=1}^k B_{\varepsilon_n R}(x_{\varepsilon_n,j})} \Big(\sum_{j=1}^k U_{\varepsilon_n, x_{\varepsilon_n,j}}\Big)^{p-1} \omega_n^2,
\end{aligned}$$

which, together with (2.2.27), implies

$$\varepsilon_n^N \le C \int_{\cup_{j=1}^k B_{\varepsilon_n R}(x_{\varepsilon_n,j})} \Big(\sum_{j=1}^k U_{\varepsilon_n,x_{\varepsilon_n,j}}\Big)^{p-1} \omega_n^2 \le C \sum_{j=1}^k \int_{B_{\varepsilon_n R}(x_{\varepsilon_n,j})} \omega_n^2. \tag{2.2.28}$$

To obtain a contradiction from (2.2.28), we only need to prove

$$\int_{B_{\varepsilon_n R}(x_{\varepsilon_n,j})} \omega_n^2 = o\big(\varepsilon_n^N\big), \quad j = 1, \cdots, k. \tag{2.2.29}$$

For this purpose, we will discuss the local behaviors of ω_n near each point $x_{\varepsilon_n,m}$. We define

$$\tilde{\omega}_{n,m}(y) = \omega_n(\varepsilon_n y + x_{\varepsilon_n,m}).$$

Then we have

$$\int_{\mathbb{R}^N} \big(|\nabla \tilde{\omega}_{n,m}|^2 + \tilde{\omega}_{n,m}^2\big) \le C.$$

Therefore, we can assume that as $n \to +\infty$,

$$\tilde{\omega}_{n,m} \rightharpoonup \omega, \quad \text{weakly in } H^1(\mathbb{R}^N),$$

and

$$\tilde{\omega}_{n,m} \to \omega, \quad \text{strongly in } L^2_{loc}(\mathbb{R}^N).$$

To prove (2.2.29), we just need to prove $\omega = 0$. For this purpose, we will find the equation for ω. From (2.2.26), we expect that ω satisfies

$$-\Delta\omega + V(p_m)\omega - pU_m^{p-1}\omega = 0, \tag{2.2.30}$$

where

$$U_m(y) = \big(V(p_m)\big)^{\frac{1}{p-1}} w\big(\sqrt{V(p_m)}\,y\big).$$

The major difficulty to prove this claim is that (2.2.26) holds just for $\phi \in E_{\varepsilon_n}$, not for all $\phi \in H^1(\mathbb{R}^N)$. Since the proof of (2.2.30) is quite lengthy, we assume it for the moment.

The non-degeneracy of the solution U_m gives

$$\omega = \sum_{i=1}^N c_i \frac{\partial U_m}{\partial y_i}. \tag{2.2.31}$$

On the other hand, for $\omega_n \in E_{\varepsilon_n}$, we have that

$$
\begin{aligned}
0 &= \Big\langle \omega_n, \frac{\partial U_{\varepsilon_n, x_{\varepsilon_n,m}}}{\partial y_i} \Big\rangle_{\varepsilon_n} \\
&= p\int_{\mathbb{R}^N} U^{p-1}_{\varepsilon_n, x_{\varepsilon_n,m}} \frac{\partial U_{\varepsilon_n, x_{\varepsilon_n,m}}}{\partial y_i}\omega_n - \int_{\mathbb{R}^N} \big(V(x_{\varepsilon_n,m}) - V(y)\big)\frac{\partial U_{\varepsilon_n, x_{\varepsilon_n,m}}}{\partial y_i}\omega_n \\
&= p\int_{\mathbb{R}^N} U^{p-1}_{\varepsilon_n, x_{\varepsilon_n,m}} \frac{\partial U_{\varepsilon_n, x_{\varepsilon_n,m}}}{\partial y_i}\omega_n + \varepsilon_n^N O\big(|\nabla V(x_{\varepsilon_n,m})| + \varepsilon_n\big),
\end{aligned}
$$

which in turn implies that

$$
\int_{\mathbb{R}^N} w^{p-1}\Big(\sqrt{V(x_{\varepsilon_n,m})}y\Big) w'\Big(\sqrt{V(x_{\varepsilon_n,m})}y\Big)\frac{y_i}{|y|}\tilde{\omega}_{n,m} = O(\varepsilon_n), \quad i{=}1,\cdots,N.
$$

Therefore, it follows that

$$
\int_{\mathbb{R}^N} U_m^{p-1}\frac{\partial U_m}{\partial y_i}\omega = 0.
$$

Hence all the constants c_i in (2.2.31) are zero. This shows that $\omega = 0$ and so (2.2.29) follows.

From (2.2.24), we find that $Q_\varepsilon L_\varepsilon$ is injective in E_ε, and $Q_\varepsilon L_\varepsilon E_\varepsilon$ is closed. This implies $Q_\varepsilon L_\varepsilon$ is surjective. In conclusion, $Q_\varepsilon L_\varepsilon$ is bijective in E_ε.

To finish the proof of this proposition, it remains to prove (2.2.30). For $\phi \in H^1(\mathbb{R}^N)$, we take

$$
Q_{\varepsilon_n}\phi = \phi - \sum_{j=1}^{k}\sum_{i=1}^{N} b_{n,i,j}\frac{\partial U_{\varepsilon_n, x_{\varepsilon_n,j}}}{\partial y_i} \in E_{\varepsilon_n}.
$$

Then, solving (2.2.18) (with $u = \phi$), we see that

$$
b_{n,h,m} = \sum_{j=1}^{k}\sum_{i=1}^{N} \alpha_{n,i,j,h,m}\Big\langle \frac{\partial U_{\varepsilon_n, x_{\varepsilon_n,j}}}{\partial y_i}, \phi\Big\rangle_{\varepsilon_n}, \tag{2.2.32}
$$

for some constants $\alpha_{n,i,j,h,m}$.

Let

$$
\tilde{r}_{n,i,j} = \Big\langle L_{\varepsilon_n}\omega_n, \frac{\partial U_{\varepsilon_n, x_{\varepsilon_n,j}}}{\partial y_i} \Big\rangle_{\varepsilon_n}.
$$

We have

$$
\begin{aligned}
&\int_{\mathbb{R}^N} \Big(\varepsilon_n^2 \nabla\omega_n\nabla\phi + V(y)\omega_n\phi - p\Big(\sum_{j=1}^{k} U_{\varepsilon_n, x_{\varepsilon_n,j}}\Big)^{p-1}\omega_n\phi\Big) \\
&= \big\langle L_{\varepsilon_n}\omega_n, \phi\big\rangle_{\varepsilon_n} = \big\langle L_{\varepsilon_n}\omega_n, Q_{\varepsilon_n}\phi\big\rangle_{\varepsilon_n} + \sum_{j=1}^{k}\sum_{i=1}^{N} b_{n,i,j}\tilde{r}_{n,i,j}.
\end{aligned} \tag{2.2.33}
$$

Moreover, by (2.2.23), we obtain

$$\begin{aligned}\langle L_{\varepsilon_n}\omega_n, Q_{\varepsilon_n}\phi\rangle_{\varepsilon_n} &= \langle Q_{\varepsilon_n}L_{\varepsilon_n}\omega_n, Q_{\varepsilon_n}\phi\rangle_{\varepsilon_n}\\ &= o(1)\|\omega_n\|_{\varepsilon_n}\|Q_{\varepsilon_n}\phi\|_{\varepsilon_n} = o(\varepsilon_n^{\frac{N}{2}})\|\phi\|_{\varepsilon_n}.\end{aligned} \tag{2.2.34}$$

Combining (2.2.33) and (2.2.34), and taking (2.2.32) into consideration, we obtain

$$\begin{aligned}&\int_{\mathbb{R}^N}\Big(\varepsilon_n^2\nabla\omega_n\nabla\phi + V(y)\omega_n\phi - p\Big(\sum_{j=1}^k U_{\varepsilon_n,x_{\varepsilon_n,j}}\Big)^{p-1}\omega_n\phi\Big)\\ &= o(\varepsilon_n^{\frac{N}{2}})\|\phi\|_{\varepsilon_n} + \sum_{j=1}^k\sum_{i=1}^N \gamma_{n,i,j}\Big\langle\frac{\partial U_{\varepsilon_n,x_{\varepsilon_n,j}}}{\partial y_i},\phi\Big\rangle_{\varepsilon_n},\end{aligned} \tag{2.2.35}$$

where $\gamma_{n,i,j}$ is some constant.

Now we can estimate $\gamma_{n,i,j}$ by taking $\phi = \frac{\partial U_{\varepsilon_n,x_{\varepsilon_n,m}}}{\partial y_h}$ in (2.2.35) to obtain

$$\begin{aligned}&\sum_{j=1}^k\sum_{i=1}^N \gamma_{n,i,j}\Big\langle\frac{\partial U_{\varepsilon_n,x_{\varepsilon_n,j}}}{\partial y_i},\frac{\partial U_{\varepsilon_n,x_{\varepsilon_n,m}}}{\partial y_h}\Big\rangle_{\varepsilon_n}\\ &= \Big\langle\omega_n,\frac{\partial U_{\varepsilon_n,x_{\varepsilon_n,m}}}{\partial y_h}\Big\rangle_{\varepsilon_n} - p\int_{\mathbb{R}^N}\Big(\sum_{j=1}^k U_{\varepsilon_n,x_{\varepsilon_n,j}}\Big)^{p-1}\omega_n\frac{\partial U_{\varepsilon_n,x_{\varepsilon_n,m}}}{\partial y_h} + o(\varepsilon_n^{N-1})\\ &= -p\int_{\mathbb{R}^N}\Big(\sum_{j=1}^k U_{\varepsilon_n,x_{\varepsilon_n,j}}\Big)^{p-1}\omega_n\frac{\partial U_{\varepsilon_n,x_{\varepsilon_n,m}}}{\partial y_h} + o(\varepsilon_n^{N-1}).\end{aligned} \tag{2.2.36}$$

In the last equality, we use the fact that $\omega_n \in E_{\varepsilon_n}$.

On the other hand, it follows from

$$-\varepsilon^2\Delta U_{\varepsilon,x_j} + V(x_j)U_{\varepsilon,x_j} = U_{\varepsilon,x_j}^p$$

that we have

$$-\varepsilon^2\Delta\frac{\partial U_{\varepsilon,x_j}}{\partial y_i} + V(x_j)\frac{\partial U_{\varepsilon,x_j}}{\partial y_i} = pU_{\varepsilon,x_j}^{p-1}\frac{\partial U_{\varepsilon,x_j}}{\partial y_i}.$$

Therefore, for $\omega_n \in E_{\varepsilon_n}$, we have

$$\begin{aligned}&p\int_{\mathbb{R}^N} U_{\varepsilon_n,x_{\varepsilon_n,j}}^{p-1}\frac{\partial U_{\varepsilon_n,x_{\varepsilon_n,j}}}{\partial y_i}\omega_n = \int_{\mathbb{R}^N}\big(V(x_{\varepsilon_n,j}) - V(y)\big)\frac{\partial U_{\varepsilon_n,x_{\varepsilon_n,j}}}{\partial y_i}\omega_n\\ &= -\int_{\mathbb{R}^N}\Big(\langle\nabla V(x_{\varepsilon_n,j}), y - x_{\varepsilon_n,j}\rangle + O(|y - x_{\varepsilon_n,j}|^2)\Big)\frac{\partial U_{\varepsilon_n,x_{\varepsilon_n,j}}}{\partial y_i}\omega_n\\ &= \varepsilon_n^N O\big(|\nabla V(x_{\varepsilon_n,j})| + \varepsilon_n\big).\end{aligned}$$

Using Lemma 6.1.1, we see that there is a constant $\sigma > 0$ such that

$$\Big(\sum_{j=1}^{k} U_{\varepsilon_n,x_{\varepsilon_n,j}}\Big)^{p-1} - U^{p-1}_{\varepsilon_n,x_{\varepsilon_n,m}} = O\Big(\sum_{j\neq m} U^{\sigma}_{\varepsilon_n,x_{\varepsilon_n,j}}\Big).$$

So, by Lemma 2.2.2, there is a $\tau > 0$ such that

$$\begin{aligned}
&p\int_{\mathbb{R}^N}\Big(\sum_{j=1}^{k} U_{\varepsilon_n,x_{\varepsilon_n,j}}\Big)^{p-1}\omega_n\frac{\partial U_{\varepsilon_n,x_{\varepsilon_n,m}}}{\partial y_h}\\
&= p\int_{\mathbb{R}^N}\Big(\Big(\sum_{j=1}^{k} U_{\varepsilon_n,x_{\varepsilon_n,j}}\Big)^{p-1} - U^{p-1}_{\varepsilon_n,x_{\varepsilon_n,m}}\Big)\omega_n\frac{\partial U_{\varepsilon_n,x_{\varepsilon_n,m}}}{\partial y_h}\\
&\quad + \varepsilon_n^N O\Big(\sum_{j=1}^{k}|\nabla V(x_{\varepsilon_n,j})| + \varepsilon_n\Big)\\
&= O\big(e^{-\frac{\tau}{\varepsilon_n}}\big)\|\omega_n\|_{\varepsilon_n} + \varepsilon_n^N O\Big(\sum_{j=1}^{k}|\nabla V(x_{\varepsilon_n,j})| + \varepsilon_n\Big) = o(\varepsilon_n^N).
\end{aligned} \tag{2.2.37}$$

Combining (2.2.36) and (2.2.37), we obtain

$$\sum_{j=1}^{k}\sum_{i=1}^{N}\gamma_{n,i,j}\Big\langle\frac{\partial U_{\varepsilon_n,x_{\varepsilon_n,j}}}{\partial y_i}, \frac{\partial U_{\varepsilon_n,x_{\varepsilon_n,m}}}{\partial y_h}\Big\rangle_{\varepsilon_n} = o(\varepsilon_n^{N-1}),$$

which, together with (2.2.20) and (2.2.21), gives

$$\gamma_{n,i,j} = o(\varepsilon_n).$$

Consequently, (2.2.35) becomes

$$\begin{aligned}
&\int_{\mathbb{R}^N}\Big(\varepsilon_n^2\nabla\omega_n\nabla\phi + V(y)\omega_n\phi - p\Big(\sum_{j=1}^{k} U_{\varepsilon_n,x_{\varepsilon_n,j}}\Big)^{p-1}\omega_n\phi\Big)\\
&= o(\varepsilon_n^{\frac{N}{2}})\|\phi\|_{\varepsilon_n}, \quad \forall\, \phi \in H^1(\mathbb{R}^N).
\end{aligned} \tag{2.2.38}$$

Now, for any $\phi \in H^1(\mathbb{R}^N)$, we let $\hat{\phi}_n(y) = \phi\big(\frac{y-x_{\varepsilon_n,m}}{\varepsilon_n}\big)$. Using (2.2.38), we obtain

$$\begin{aligned}
&\int_{\mathbb{R}^N}\Big(\nabla\tilde{\omega}_{n,m}\nabla\phi + V(\varepsilon_n y + x_{\varepsilon_n,m})\tilde{\omega}_{n,m}\phi\\
&\quad - p\Big(\sum_{j=1}^{k} U_{\varepsilon_n,x_{\varepsilon_n,j}}(\varepsilon_n y + x_{\varepsilon_n,m})\Big)^{p-1}\tilde{\omega}_{n,m}\phi\Big)
\end{aligned}$$

$$= \varepsilon_n^{-N} \int_{\mathbb{R}^N} \Big(\varepsilon_n^2 \nabla \omega_n \nabla \hat{\phi}_n + V(y) \omega_n \hat{\phi}_n - p\Big(\sum_{j=1}^{k} U_{\varepsilon_n, x_{\varepsilon_n,j}}\Big)^{p-1} \omega_n \hat{\phi}_n \Big)$$

$$= o(\varepsilon_n^{-\frac{N}{2}}) \|\hat{\phi}_n\|_{\varepsilon_n} = o(1). \tag{2.2.39}$$

Letting $n \to +\infty$ in (2.2.39), we see that ω satisfies

$$-\Delta \omega + V(p_m)\omega - pU_m^{p-1}\omega = 0.$$

Hence, the result follows. □

We are now ready to carry out the reduction for (2.2.13). That is, we consider the following problem

$$Q_\varepsilon L_\varepsilon \omega = Q_\varepsilon l_\varepsilon + Q_\varepsilon R_\varepsilon \omega, \quad \omega \in E_\varepsilon. \tag{2.2.40}$$

Note that this problem is a perturbation of the following problem:

$$\begin{cases} Q_\varepsilon L_\varepsilon \omega_\varepsilon = Q_\varepsilon l_\varepsilon, & x \in \mathbb{R}^N, \\ \omega_\varepsilon \in H^1(\mathbb{R}^N). \end{cases}$$

Similar to the discussion for the implicit function theorem, we can expect from Proposition 2.2.3 that (2.2.40) has a solution ω_ε satisfying $\|\omega_\varepsilon\|_\varepsilon \le C\|l_\varepsilon\|_\varepsilon$. In order to apply the contraction mapping theorem to carry out the reduction, we need to now estimate $\|l_\varepsilon\|_\varepsilon$ and $\|R_\varepsilon(\omega)\|_\varepsilon$.

Lemma 2.2.4 *We have*

$$\|l_\varepsilon\|_\varepsilon \le C\varepsilon^{\frac{N}{2}} \Big(\varepsilon \Big| \sum_{j=1}^{k} \nabla V(x_j) \Big| + \varepsilon^2 \Big).$$

Proof. Recall that

$$\langle l_\varepsilon, \eta \rangle_\varepsilon = \int_{\mathbb{R}^N} \Big(\sum_{j=1}^{k} (V(x_j) - V(y)) U_{\varepsilon, x_j} + \Big(\sum_{j=1}^{k} U_{\varepsilon, x_j} \Big)^p - \sum_{j=1}^{k} U_{\varepsilon, x_j}^p \Big) \eta.$$

One has

$$\begin{aligned} & \Big| \int_{\mathbb{R}^N} (V(y) - V(x_j)) U_{\varepsilon, x_j} \eta \Big| \\ & \le C \Big(\int_{\mathbb{R}^N} (V(y) - V(x_j))^2 U_{\varepsilon, x_j}^2 \Big)^{\frac{1}{2}} \|\eta\|_\varepsilon \\ & = C \Big(\varepsilon^N \int_{\mathbb{R}^N} (V(\varepsilon y + x_j) - V(x_j))^2 U_{\varepsilon, x_j}^2 (\varepsilon y + x_j) \Big)^{\frac{1}{2}} \|\eta\|_\varepsilon \end{aligned}$$

$$
\begin{aligned}
&\leq C\varepsilon^{\frac{N}{2}}\Big(\int_{\mathbb{R}^N}\big(\varepsilon|\nabla V(x_j)||y|+\varepsilon^2|y|^2\big)^2 U_{\varepsilon,x_j}^2(\varepsilon y+x_j)\Big)^{\frac{1}{2}}\|\eta\|_\varepsilon \\
&\leq C\varepsilon^{\frac{N}{2}}\big(\varepsilon|\nabla V(x_j)|+\varepsilon^2\big)\|\eta\|_\varepsilon.
\end{aligned}
\tag{2.2.41}
$$

On the other hand, we have

$$
|U_{\varepsilon,x_j}|\leq Ce^{-\frac{\sqrt{\alpha}|x-x_j|}{\varepsilon}}
$$

for any $\alpha\in(0,\min_{x\in\mathbb{R}^N}V(x))$. Using Lemma 6.1.1, we can deduce that if $p\in(1,2]$, then

$$
\begin{aligned}
\Big(\sum_{j=1}^k U_{\varepsilon,x_j}\Big)^p-\sum_{j=1}^k U_{\varepsilon,x_j}^p &= p\sum_{j\neq i}U_{\varepsilon,x_i}^{p-1}U_{\varepsilon,x_j}+O\Big(\sum_{j\neq i}U_{\varepsilon,x_j}^p\Big)\\
&=O\Big(\sum_{j\neq i}e^{-\frac{\sqrt{\alpha}(p-1)|x-x_j|}{\varepsilon}}\Big),\quad \forall\, x\in B_\theta(x_i),
\end{aligned}
\tag{2.2.42}
$$

where $\theta>0$ is a small constant, while for $p>2$, we may apply (6.1.1) to obtain

$$
\begin{aligned}
\Big(\sum_{j=1}^k U_{\varepsilon,x_j}\Big)^p-\sum_{j=1}^k U_{\varepsilon,x_j}^p &= p\sum_{j\neq i}U_{\varepsilon,x_i}^{p-1}U_{\varepsilon,x_j}+O\Big(\sum_{j\neq i}U_{\varepsilon,x_i}^{p-2}U_{\varepsilon,x_j}^2+\sum_{j\neq i}U_{\varepsilon,x_j}^p\Big)\\
&=O\Big(\sum_{j\neq i}e^{-\frac{\sqrt{\alpha}|x-x_j|}{\varepsilon}}\Big),\quad \forall\, x\in B_\theta(x_i).
\end{aligned}
\tag{2.2.43}
$$

From (2.2.42) and (2.2.43), we have

$$
\begin{aligned}
&\Big|\int_{\mathbb{R}^N}\Big(\Big(\sum_{j=1}^k U_{\varepsilon,x_j}\Big)^p-\sum_{j=1}^k U_{\varepsilon,x_j}^p\Big)\eta\Big|\\
&=\Big|\int_{\bigcup_{\ell=1}^k B_\theta(x_\ell)}\Big(\Big(\sum_{j=1}^k U_{\varepsilon,x_j}\Big)^p-\sum_{j=1}^k U_{\varepsilon,x_j}^p\Big)\eta\Big|\\
&\quad+\Big|\int_{\mathbb{R}^N\setminus\bigcup_{\ell=1}^k B_\theta(x_\ell)}\Big(\Big(\sum_{j=1}^k U_{\varepsilon,x_j}\Big)^p-\sum_{j=1}^k U_{\varepsilon,x_j}^p\Big)\eta\Big|\\
&\leq C\varepsilon^{\frac{N}{2}}e^{-\frac{\tau}{\varepsilon}}\|\eta\|_\varepsilon
\end{aligned}
\tag{2.2.44}
$$

for some small $\tau>0$. The conclusion of the lemma now follows from (2.2.41) and (2.2.44). □

Lemma 2.2.5 *We have*

$$\|R_\varepsilon(\omega)\|_\varepsilon \le C\varepsilon^{N(1-\frac{\min(p,2)+1}{2})}\|\omega\|_\varepsilon^{\min(p,2)}.$$

Proof. Suppose that $p \le 2$. Then by (6.1.2), one has

$$\begin{aligned}
R_\varepsilon(\omega) &= \Big(\sum_{j=1}^k U_{\varepsilon,x_j} + \omega\Big)_+^p - \Big(\sum_{j=1}^k U_{\varepsilon,x_j}\Big)^p - p\Big(\sum_{j=1}^k U_{\varepsilon,x_j}\Big)^{p-1}\omega \\
&= p\Big(\sum_{j=1}^k U_{\varepsilon,x_j} + t\omega\Big)_+^{p-1}\omega - p\Big(\sum_{j=1}^k U_{\varepsilon,x_j}\Big)^{p-1}\omega \\
&= O\big(|\omega|^p\big),
\end{aligned}$$

where $t \in (0, 1)$. Consequently, we obtain

$$\Big|\int_{\mathbb{R}^N} R_\varepsilon(\omega)\eta\Big| \le C\int_{\mathbb{R}^N}|\omega|^p|\eta| \le C\Big(\int_{\mathbb{R}^N}|\omega|^{p+1}\Big)^{\frac{p}{p+1}}\Big(\int_{\mathbb{R}^N}|\eta|^{p+1}\Big)^{\frac{1}{p+1}}. \tag{2.2.45}$$

For any ξ, let $\tilde{\xi}(y) = \xi(\varepsilon y)$. We then have

$$\begin{aligned}
\int_{\mathbb{R}^N}|\xi|^{p+1} &= \varepsilon^N\int_{\mathbb{R}^N}|\tilde{\xi}|^{p+1} \le C\varepsilon^N\Big(\int_{\mathbb{R}^N}\big(|\nabla\tilde{\xi}|^2 + \tilde{\xi}^2\big)\Big)^{\frac{p+1}{2}} \\
&= C\varepsilon^{N(1-\frac{p+1}{2})}\Big(\int_{\mathbb{R}^N}\big(\varepsilon^2|\nabla\xi|^2 + \xi^2\big)\Big)^{\frac{p+1}{2}}.
\end{aligned} \tag{2.2.46}$$

Now, by (2.2.45), we obtain

$$\Big|\int_{\mathbb{R}^N} R_\varepsilon(\omega)\eta\Big| \le C\int_{\mathbb{R}^N}|\omega|^p|\eta| \le C\varepsilon^{N(1-\frac{p+1}{2})}\|\omega\|_\varepsilon^p\|\eta\|_\varepsilon.$$

Thus, this lemma is proved in the case $p \le 2$.

Now, we consider the case $p > 2$. Then it follows from (6.1.1) that

$$|R_\varepsilon(\omega)| \le C\Big(\sum_{j=1}^k U_{\varepsilon,x_j}\Big)^{p-2}\omega^2 + C|\omega|^p.$$

On the other hand, by the Hölder inequality and (2.2.46), we have

$$\begin{aligned}
&\int_{\mathbb{R}^N}\Big(\sum_{j=1}^k U_{\varepsilon,x_j}\Big)^{p-2}\omega^2|\eta| \\
&\le \Big(\int_{\mathbb{R}^N}\Big(\sum_{j=1}^k U_{\varepsilon,x_j}\Big)^{p+1}\Big)^{\frac{p-2}{p+1}}\Big(\int_{\mathbb{R}^N}|\omega|^{p+1}\Big)^{\frac{2}{p+1}}\Big(\int_{\mathbb{R}^N}|\eta|^{p+1}\Big)^{\frac{1}{p+1}}
\end{aligned}$$

$$\le C\varepsilon^{\frac{p-2}{p+1}N+N(1-\frac{p+1}{2})\frac{3}{p+1}}\|\omega\|_\varepsilon^2\|\eta\|_\varepsilon$$
$$= C\varepsilon^{-\frac{N}{2}}\|\omega\|_\varepsilon^2\|\eta\|_\varepsilon.$$

This proves the lemma for the case $p > 2$. □

We are now ready to carry out the reduction argument for (2.2.40).

Proposition 2.2.6 *Let $\theta > 0$ be small so that $B_\theta(p_i) \cap B_\theta(p_j) = \emptyset$, whenever $i \neq j$ and where $i, j = 1, \cdots, k$. Then, there exists $\varepsilon_0 > 0$ such that for any $\varepsilon \in (0, \varepsilon_0]$, $x_j \in B_\theta(p_j)$, there is a unique $\omega_\varepsilon \in E_\varepsilon$ satisfying (2.2.40) and*

$$\|\omega_\varepsilon\|_\varepsilon \le C\|l_\varepsilon\|_\varepsilon \le C\varepsilon^{\frac{N}{2}}\Big(\varepsilon\sum_{j=1}^{k}|\nabla V(x_j)| + \varepsilon^2\Big).$$

Proof. In view of Proposition 2.2.3, we can rewrite (2.2.40) as

$$\omega = B\omega := (Q_\varepsilon L_\varepsilon)^{-1}l_\varepsilon + (Q_\varepsilon L_\varepsilon)^{-1}R_\varepsilon(\omega).$$

It follows from Proposition 2.2.3 and Lemma 2.2.4 that

$$\|(Q_\varepsilon L_\varepsilon)^{-1}l_\varepsilon\|_\varepsilon \le C\|l_\varepsilon\|_\varepsilon \le C\varepsilon^{\frac{N}{2}+1}.$$

Now we will apply the contraction mapping theorem in a ball whose radius is slightly bigger than $C\|l_\varepsilon\|_\varepsilon$. Let

$$S := \Big\{\omega : \omega \in E_\varepsilon,\ \|\omega_\varepsilon\|_\varepsilon \le \varepsilon^{\frac{N}{2}+1-\tau}\Big\},$$

where $\tau > 0$ is a fixed small constant.

(i) B maps S to S. In fact, for any $\omega \in S$, it follows from Lemmas 2.2.4 and 2.2.5 that

$$\begin{aligned}\|B\omega\|_\varepsilon &\le C\|l_\varepsilon\|_\varepsilon + C\|R_\varepsilon(\omega)\|_\varepsilon\\ &\le C\varepsilon^{\frac{N}{2}+1} + C\varepsilon^{N(1-\frac{\min(p,2)+1}{2})}\|\omega\|_\varepsilon^{\min(p,2)} \qquad (2.2.47)\\ &\le C\varepsilon^{\frac{N}{2}+1} + C\varepsilon^{N(1-\frac{\min(p,2)+1}{2})}\big(\varepsilon^{\frac{N}{2}+1-\tau}\big)^{\min(p,2)} \le \varepsilon^{\frac{N}{2}+1-\tau}.\end{aligned}$$

(ii) B is a contraction map. To prove this claim, for any $\omega_1,\ \omega_2 \in S$, we have

$$\|B\omega_1 - B\omega_2\|_\varepsilon \le C\|R_\varepsilon(\omega_1) - R_\varepsilon(\omega_2)\|_\varepsilon.$$

On the other hand, one has

$$
\begin{aligned}
&R_\varepsilon(\omega_1) - R_\varepsilon(\omega_2)\\
&= \Big(\sum_{j=1}^k U_{\varepsilon,x_j} + \omega_1\Big)_+^p - \Big(\sum_{j=1}^k U_{\varepsilon,x_j} + \omega_2\Big)_+^p - p\Big(\sum_{j=1}^k U_{\varepsilon,x_j}\Big)^{p-1}(\omega_1 - \omega_2)\\
&= p\bigg(\Big(\sum_{j=1}^k U_{\varepsilon,x_j} + \omega_1 + t(\omega_1 - \omega_2)\Big)_+^{p-1} - \Big(\sum_{j=1}^k U_{\varepsilon,x_j}\Big)^{p-1}\bigg)(\omega_1 - \omega_2),
\end{aligned}
$$

where $t \in (0, 1)$.

If $1 < p \le 2$, it then follows from (6.1.2) that

$$
|R_\varepsilon(\omega_1) - R_\varepsilon(\omega_2)| \le C\big(|\omega_1|^{p-1} + |\omega_2|^{p-1}\big)|\omega_1 - \omega_2|.
$$

Hence we have

$$
\begin{aligned}
&\Big|\int_{\mathbb{R}^N} \big|R_\varepsilon(\omega_1) - R_\varepsilon(\omega_2)\big|\eta\Big|\\
&\le \Big(\int_{\mathbb{R}^N} \big(|\omega_1| + |\omega_2|\big)^{p+1}\Big)^{\frac{p-1}{p+1}} \Big(\int_{\mathbb{R}^N} |\omega_1 - \omega_2|^{p+1}\Big)^{\frac{1}{p+1}} \Big(\int_{\mathbb{R}^N} |\eta|^{p+1}\Big)^{\frac{1}{p+1}}\\
&\le C\varepsilon^{-\frac{(p-1)N}{2}}\big(\|\omega_1\|_\varepsilon^{p-1} + \|\omega_2\|_\varepsilon^{p-1}\big)\|\omega_1 - \omega_2\|_\varepsilon \|\eta\|_\varepsilon\\
&\le C\varepsilon^{(p-1)(1-\tau)}\|\omega_1 - \omega_2\|_\varepsilon\|\eta\|_\varepsilon \le \frac{1}{2}\|\omega_1 - \omega_2\|_\varepsilon\|\eta\|_\varepsilon.
\end{aligned}
$$

If $p > 2$, then by (6.1.1), one has

$$
\begin{aligned}
|R_\varepsilon(\omega_1) - R_\varepsilon(\omega_2)| \le{}& C\Big(\sum_{j=1}^k U_{\varepsilon,x_j}\Big)^{p-2}\big(|\omega_1| + |\omega_2|\big)|\omega_1 - \omega_2|\\
&+ C\big(|\omega_1|^{p-1} + |\omega_2|^{p-1}\big)|\omega_1 - \omega_2|.
\end{aligned}
$$

We also have

$$
\begin{aligned}
&\Big|\int_{\mathbb{R}^N} \Big(\sum_{j=1}^k U_{\varepsilon,x_j}\Big)^{p-2}\big(|\omega_1| + |\omega_2|\big)|\omega_1 - \omega_2|\eta\Big|\\
&\le \bigg(\int_{\mathbb{R}^N} \Big(\sum_{j=1}^k U_{\varepsilon,x_j}\Big)^{p+1}\bigg)^{\frac{p-2}{p+1}} \Big(\int_{\mathbb{R}^N} \big(|\omega_1| + |\omega_2|\big)^{p+1}\Big)^{\frac{1}{p+1}}\\
&\quad \times \Big(\int_{\mathbb{R}^N} |\omega_1 - \omega_2|^{p+1}\Big)^{\frac{1}{p+1}} \Big(\int_{\mathbb{R}^N} |\eta|^{p+1}\Big)^{\frac{1}{p+1}}\\
&\le C\varepsilon^{-\frac{N}{2}}\big(\|\omega_1\|_\varepsilon + \|\omega_2\|_\varepsilon\big)\|\omega_1 - \omega_2\|_\varepsilon\|\eta\|_\varepsilon\\
&\le C\varepsilon^{1-\tau}\|\omega_1 - \omega_2\|_\varepsilon\|\eta\|_\varepsilon.
\end{aligned}
$$

In both cases, we have proved that

$$\|B\omega_1 - B\omega_2\|_\varepsilon \le \frac{1}{2}\|\omega_1 - \omega_2\|_\varepsilon, \quad \forall\, \omega_1,\ \omega_2 \in S.$$

By the contraction mapping theorem, we conclude that for any $\varepsilon \in (0, \varepsilon_0]$, $x_j \in B_\theta(p_j)$, there is $\omega_\varepsilon \in E_\varepsilon$, depending on x_j and ε, satisfying

$$\omega_\varepsilon = B\omega_\varepsilon.$$

Similar to (2.2.47), we obtain

$$\begin{aligned}
\|\omega\|_\varepsilon = \|B\omega\|_\varepsilon &\le C\|l_\varepsilon\|_\varepsilon + C\|R_\varepsilon(\omega)\|_\varepsilon \\
&\le C\|l_\varepsilon\|_\varepsilon + C\varepsilon^{N(1-\frac{\min(p,2)+1}{2})}\|\omega\|_\varepsilon^{\min(p,2)} \\
&\le C\|l_\varepsilon\|_\varepsilon + C\varepsilon^{N(1-\frac{\min(p,2)+1}{2})}\left(\varepsilon^{\frac{N}{2}+1-\tau}\right)^{\min(p,2)-1}\|\omega\|_\varepsilon \\
&\le C\|l_\varepsilon\|_\varepsilon + \varepsilon^{(1-\tau)[\min(p,2)-1]}\|\omega\|_\varepsilon,
\end{aligned}$$

which yields

$$\|\omega\|_\varepsilon \le C\|l_\varepsilon\|_\varepsilon \le C\varepsilon^{\frac{N}{2}}\Big(\varepsilon\sum_{j=1}^{k}|\nabla V(x_j)| + \varepsilon^2\Big). \qquad \square$$

Proposition 2.2.6 implies that

$$L_\varepsilon\omega_\varepsilon - l_\varepsilon - R_\varepsilon(\omega_\varepsilon) = \sum_{j=1}^{k}\sum_{i=1}^{N} a_{\varepsilon,i,j}\frac{\partial U_{\varepsilon,x_j}}{\partial y_i}, \tag{2.2.48}$$

for some constants $a_{\varepsilon,i,j}$.

The second step is to choose x_j suitably, such that $a_{\varepsilon,i,j} = 0$, $i = 1, \cdots, N$, $j = 1, \cdots, k$.

The function in the right-hand side of (2.2.48) belongs to

$$E_\varepsilon^\perp = span\Big\{\frac{\partial U_{\varepsilon,x_j}}{\partial y_i},\ i = 1, \cdots, N,\ j = 1, \cdots, k\Big\}.$$

Therefore, if the left-hand side of (2.2.48) belongs to E_ε, then the function in the right-hand side of (2.2.48) must be zero.

We use the notation

$$u_\varepsilon = \sum_{j=1}^{k} U_{\varepsilon,x_j} + \omega_\varepsilon.$$

Then one has

$$\begin{aligned}
&\big\langle L_\varepsilon\omega_\varepsilon - l_\varepsilon - R_\varepsilon(\omega_\varepsilon), \psi\big\rangle_\varepsilon \\
&= \int_{\mathbb{R}^N}\big(\varepsilon^2\nabla u_\varepsilon\nabla\psi + V(y)u_\varepsilon\psi - (u_\varepsilon)_+^p\psi\big), \forall\ \psi \in H^1(\mathbb{R}^N).
\end{aligned}$$

We now prove the following:

Lemma 2.2.7 *Suppose that $x_{\varepsilon,j}$ satisfies*

$$\int_{\mathbb{R}^N}\Big(\varepsilon^2\nabla u_\varepsilon\nabla\frac{\partial U_{\varepsilon,x_{\varepsilon,j}}}{\partial y_i}+V(y)u_\varepsilon\frac{\partial U_{\varepsilon,x_{\varepsilon,j}}}{\partial y_i}-(u_\varepsilon)_+^p\frac{\partial U_{\varepsilon,x_{\varepsilon,j}}}{\partial y_i}\Big)=0,$$
$$i=1,\cdots,N,\ j=1,\cdots,k, \tag{2.2.49}$$

then $a_{\varepsilon,i,j}=0$, $i=1,\cdots,N$, $j=1,\cdots,k$.

Proof. If (2.2.49) holds, then

$$\sum_{m=1}^k\sum_{h=1}^N a_{\varepsilon,h,m}\Big\langle\frac{\partial U_{\varepsilon,x_{\varepsilon,j}}}{\partial y_i},\frac{\partial U_{\varepsilon,x_{\varepsilon,m}}}{\partial y_h}\Big\rangle_\varepsilon=0,\quad j=1,\cdots,k,\ i=1,\cdots,N.$$

Using (2.2.20) and (2.2.21), we conclude that

$$a_{\varepsilon,i,j}=0,\ i=1,\cdots,N,\ j=1,\cdots,k. \qquad \square$$

Next, we want to solve the algebraic equations (2.2.49). This is a finite dimensional problem. We can use the degree theorem to solve it. For this purpose, we need to prove that the function in the left-hand side of (2.2.49) is continuous. Before we can solve (2.2.49), we need the following result:

Proposition 2.2.8 *Let ω_ε be the map obtained in Proposition 2.2.6. Then ω_ε is continuous as a map from $B_\theta(p_1)\times\cdots\times B_\theta(p_k)$ to $H^1(\mathbb{R}^N)$.*

Proof. Before we prove this proposition, we point out that the continuity can be derived from the uniqueness of solution for (2.2.40). The minor difficulty in the proof of this proposition comes from the fact that the subspace E_ε also depends on x_j, $j=1,\cdots,k$. To avoid any confusion, we use $E_{\varepsilon,x}$ and $\omega_{\varepsilon,x}$ to denote this subspace and the function ω_ε, respectively. For the same reason, we also use $L_{\varepsilon,x}$ and $Q_{\varepsilon,x}$ to denote the linear operators defined in (2.2.14) and (2.2.18), respectively.

Note that in this lemma, ε is fixed. Let

$$W_{\varepsilon,x}=\sum_{j=1}^k U_{\varepsilon,x_j}.$$

Suppose that

$$x^{(n)}=(x_1^{(n)},\cdots,x_k^{(n)})\to x^\infty=(x_1^\infty,\cdots,x_k^\infty).$$

For any $\eta \in E_{\varepsilon,x^{(n)}}$, one has

$$\int_{\mathbb{R}^N} \Big(\varepsilon^2 \nabla (W_{\varepsilon,x^{(n)}} + \omega_{\varepsilon,x^{(n)}}) \nabla \eta + V(y)(W_{\varepsilon,x^{(n)}} + \omega_{\varepsilon,x^{(n)}})\eta - (W_{\varepsilon,x^{(n)}} + \omega_{\varepsilon,x^{(n)}})^p \eta\Big) = 0. \tag{2.2.50}$$

For n and m large, and $\eta \in E_{\varepsilon,x^{(n)}}$, we decompose

$$\eta_{n,m} := \eta - \sum_{j=1}^{k}\sum_{i=1}^{N} \bar{c}_{m,i,j} \frac{\partial U_{\varepsilon,x_j^{(m)}}}{\partial y_i} \in E_{\varepsilon,x^{(m)}}.$$

Noting that

$$\Big\langle \frac{\partial U_{\varepsilon,x_j^{(m)}}}{\partial y_i}, \eta\Big\rangle_\varepsilon = \Big\langle \frac{\partial U_{\varepsilon,x_j^{(m)}}}{\partial y_i} - \frac{\partial U_{\varepsilon,x_j^{(n)}}}{\partial y_i}, \eta\Big\rangle_\varepsilon = o(1)\|\eta\|_\varepsilon, \quad \text{as } n, m \to +\infty,$$

we can prove that $\bar{c}_{m,i,j} = o(1)\|\eta\|_\varepsilon$ as $n \to +\infty$.

Therefore, it follows from (2.2.50) that for any $\eta \in E_{\varepsilon,x^{(n)}}$, we have

$$\int_{\mathbb{R}^N} \Big(\varepsilon^2 \nabla (W_{\varepsilon,x^{(m)}} + \omega_{\varepsilon,x^{(m)}}) \nabla \eta + V(y)(W_{\varepsilon,x^{(m)}} + \omega_{\varepsilon,x^{(m)}})\eta\Big) - \int_{\mathbb{R}^N} (W_{\varepsilon,x^{(m)}} + \omega_{\varepsilon,x^{(m)}})^p \eta = o(1)\|\eta\|_\varepsilon,$$

which, together with

$$\int_{\mathbb{R}^N} \Big(\varepsilon^2 \nabla (W_{\varepsilon,x^{(n)}} + \omega_{\varepsilon,x^{(n)}}) \nabla \eta + V(y)(W_{\varepsilon,x^{(n)}} + \omega_{\varepsilon,x^{(n)}})\eta\Big) - \int_{\mathbb{R}^N} (W_{\varepsilon,x^{(n)}} + \omega_{\varepsilon,x^{(n)}})^p \eta = 0,$$

gives

$$\begin{aligned}
&\Big|\big\langle Q_{\varepsilon,x^{(n)}} L_{\varepsilon,x^{(n)}}(\omega_{\varepsilon,x^{(m)}} - \omega_{\varepsilon,x^{(n)}}), \eta\big\rangle_\varepsilon\Big| \\
&= \Big|\big\langle L_{\varepsilon,x^{(n)}}(\omega_{\varepsilon,x^{(m)}} - \omega_{\varepsilon,x^{(n)}}), \eta\big\rangle_\varepsilon\Big| \\
&= \Big|\int_{\mathbb{R}^N} \Big((W_{\varepsilon,x^{(m)}} + \omega_{\varepsilon,x_\varepsilon^{(m)}})^p - (W_{\varepsilon,x^{(n)}} + \omega_{\varepsilon,x^{(n)}})^p\Big)\eta \\
&\quad - p\int_{\mathbb{R}^N} W^{p-1}_{\varepsilon,x^{(n)}}(\omega_{\varepsilon,x^{(m)}} - \omega_{\varepsilon,x^{(n)}})\eta\Big| + o(1)\|\eta\|_\varepsilon \\
&\le C\int_{\mathbb{R}^N} \big(|\omega_{\varepsilon,x^{(n)}}|^{p-1} + |\omega_{\varepsilon,x^{(m)}}|^{p-1}\big)|\omega_{\varepsilon,x^{(n)}} - \omega_{\varepsilon,x^{(m)}}||\eta| + o(1)\|\eta\|_\varepsilon \\
&\le C\varepsilon^{-\frac{N(p-1)}{2}} \big(\|\omega_{\varepsilon,x^{(n)}}\|_\varepsilon^{p-1} + \|\omega_{\varepsilon,x^{(m)}}\|_\varepsilon^{p-1}\big)\|\omega_{\varepsilon,x^{(n)}} - \omega_{\varepsilon,x^{(m)}}\|_\varepsilon\|\eta\|_\varepsilon + o(1)\|\eta\|_\varepsilon \\
&\le C\varepsilon^{p-1}\|\omega_{\varepsilon,x^{(n)}} - \omega_{\varepsilon,x^{(m)}}\|_\varepsilon\|\eta\|_\varepsilon + o(1)\|\eta\|_\varepsilon,
\end{aligned} \tag{2.2.51}$$

if $p \le 2$. Similarly, if $p > 2$, we have

$$\begin{aligned}
&\left|\left\langle Q_{\varepsilon,x^{(n)}} L_{\varepsilon,x^{(n)}}(\omega_{\varepsilon,x^{(m)}} - \omega_{\varepsilon,x^{(n)}}), \eta\right\rangle_\varepsilon\right| \\
&\le C\int_{\mathbb{R}^N} (W_{\varepsilon,x^{(m)}} + W_{\varepsilon,x^{(n)}})^{p-2}\big(|\omega_{\varepsilon,x^{(n)}}| + |\omega_{\varepsilon,x^{(m)}}|\big)|\omega_{\varepsilon,x^{(n)}} - \omega_{\varepsilon,x^{(m)}}||\eta| \\
&\quad + C\int_{\mathbb{R}^N} \big(|\omega_{\varepsilon,x^{(n)}}|^{p-1} + |\omega_{\varepsilon,x^{(m)}}|^{p-1}\big)|\omega_{\varepsilon,x^{(n)}} - \omega_{\varepsilon,x^{(m)}}||\eta| \\
&\quad + o(1)\|\eta\|_\varepsilon \le C\varepsilon\|\omega_{\varepsilon,x^{(n)}} - \omega_{\varepsilon,x^{(m)}}\|_\varepsilon\|\eta\|_\varepsilon + o(1)\|\eta\|_\varepsilon.
\end{aligned} \tag{2.2.52}$$

On the other hand, for n and m large, we decompose

$$\omega_{n,m} := \omega_{\varepsilon,x^{(n)}} - \omega_{\varepsilon,x^{(m)}} - \sum_{j=1}^{k}\sum_{i=1}^{N} c_{m,n,i,j}\frac{\partial U_{\varepsilon,x_j^{(n)}}}{\partial y_i} \in E_{\varepsilon,x^{(n)}}. \tag{2.2.53}$$

Since

$$\left\langle \frac{\partial U_{\varepsilon,x_j^{(n)}}}{\partial y_i}, \omega_{\varepsilon,x^{(n)}} - \omega_{\varepsilon,x^{(m)}}\right\rangle_\varepsilon = -\left\langle \frac{\partial U_{\varepsilon,x_j^{(n)}}}{\partial y_i} - \frac{\partial U_{\varepsilon,x_j^{(m)}}}{\partial y_i}, \omega_{\varepsilon,x^{(m)}}\right\rangle_\varepsilon \to 0,$$
$$\text{as } n, m \to +\infty,$$

it follows that $c_{m,n,i,j} = o(1)$ as $n, m \to +\infty$.

If $p \le 2$, inserting (2.2.53) into (2.2.51), we obtain

$$\left|\left\langle Q_\varepsilon L_\varepsilon \omega_{m,n}, \eta\right\rangle_\varepsilon\right| \le C\varepsilon^{p-1}\|\omega_{m,n}\|_\varepsilon\|\eta\|_\varepsilon + o(1)\|\eta\|_\varepsilon, \quad \forall\, \eta \in E_{\varepsilon,x^{(n)}},$$

which in turn implies that

$$\|Q_\varepsilon L_\varepsilon \omega_{m,n}\|_\varepsilon \le C\varepsilon^{p-1}\|\omega_{m,n}\|_\varepsilon + o(1). \tag{2.2.54}$$

By Proposition 2.2.3, (2.2.54) then gives

$$\|\omega_{m,n}\|_\varepsilon \le C\varepsilon^{p-1}\|\omega_{m,n}\|_\varepsilon + o(1).$$

Thus, $\|\omega_{m,n}\|_\varepsilon \to 0$ as $m, n \to +\infty$. If $p > 2$, using (2.2.52), we also have $\|\omega_{m,n}\|_\varepsilon \to 0$ as $m, n \to +\infty$. It then follows that $\|\omega_{\varepsilon,x^{(m)}} - \omega_{\varepsilon,x^{(n)}}\|_\varepsilon \to 0$ as $m, n \to +\infty$. This shows that $\omega_{\varepsilon,x^{(n)}} \to \omega$ strongly in $H^1(\mathbb{R}^N)$. Moreover, by the uniqueness of $\omega_{\varepsilon,x^\infty}$ in Proposition 2.2.6, we conclude that $\omega = \omega_{\varepsilon,x^\infty}$. □

Remark 2.2.9 With more careful analysis, we can prove that ω_ε as a map from $B_\theta(p_1) \times \cdots \times B_\theta(p_k)$ to $H^1(\mathbb{R}^N)$ is C^1. Here, we will omit the proof.

Theorem 2.2.10 (Y. Y. Li [99], 1997; Cao–Noussair–Yan [35], 1999) *Suppose that p_j is a critical point of $V(y)$ satisfying $deg(\nabla V, B_\delta(p_j), 0) \neq 0$,*

$j = 1, \cdots, k$. Then there exists $\varepsilon_0 > 0$ such that for any $\varepsilon \in (0, \varepsilon_0]$, (2.2.11) has a solution of the form

$$u_\varepsilon = \sum_{j=1}^{k} U_{\varepsilon,x_{\varepsilon,j}} + \omega_\varepsilon,$$

for some $x_{\varepsilon,j} \in B_\delta(p_j)$, and $\|\omega_\varepsilon\|_\varepsilon = O(\varepsilon^{\frac{N}{2}+1})$.

Proof. We only need to solve (2.2.49). The main task is to find the main term for the function in the left-hand side of (2.2.49). We first estimate the left-hand side of (2.2.49) with $\omega_\varepsilon = 0$. Then we show that the contribution of the error term ω_ε to the function in the left-hand side of (2.2.49) is negligible.

Now, let

$$W_{\varepsilon,x} = \sum_{j=1}^{k} U_{\varepsilon,x_j}.$$

Applying the symmetry of U_{ε,x_j} and Lemma 2.2.2, we have

$$\begin{aligned}\int_{\mathbb{R}^N} \varepsilon^2 \nabla W_{\varepsilon,x} \nabla \frac{\partial U_{\varepsilon,x_j}}{\partial y_i} &= \int_{\mathbb{R}^N} \Big(pU_{\varepsilon,x_j}^{p-1} - V(x_j)\Big) W_{\varepsilon,x} \frac{\partial U_{\varepsilon,x_j}}{\partial y_i} \\ &= \sum_{m \neq j} \int_{\mathbb{R}^N} \Big(pU_{\varepsilon,x_{\varepsilon,j}}^{p-1} - V(x_j)\Big) U_{\varepsilon,x_m} \frac{\partial U_{\varepsilon,x_j}}{\partial y_i} \\ &= O\big(e^{-\tau/\varepsilon}\big),\end{aligned} \tag{2.2.55}$$

for some $\tau > 0$. Similarly, by Lemma 6.1.1, we have

$$\begin{aligned}\int_{\mathbb{R}^N} W_{\varepsilon,x}^p \frac{\partial U_{\varepsilon,x_j}}{\partial y_i} &= \int_{\mathbb{R}^N} \big(W_{\varepsilon,x}^p - U_{\varepsilon,x_j}^p\big) \frac{\partial U_{\varepsilon,x_j}}{\partial y_i} \\ &= O\Big(\sum_{m \neq j} \int_{\mathbb{R}^N} U_{\varepsilon,x_m}^{p-1} U_{\varepsilon,x_j} \Big| \frac{\partial U_{\varepsilon,x_j}}{\partial y_i} \Big|\Big) = O\big(e^{-\tau/\varepsilon}\big).\end{aligned} \tag{2.2.56}$$

Moreover, it follows from Lemma 2.2.2 that

$$\begin{aligned}\int_{\mathbb{R}^N} V(y) W_{\varepsilon,x} \frac{\partial U_{\varepsilon,x_j}}{\partial y_i} &= \int_{\mathbb{R}^N} V(y) U_{\varepsilon,x_j} \frac{\partial U_{\varepsilon,x_j}}{\partial y_i} + O\big(e^{-\tau/\varepsilon}\big) \\ &= \frac{1}{2} \int_{\mathbb{R}^N} V(y) \frac{\partial U_{\varepsilon,x_j}^2}{\partial y_i} + O\big(e^{-\tau/\varepsilon}\big) \\ &= -\frac{1}{2} \int_{\mathbb{R}^N} \frac{\partial V(y)}{\partial y_i} U_{\varepsilon,x_j}^2 + O\big(e^{-\tau/\varepsilon}\big)\end{aligned}$$

$$
\begin{aligned}
&= -\frac{1}{2}\varepsilon^N \int_{\mathbb{R}^N} \frac{\partial V(\varepsilon y + x_j)}{\partial y_i} U^2_{\varepsilon,x_j}(\varepsilon y + x_j) + O\big(e^{-\tau/\varepsilon}\big) \\
&= -\frac{1}{2}\varepsilon^N \frac{\partial V(x_j)}{\partial x_i} \int_{\mathbb{R}^N} U^2_{\varepsilon,x_j}(\varepsilon y + x_j) + O\big(\varepsilon^{N+1}\big).
\end{aligned} \tag{2.2.57}
$$

Combining (2.2.55)–(2.2.2), we obtain

$$
\begin{aligned}
&\int_{\mathbb{R}^N} \Big(\varepsilon^2 \nabla W_{\varepsilon,x} \nabla \frac{\partial U_{\varepsilon,x_j}}{\partial y_i} + V(y) W_{\varepsilon,x} \frac{\partial U_{\varepsilon,x_j}}{\partial y_i} - W^p_{\varepsilon,x} \frac{\partial U_{\varepsilon,x_j}}{\partial y_i}\Big) \\
&= -\frac{1}{2}\varepsilon^N \frac{\partial V(x_j)}{\partial x_i} \int_{\mathbb{R}^N} U^2_{\varepsilon,x_j}(\varepsilon y + x_j) + O\big(\varepsilon^{N+1}\big).
\end{aligned} \tag{2.2.58}
$$

We will now show that the contribution of the error term ω_ε to the function in the left-hand side of (2.2.49) is negligible. For $\omega_\varepsilon \in E_\varepsilon$, we have that

$$
\begin{aligned}
&\int_{\mathbb{R}^N} \Big(\varepsilon^2 \nabla \Big(\sum_{m=1}^k U_{\varepsilon,x_m} + \omega_\varepsilon\Big) \nabla \frac{\partial U_{\varepsilon,x_j}}{\partial y_i} + V(y)\Big(\sum_{m=1}^k U_{\varepsilon,x_m} + \omega_\varepsilon\Big) \frac{\partial U_{\varepsilon,x_j}}{\partial y_i}\Big) \\
&= \int_{\mathbb{R}^N} \Big(\varepsilon^2 \nabla \sum_{m=1}^k U_{\varepsilon,x_m} \nabla \frac{\partial U_{\varepsilon,x_j}}{\partial y_i} + V(y) \sum_{m=1}^k U_{\varepsilon,x_m} \frac{\partial U_{\varepsilon,x_j}}{\partial y_i}\Big), \\
&\quad i = 1, \cdots, N, \ j = 1, \cdots, k.
\end{aligned}
$$

On the other hand, it follows from Lemma 6.1.1 that

$$
\begin{aligned}
&\int_{\mathbb{R}^N} \Big(\sum_{m=1}^k U_{\varepsilon,x_m} + \omega_\varepsilon\Big)_+^p \frac{\partial U_{\varepsilon,x_j}}{\partial y_i} \\
&= \int_{\mathbb{R}^N} \Big(\sum_{m=1}^k U_{\varepsilon,x_m}\Big)^p \frac{\partial U_{\varepsilon,x_j}}{\partial y_i} - p \int_{\mathbb{R}^N} \Big(\sum_{m=1}^k U_{\varepsilon,x_m}\Big)^{p-1} \frac{\partial U_{\varepsilon,x_j}}{\partial y_i} \omega_\varepsilon \\
&\quad + \begin{cases} O\Big(\displaystyle\int_{\mathbb{R}^N} \Big|\frac{\partial U_{\varepsilon,x_j}}{\partial y_i}\Big| |\omega_\varepsilon|^p\Big), & p \le 2, \\ O\Big(\displaystyle\int_{\mathbb{R}^N} \Big|\frac{\partial U_{\varepsilon,x_j}}{\partial y_i}\Big| |\omega_\varepsilon|^p + \int_{\mathbb{R}^N} \Big(\sum_{m=1}^k U_{\varepsilon,x_m}\Big)^{p-2} \Big|\frac{\partial U_{\varepsilon,x_j}}{\partial y_i}\Big| |\omega_\varepsilon|^2\Big), & p > 2. \end{cases}
\end{aligned}
$$

Now we observe that

$$
\begin{aligned}
&\int_{\mathbb{R}^N} \Big(\sum_{m=1}^k U_{\varepsilon,x_m}\Big)^{p-1} \frac{\partial U_{\varepsilon,x_j}}{\partial y_i} \omega_\varepsilon \\
&= \int_{\mathbb{R}^N} \Big(\Big(\sum_{m=1}^k U_{\varepsilon,x_m}\Big)^{p-1} - U^{p-1}_{\varepsilon,x_j}\Big) \frac{\partial U_{\varepsilon,x_j}}{\partial y_i} \omega_\varepsilon + \int_{\mathbb{R}^N} U^{p-1}_{\varepsilon,x_j} \frac{\partial U_{\varepsilon,x_j}}{\partial y_i} \omega_\varepsilon
\end{aligned}
$$

$$
\begin{aligned}
&= O\big(e^{-\frac{\tau}{\varepsilon}}\big) + \varepsilon^N O\Big(\sum_{j=1}^{k} |\nabla V(x_j)|\varepsilon + \varepsilon^2\Big)\\
&= \varepsilon^N O\Big(\sum_{j=1}^{k} |\nabla V(x_j)|\varepsilon + \varepsilon^2\Big),
\end{aligned}
$$

since

$$
\begin{aligned}
& p\int_{\mathbb{R}^N} U_{\varepsilon,x_j}^{p-1}\frac{\partial U_{\varepsilon,x_j}}{\partial y_i}\omega_\varepsilon\\
&= \int_{\mathbb{R}^N}\Big(\varepsilon^2\nabla\frac{\partial U_{\varepsilon,x_j}}{\partial y_j}\nabla\omega_\varepsilon + V(y)\frac{\partial U_{\varepsilon,x_j}}{\partial y_i}\omega_\varepsilon\Big) + \int_{\mathbb{R}^N}\big(V(x_j) - V(y)\big)\frac{\partial U_{\varepsilon,x_j}}{\partial y_i}\omega_\varepsilon\\
&= \int_{\mathbb{R}^N}\big(V(x_j) - V(y)\big)\frac{\partial U_{\varepsilon,x_j}}{\partial y_i}\omega_\varepsilon\\
&= O\bigg(\Big(\int_{\mathbb{R}^N}\big(V(x_j) - V(y)\big)^2\Big(\frac{\partial U_{\varepsilon,x_j}}{\partial y_i}\Big)^2\Big)^{\frac{1}{2}}\bigg)\|\omega_\varepsilon\|_\varepsilon\\
&= \varepsilon^{\frac{N}{2}} O\Big(|\nabla V(x_j)| + \varepsilon\Big)\|\omega_\varepsilon\|_\varepsilon\\
&= \varepsilon^N O\Big(\sum_{j=1}^{k} |\nabla V(x_j)|\varepsilon + \varepsilon^2\Big).
\end{aligned}
$$

For the other terms, we have the estimates

$$
\begin{aligned}
\int_{\mathbb{R}^N}\Big|\frac{\partial U_{\varepsilon,x_j}}{\partial y_i}\Big||\omega_\varepsilon|^p &\le C\varepsilon^{\frac{N}{p+1}-1}\Big(\int_{\mathbb{R}^N}|\omega_\varepsilon|^{p+1}\Big)^{\frac{p}{p+1}}\\
&\le C\varepsilon^{N-\frac{pN}{2}-1}\|\omega_\varepsilon\|_\varepsilon^p \le C\varepsilon^{N+p-1}
\end{aligned}
$$

and

$$
\begin{aligned}
&\int_{\mathbb{R}^N}\Big(\sum_{m=1}^{k} U_{\varepsilon,x_m}\Big)^{p-2}\Big|\frac{\partial U_{\varepsilon,x_j}}{\partial y_i}\Big||\omega_\varepsilon|^2\\
&\quad\le C\int_{\mathbb{R}^N} U_{\varepsilon,x_j}^{p-2}\Big|\frac{\partial U_{\varepsilon,x_j}}{\partial y_i}\Big||\omega_\varepsilon|^2 + O\big(e^{-\tau/\varepsilon}\big)\\
&\quad\le \varepsilon^{-1}\int_{\mathbb{R}^N} U_{\varepsilon,x_j}^{p-1}|\omega_\varepsilon|^2 + O(e^{-\tau/\varepsilon}) \le C\varepsilon^{\frac{N(p-1)}{p+1}-1}\Big(\int_{\mathbb{R}^N}|\omega_\varepsilon|^{p+1}\Big)^{\frac{2}{p+1}}\\
&\quad\quad + O(e^{-\tau/\varepsilon}) \le C\varepsilon^{-1}\|\omega_\varepsilon\|_\varepsilon^2 + O(e^{-\tau/\varepsilon}) \le C\varepsilon^{N+1}.
\end{aligned}
$$

In conclusion, we have proved

$$
\begin{aligned}
&\int_{\mathbb{R}^N}\Big(\varepsilon^2\nabla u_\varepsilon\nabla\frac{\partial U_{\varepsilon,x_j}}{\partial y_i} + V(y)u_\varepsilon\frac{\partial U_{\varepsilon,x_j}}{\partial y_i} - (u_\varepsilon)_+^p\frac{\partial U_{\varepsilon,x_j}}{\partial y_i}\Big)\\
&\quad= \int_{\mathbb{R}^N}\Big(\varepsilon^2\nabla W_{\varepsilon,x}\nabla\frac{\partial U_{\varepsilon,x_j}}{\partial y_i} + V(y)W_{\varepsilon,x}\frac{\partial U_{\varepsilon,x_j}}{\partial y_i} - W_{\varepsilon,x}^p\frac{\partial U_{\varepsilon,x_j}}{\partial y_i}\Big)\\
&\quad\quad + \varepsilon^N O\big(\varepsilon^{\min(1,p-1)}\big).
\end{aligned}
$$

Therefore, (2.2.49) is equivalent to

$$\nabla V(x_j) = O\big(\varepsilon^{\min(1,p-1)}\big), \quad j = 1, \cdots, k. \tag{2.2.59}$$

By the assumption that $\deg(\nabla V, B_\delta(p_j), 0) \neq 0$, we deduce that (2.2.59) has a solution $x_{\varepsilon,j} \in B_\theta(p_j)$, and $|x_{\varepsilon,j} - p_j| = O\big(\varepsilon^{\min(1,p-1)}\big)$, $j = 1, \cdots, k$. Hence, we complete the proof of this theorem. □

For some properties of the solution u_ε, we have the following result:

Proposition 2.2.11 *Let u_ε be the solution constructed in Theorem 2.2.10. Then the following statements hold.*

(i) *u_ε has k local maximum points $x_{\varepsilon,j} \in \mathbb{R}^N$ such that $x_{\varepsilon,j} \to p_j \in \mathbb{R}^N$ as $\varepsilon \to 0$, and $p_i \neq p_j$ for $i \neq j$. Moreover, each p_j is a critical point of $V(y)$.*

(ii) *For any given small $\tau > 0$, there exists a large constant $R > 0$, such that*

$$|u_\varepsilon(x)| \le \tau, \quad \forall x \in \mathbb{R}^N \setminus \cup_{j=1}^k B_{R\varepsilon}(x_{\varepsilon,j}).$$

In other words, $u_\varepsilon(x)$ is a k-peak solution.

(iii) *There exists $C > 0$ such that*

$$\|u_\varepsilon\|_\varepsilon^2 := \int_{\mathbb{R}^N} \big(\varepsilon^2|\nabla u_\varepsilon|^2 + V(y)u_\varepsilon^2\big) \le C\varepsilon^N.$$

In particular, (2.1.6) *and* (2.1.7) *hold.*

Proof. From Theorem 2.2.10, u_ε satisfies (i) and (iii). To show (ii), we just need to show that as $\varepsilon \to 0$, $\|\omega_\varepsilon\|_{L^\infty(\mathbb{R}^N)} \to 0$. For this purpose, we let $\tilde{\omega}_\varepsilon(y) = \omega_\varepsilon(\varepsilon y)$. Then,

$$\begin{aligned}
-\Delta\tilde{\omega}_\varepsilon + V(\varepsilon y)\tilde{\omega}_\varepsilon &= \Big(\sum_{j=1}^k U_{\varepsilon,x_{\varepsilon,j}}(\varepsilon y) + \tilde{\omega}_\varepsilon\Big)^p - \sum_{j=1}^k U^p_{\varepsilon,x_{\varepsilon,j}}(\varepsilon y) \\
&\quad + \sum_{j=1}^k \big(V(x_j) - V(\varepsilon y)\big)U_{\varepsilon,x_{\varepsilon,j}}(\varepsilon y) \\
&= \Big(\sum_{j=1}^k U_{\varepsilon,x_{\varepsilon,j}}(\varepsilon y) + \tilde{\omega}_\varepsilon\Big)^p - \Big(\sum_{j=1}^k U_{\varepsilon,x_{\varepsilon,j}}(\varepsilon y)\Big)^p
\end{aligned}$$

$$
\begin{aligned}
& + \Big(\sum_{j=1}^{k} U_{\varepsilon,x_{\varepsilon,j}}(\varepsilon y)\Big)^p - \sum_{j=1}^{k} U^p_{\varepsilon,x_{\varepsilon,j}}(\varepsilon y) + o(1) \\
& = O\big(|\tilde{\omega}_\varepsilon| + |\tilde{\omega}_\varepsilon|^p\big) + o(1),
\end{aligned}
$$

and

$$\|\tilde{\omega}_\varepsilon\|_{H^1(\mathbb{R}^N)} = o(1).$$

We can prove via the Moser iteration that

$$\|\omega_\varepsilon\|_{L^\infty(B_1(x_0))} \le C\|\tilde{\omega}_\varepsilon\|_{L^2(B_2(x_0))} + o(1) = o(1).$$

□

Let us summarize the general procedure to use the reduction argument to construct a peak solution.

Step 1. First, we need to determine the corresponding limit problem, so that we can find an approximate solution for the problem. For the nonlinear Schrödinger equation (2.2.11), the corresponding limit problem is (2.2.2) if the local maximum point of the peak solution u_ε is x_ε. For (2.2.11), we use the solution $U_{\varepsilon,x_\varepsilon}$ as an approximate solution. For other problem, we may need to modify it.

Step 2. To carry out the reduction argument, what we really need is the corresponding limit problem has a radial solution, which is non-degenerate in the sense that the kernel of linearized operator for this radial solution is generated exactly by the invariance of the limit problem. For (2.2.2), it is invariant under the translation. So what we need is (2.2.15).

Step 3. In order to solve the corresponding finite dimensional problem, we first need to check that the error term obtained in the reduction argument is negligible. Note that the size of the error term ω_ε is controlled by the approximate solution. See (2.2.2) and Proposition 2.2.6. If the size of the error term is not small enough, we should modify the approximate solution so that the term l_ε becomes smaller.

Generally speaking, to determine whether the approximate solution W_ε is good or not, we should first calculate the corresponding functions without the error term and determine the order of the main term (see, for example, (2.2.58)). To prove that the error term is negligible, what we need is that $\|\omega_\varepsilon\|_\varepsilon^2$ is smaller than the main term. Thus, the approximate solution W_ε is good if the term $\|l_\varepsilon\|_\varepsilon^2$ is smaller than the main term.

Before we end this subsection, let us revisit the finite dimensional problem (2.2.49). This will be useful in the construction of solutions with clustering peaks in the next section.

Similar to Lemma 2.2.7, we have the following:

Lemma 2.2.12 *Suppose that $x_{\varepsilon,j}$ satisfies*

$$\begin{aligned}&\int_{\mathbb{R}^N}\Big(\varepsilon^2\nabla u_\varepsilon\nabla\frac{\partial(U_{\varepsilon,x_{\varepsilon,j}}+\omega_\varepsilon)}{\partial x_{ji}}+V(y)u_\varepsilon\frac{\partial(U_{\varepsilon,x_{\varepsilon,j}}+\omega_\varepsilon)}{\partial x_{ji}}\\&-(u_\varepsilon)_+^p\frac{\partial(U_{\varepsilon,x_{\varepsilon,j}}+\omega_\varepsilon)}{\partial x_{ji}}\Big)=0,\, i=1,\cdots,N,\ j=1,\cdots,k,\end{aligned}\tag{2.2.60}$$

then $a_{\varepsilon,i,j}=0$, $i=1,\cdots,N$, $j=1,\cdots,k$.

Proof. If (2.2.60) holds, then

$$\sum_{m=1}^{k}\sum_{h=1}^{N}a_{\varepsilon,h,m}\Big\langle\frac{\partial(U_{\varepsilon,x_{\varepsilon,j}}+\omega_\varepsilon)}{\partial x_{ji}},\frac{\partial U_{\varepsilon,x_{\varepsilon,m}}}{\partial y_h}\Big\rangle_\varepsilon=0,\quad j=1,\cdots,k,\ i=1,\cdots,N.$$

Noting that

$$\Big\langle\frac{\partial U_{\varepsilon,x_{\varepsilon,m}}}{\partial y_h},\omega_\varepsilon\Big\rangle_\varepsilon=0,$$

for any $x_{\varepsilon,m}$, one then has

$$\Big\langle\frac{\partial U_{\varepsilon,x_{\varepsilon,m}}}{\partial y_h},\frac{\partial\omega_\varepsilon}{\partial x_{j,i}}\Big\rangle_\varepsilon=-\Big\langle\frac{\partial^2 U_{\varepsilon,x_{\varepsilon,j}}}{\partial x_{j,i}\partial y_h},\omega_\varepsilon\Big\rangle_\varepsilon.$$

By direct calculations, we find that

$$\begin{aligned}&\frac{\partial U_{\varepsilon,x_j}(y)}{\partial x_{ji}}\\&=\frac{\partial\big(V(x_j)\big)^{\frac{1}{p-1}}}{\partial x_{ji}}w\left(\frac{\sqrt{V(x_j)}}{\varepsilon}|y-x_j|\right)\\&\quad+\big(V(x_j)\big)^{\frac{1}{p-1}}w'\left(\frac{\sqrt{V(x_j)}}{\varepsilon}|y-x_j|\right)\\&\quad\times\left(\frac{\sqrt{V(x_j)}}{\varepsilon}\frac{x_{ji}-y_i}{|y-x_j|}+\frac{|y-x_j|}{\varepsilon}\frac{\partial\sqrt{V(x_j)}}{\partial x_{ji}}\right).\end{aligned}$$

Consequently, we obtain

$$\begin{aligned}\Big\langle\frac{\partial U_{\varepsilon,x_{\varepsilon,m}}}{\partial y_h},\frac{\partial\omega_\varepsilon}{\partial x_{j,i}}\Big\rangle_\varepsilon&=-\Big\langle\frac{\partial^2 U_{\varepsilon,x_{\varepsilon,j}}}{\partial x_{j,i}\partial y_h},\omega_\varepsilon\Big\rangle_\varepsilon=O\big(\varepsilon^{\frac{N}{2}-2}\big)\|\omega_\varepsilon\|_\varepsilon\\&=O\big(\varepsilon^{N-2+\min(p-1,1)}\big),\end{aligned}\tag{2.2.61}$$

and

$$\Big\langle\frac{\partial U_{\varepsilon,x_{\varepsilon,j}}}{\partial x_{ji}},\frac{\partial U_{\varepsilon,x_{\varepsilon,m}}}{\partial y_h}\Big\rangle_\varepsilon=-\Big\langle\frac{\partial U_{\varepsilon,x_{\varepsilon,j}}}{\partial y_i},\frac{\partial U_{\varepsilon,x_{\varepsilon,m}}}{\partial y_h}\Big\rangle_\varepsilon+O(\varepsilon^{N-1}).$$

Using (2.2.61), (2.2.20) and (2.2.21), we conclude that $a_{\varepsilon,i,j}=0$ for $i=1,\cdots,N$, $j=1,\cdots,k$. □

We define the function

$$K(\mathbf{x}) = I\Big(\sum_{j=1}^{k} U_{\varepsilon,x_j} + \omega_\varepsilon\Big),$$

where

$$I(u) = \frac{1}{2}\int_{\mathbb{R}^N} \big(\varepsilon^2|\nabla u|^2 + V(y)u^2\big) - \frac{1}{p+1}\int_{\mathbb{R}^N} u_+^{p+1},$$

and ω_ε is the function obtained in Proposition 2.2.6.

Note that the left-hand side of (2.2.60) is exactly the derivative of $K(\mathbf{x})$. Hence we have proved the following result.

Lemma 2.2.13 *Suppose that $\mathbf{x}_\varepsilon$ is a critical point of $K(\mathbf{x})$. Then $a_{\varepsilon,i,j} = 0$ for $i = 1, \cdots, N$ and $j = 1, \cdots, k$.*

2.3 Existence of Clustering Peak Solutions

In the last section, we construct a solution for the nonlinear Schrödinger equation (2.2.11) with k peaks separating from each other. In this section, we will consider peak solutions whose peaks cluster together at one point. This is not always possible for (2.2.11). For example, if the function $V(x)$ is radially symmetric and $V'(r) > 0$ for all $r > 0$, then any solution u_ε of (2.2.11) is radial and is decreasing. Therefore, u_ε has exactly one local maximum point.

Let $x_{\varepsilon,j}$ be a local maximum point of a solution u_ε with the property that $x_{\varepsilon,j} \to x_0$ as $\varepsilon \to 0$. Then for any $R > 0$,

$$u_\varepsilon(\varepsilon x + x_{\varepsilon,j}) \to \big(V(x_0)\big)^{\frac{1}{p-1}} w\big(\sqrt{V(x_0)}|x|\big) \quad \text{in } C^1(B_R(0)).$$

Since $w'(r) < 0$, it follows that $u_\varepsilon(\varepsilon x + x_{\varepsilon,j})$ has no any other local maximum point point in $B_R(0)$ if $\varepsilon > 0$ is small. This shows that if $x_{\varepsilon,j}$ and $x_{\varepsilon,m}$ are two local maximum points of u_ε, then $\frac{|x_{\varepsilon,j}-x_{\varepsilon,m}|}{\varepsilon} \to +\infty$ as $\varepsilon \to 0$.

In this section, we will prove that if x_0 is an isolated local maximum point of $V(x)$, then for any integer $k > 0$, (2.2.11) has a k-peak solution, whose peaks cluster at x_0. In other words, we want to construct a solution for (2.2.11) of the form

$$u_\varepsilon(x) = \sum_{j=1}^{k} U_{\varepsilon,x_{\varepsilon,j}} + \omega_\varepsilon,$$

such that as $\varepsilon \to 0$, $x_{\varepsilon,j} \to x_0$, $\frac{|x_{\varepsilon,j}-x_{\varepsilon,m}|}{\varepsilon} \to +\infty$ if $m \neq j$, and $\|\omega_\varepsilon\|_\varepsilon^2 = o(\varepsilon^N)$.

In the following, we always assume that each x_j is close to x_0, $j=1,\cdots,k$, while $\frac{|x_j-x_m|}{\varepsilon}$ is large for any $j\neq m$.

In view of Lemma 2.2.13, we can study the existence of critical points of the function $K(\mathbf{x})$. In the same spirit as in the last subsection, we proceed as follows.

(1) Estimate $I\big(\sum_{j=1}^k U_{\varepsilon,x_j}\big)$ and find the main terms for this function.
(2) Show that the error term ω_ε is negligible.

We begin by proving the following result.

Proposition 2.3.1 *We have the following estimate*

$$
\begin{aligned}
&I\Big(\sum_{j=1}^k U_{\varepsilon,x_j}\Big)\\
=&A\sum_{j=1}^k V(x_j)^{\frac{p+1}{p-1}-\frac{N}{2}}\varepsilon^N-(a_0+o(1))\varepsilon^N\\
&\times\sum_{j>m}V(x_0)^{\frac{p+1}{p-1}-\frac{N}{2}}e^{-\frac{(\sqrt{V(x_0)}+o(1))|x_m-x_j|}{\varepsilon}}\Big(\frac{\varepsilon}{\sqrt{V(x_0)}|x_m-x_j|}\Big)^{\frac{N-1}{2}}\\
&+\varepsilon^N O\Big(\varepsilon+\sum_{m\neq j}e^{-\frac{(1+\tau)\sqrt{V(x_0)}|x_m-x_j|}{\varepsilon}}\Big),
\end{aligned}
$$

where $\tau>0$ is a small constant, $a_0>0$ is some constant and

$$
A=\Big(\frac{1}{2}-\frac{1}{p+1}\Big)\int_{\mathbb{R}^N}w^{p+1}>0.
$$

Proof. We have

$$
\begin{aligned}
&\frac{1}{2}\int_{\mathbb{R}^N}\Big(\varepsilon^2|\nabla\sum_{j=1}^k U_{\varepsilon,x_j}|^2+V(y)\Big(\sum_{j=1}^k U_{\varepsilon,x_j}\Big)^2\Big)\\
&=\frac{1}{2}\sum_{j=1}^k\int_{\mathbb{R}^N}(\varepsilon^2|\nabla U_{\varepsilon,x_j}|^2+V(y)U_{\varepsilon,x_j}^2)\\
&\quad+\sum_{j=1}^k\sum_{j>m}\int_{\mathbb{R}^N}(\varepsilon^2\nabla U_{\varepsilon,x_j}\nabla U_{\varepsilon,x_m}+V(y)U_{\varepsilon,x_j}U_{\varepsilon,x_m}).
\end{aligned}
$$

From

$$
\int_{\mathbb{R}^N}V(y)U_{\varepsilon,x_j}U_{\varepsilon,x_m}=V(x_j)\int_{\mathbb{R}^N}U_{\varepsilon,x_j}U_{\varepsilon,x_m}+O(\varepsilon^{N+1}),
$$

and

$$
-\varepsilon^2\Delta U_{\varepsilon,x_j}+V(x_j)U_{\varepsilon,x_j}=U_{\varepsilon,x_j}^p,
$$

we obtain

$$
\begin{aligned}
&\frac{1}{2}\int_{\mathbb{R}^N}\Big(\varepsilon^2\Big|\nabla\sum_{j=1}^k U_{\varepsilon,x_j}\Big|^2+V(y)\Big(\sum_{j=1}^k U_{\varepsilon,x_j}\Big)^2\Big)\\
&=\frac{1}{2}\sum_{j=1}^k\int_{\mathbb{R}^N}\big(\varepsilon^2|\nabla U_{\varepsilon,x_j}|^2+V(x_j)U_{\varepsilon,x_j}^2\big)\\
&\quad+\sum_{j=1}^k\sum_{j>m}\int_{\mathbb{R}^N}\big(\varepsilon^2\nabla U_{\varepsilon,x_j}\nabla U_{\varepsilon,x_m}+V(x_j)U_{\varepsilon,x_j}U_{\varepsilon,x_m}\big)+O\big(\varepsilon^{N+1}\big)\\
&=\frac{1}{2}\sum_{j=1}^k\int_{\mathbb{R}^N}\big(\varepsilon^2|\nabla U_{\varepsilon,x_j}|^2+V(x_j)U_{\varepsilon,x_j}^2\big)\\
&\quad+\sum_{j=1}^k\sum_{j>m}\int_{\mathbb{R}^N}U_{\varepsilon,x_j}^p U_{\varepsilon,x_m}+O\big(\varepsilon^{N+1}\big).
\end{aligned}
$$

We also have

$$
\begin{aligned}
&\frac{1}{p+1}\int_{\mathbb{R}^N}\Big(\sum_{j=1}^k U_{\varepsilon,x_j}\Big)^{p+1}\\
&=\frac{1}{p+1}\sum_{j=1}^k\int_{\mathbb{R}^N}U_{\varepsilon,x_j}^{p+1}+\frac{1}{p+1}\int_{\mathbb{R}^N}\Big(\Big(\sum_{j=1}^k U_{\varepsilon,x_j}\Big)^{p+1}-\sum_{j=1}^k U_{\varepsilon,x_j}^{p+1}\Big).
\end{aligned}
$$

Thus, it follows that

$$
\begin{aligned}
I\Big(\sum_{j=1}^k U_{\varepsilon,x_j}\Big)=k\sum_{j=1}^k A_j-\frac{1}{p+1}\int_{\mathbb{R}^N}\Big(\Big(\sum_{j=1}^k U_{\varepsilon,x_j}\Big)^{p+1}-\sum_{j=1}^k U_{\varepsilon,x_j}^{p+1}\Big)\\
+\sum_{j=1}^k\sum_{j>m}\int_{\mathbb{R}^N}U_{\varepsilon,x_j}^p U_{\varepsilon,x_m}+O\big(\varepsilon^{N+1}\big),
\end{aligned}
\tag{2.3.1}
$$

where

$$
\begin{aligned}
A_j&=\frac{1}{2}\int_{\mathbb{R}^N}\big(\varepsilon^2|\nabla U_{\varepsilon,x_j}|^2+V(x_j)U_{\varepsilon,x_j}^2\big)-\frac{1}{p+1}\int_{\mathbb{R}^N}U_{\varepsilon,x_j}^{p+1}\\
&=\Big(\frac{1}{2}-\frac{1}{p+1}\Big)\int_{\mathbb{R}^N}U_{\varepsilon,x_j}^{p+1}
\end{aligned}
\tag{2.3.2}
$$

$$
\begin{aligned}
&= \Big(\frac{1}{2} - \frac{1}{p+1}\Big) \int_{\mathbb{R}^N} \Big(V(x_j)^{\frac{1}{p-1}} w\Big(\frac{\sqrt{V(x_j)}(y-x_j)}{\varepsilon}\Big)\Big)^{p+1} \\
&= \Big(\frac{1}{2} - \frac{1}{p+1}\Big) V(x_j)^{\frac{p+1}{p-1}-\frac{N}{2}} \varepsilon^N \int_{\mathbb{R}^N} w^{p+1}.
\end{aligned}
$$

On the other hand, using Lemma 6.1.2, we have that

$$
\begin{aligned}
&\int_{\mathbb{R}^N} \Big(\Big(\sum_{j=1}^k U_{\varepsilon,x_j}\Big)^{p+1} - \sum_{j=1}^k U_{\varepsilon,x_j}^{p+1}\Big) \\
&= (p+1)\sum_{j=1}^k \sum_{j>m} \int_{\mathbb{R}^N} U_{\varepsilon,x_j}^p U_{\varepsilon,x_m} + (p+1)\sum_{j=1}^k \int_{\mathbb{R}^N} U_{\varepsilon,x_j}\Big(\sum_{m>j} U_{\varepsilon,x_m}\Big)^p \\
&\quad + \varepsilon^N O\Big(\sum_{m\neq j} e^{-\frac{(1+\tau)\sqrt{V(x_0)}|x_m-x_j|}{\varepsilon}}\Big) \\
&= (p+1)\sum_{j=1}^k \sum_{j>m} \int_{\mathbb{R}^N} U_{\varepsilon,x_j}^p U_{\varepsilon,x_m} + (p+1)\sum_{j=1}^k \int_{\mathbb{R}^N} U_{\varepsilon,x_j} \sum_{m>j} U_{\varepsilon,x_m}^p \\
&\quad + \varepsilon^N O\Big(\sum_{m\neq j} e^{-\frac{(1+\tau)\sqrt{V(x_0)}|x_m-x_j|}{\varepsilon}}\Big).
\end{aligned} \tag{2.3.3}
$$

Combining (2.3.1), (2.3.2) and (2.3.3), we are led to the following estimate:

$$
\begin{aligned}
I\Big(\sum_{j=1}^k U_{\varepsilon,x_j}\Big) &= A\varepsilon^N \sum_{j=1}^k V(x_j)^{\frac{p+1}{p-1}-\frac{N}{2}} - \sum_{j=1}^k \int_{\mathbb{R}^N} U_{\varepsilon,x_j} \sum_{m>j} U_{\varepsilon,x_m}^p \\
&\quad + \varepsilon^N O\Big(\varepsilon + \sum_{m\neq j} e^{-\frac{(1+\tau)\sqrt{V(x_0)}|x_m-x_j|}{\varepsilon}}\Big).
\end{aligned}
$$

By Lemma 6.1.4 and

$$
U_{\varepsilon,x_j}(y) = \big(V(x_j)\big)^{\frac{1}{p-1}} w\left(\frac{\sqrt{V(x_j)}}{\varepsilon}(y-x_j)\right),
$$

we find that for $j \neq m$,

$$
\begin{aligned}
&\int_{\mathbb{R}^N} U_{\varepsilon,x_j}^p U_{\varepsilon,x_m} \\
&= \varepsilon^N V(x_j)^{\frac{p}{p-1}-\frac{N}{2}} V(x_m)^{\frac{1}{p-1}} \int_{\mathbb{R}^N} w^p(x) w\Big(\frac{\sqrt{V(x_m)}}{\sqrt{V(x_j)}}\Big(x - \frac{\sqrt{V(x_j)}|x_m-x_j|}{\varepsilon}\Big)\Big) \\
&= \varepsilon^N V(x_0)^{\frac{p+1}{p-1}-\frac{N}{2}} e^{-\frac{(\sqrt{V(x_0)}+o(1))|x_m-x_j|}{\varepsilon}} \Big(\frac{\varepsilon}{\sqrt{V(x_0)}|x_m-x_j|}\Big)^{\frac{N-1}{2}} \big(a_0 + o(1)\big).
\end{aligned}
$$

Hence the proposition follows. □

From Proposition 2.3.1, the main term for $I\big(\sum_{j=1}^{k} U_{\varepsilon,x_j}\big)$ is

$$
\begin{aligned}
F(\mathbf{x}) = {} & A\sum_{j=1}^{k} V(x_j)^{\frac{p+1}{p-2}-\frac{N}{2}}\varepsilon^N \\
& - (a_0+o(1))\varepsilon^N \sum_{j>m} V(x_0)^{\frac{p+1}{p-1}-\frac{N}{2}} e^{-\frac{(\sqrt{V(x_0)}+o(1))|x_m-x_j|}{\varepsilon}} \\
& \times \Big(\frac{\varepsilon}{\sqrt{V(x_0)}|x_m-x_j|}\Big)^{\frac{N-1}{2}}.
\end{aligned}
$$

Note that $\frac{p+1}{p-2}-\frac{N}{2}>0$ since $p<\frac{N+2}{N-2}$. If x_0 is a local maximum point of $V(x)$, then the function $F(\mathbf{x})$ is smaller if x_j moves away from x_0. On the other hand, from the second term in x_j, we find that $F(\mathbf{x})$ is smaller if x_j moves toward to x_m. Hence $F(\mathbf{x})$ attains its maximum in the interior of the following set

$$
D=\Big\{\mathbf{x}: x_j\in \overline{B_\theta(x_0)},\ j=1,\cdots,k,\ \ |x_m-x_j|\ge \theta\varepsilon\ln\frac{1}{\varepsilon},\ \ m\ne j\Big\}, \tag{2.3.4}
$$

where $\theta>0$ is a fixed small constant. If we can prove that the error term ω_ε is negligible, then $K(\mathbf{x})$ has a critical point in D. We make the following expansion:

$$
\begin{aligned}
K(\mathbf{x}) &= I\Big(\sum_{j=1}^{k} U_{\varepsilon,x_j}+\omega_\varepsilon\Big) \\
&= I\Big(\sum_{j=1}^{k} U_{\varepsilon,x_j}\Big) + \big\langle l_\varepsilon,\omega_\varepsilon\big\rangle_\varepsilon + O\Big(\|\omega_\varepsilon\|_\varepsilon^2+\varepsilon^{N(1-\frac{p+1}{2})}\|\omega_\varepsilon\|_\varepsilon^{p+1}\Big) \\
&= I\Big(\sum_{j=1}^{k} U_{\varepsilon,x_j}\Big) + O\Big(\|l_\varepsilon\|_\varepsilon\|\omega_\varepsilon\|_\varepsilon+\|\omega_\varepsilon\|_\varepsilon^2+\varepsilon^{N(1-\frac{p+1}{2})}\|\omega_\varepsilon\|_\varepsilon^{p+1}\Big).
\end{aligned} \tag{2.3.5}
$$

As in the proof of Lemma 2.2.4, using (6.1.4), we can deduce

$$
\|l_\varepsilon\|_\varepsilon \le C\varepsilon^{\frac{N}{2}}\Big(\varepsilon+\sum_{m\ne j} e^{-\frac{(1+\tau)\sqrt{V(x_0)}|x_m-x_j|}{2\varepsilon}}\Big),
$$

where $\tau>0$ is a small constant. See the estimates in (2.2.41), (2.2.42) and (2.2.43). As a result, one has

$$
\|\omega_\varepsilon\|_\varepsilon \le C\|l_\varepsilon\|_\varepsilon \le C\varepsilon^{\frac{N}{2}}\Big(\varepsilon+\sum_{m\ne j} e^{-\frac{(1+\tau)\sqrt{V(x_0)}|x_m-x_j|}{2\varepsilon}}\Big). \tag{2.3.6}
$$

Combining Proposition 2.3.1, (2.3.5) and (2.3.6), we obtain

$$
\begin{aligned}
K(\mathbf{x}) &= I\Big(\sum_{j=1}^{k} U_{\varepsilon,x_j} + \omega_\varepsilon\Big) = A\sum_{j=1}^{k} V(x_j)^{\frac{p+1}{p-2}-\frac{N}{2}}\varepsilon^N \\
&- (a_0 + o(1))\varepsilon^N \sum_{j>m} V(x_0)^{\frac{p+1}{p-1}-\frac{N}{2}} e^{-\frac{(\sqrt{V(x_0)}+o(1))|x_m - x_j|}{\varepsilon}} \\
&\times \Big(\frac{\varepsilon}{\sqrt{V(x_0)}|x_m - x_j|}\Big)^{\frac{N-1}{2}} \\
&+ \varepsilon^N O\Big(\varepsilon + \sum_{m\neq j} e^{-\frac{(1+\tau)\sqrt{V(x_0)}|x_m - x_j|}{\varepsilon}}\Big),
\end{aligned} \tag{2.3.7}
$$

where $\tau > 0$ is a small constant, $A > 0$ and $B > 0$ are some constants.

We are now ready to prove the following result on the existence of clustering peak solutions for (2.2.11).

Theorem 2.3.2 (Kang–Wei [90], 2000; Noussair–Yan [116], 2000) *Suppose x_0 is an isolated strict local maximum point of $V(x)$. Then, for any integer $k > 0$, there exists $\varepsilon_0 > 0$, such that for any $\varepsilon \in (0, \varepsilon_0]$, (2.2.11) has a solution of the form*

$$
u_\varepsilon(x) = \sum_{j=1}^{k} U_{\varepsilon,x_{\varepsilon,j}} + \omega_\varepsilon,
$$

for some $x_{\varepsilon,j} \in B_\delta(x_0)$, and $\|\omega_\varepsilon\|_\varepsilon = o(\varepsilon^{\frac{N}{2}})$. Moreover, as $\varepsilon \to 0$, $x_{\varepsilon,j} \to x_0$, $\frac{|x_{\varepsilon,j}-x_{\varepsilon,m}|}{\varepsilon} \to +\infty$ if $m \neq j$.

Proof. First, we can use a similar argument to that in Proposition 2.2.6 to prove that for any $\mathbf{x} \in D$, there exists $\omega_\varepsilon \in E_\varepsilon$ satisfying (2.2.40) and (2.3.6).

Next, we need to choose $\mathbf{x} \in D$ suitable, so that $\sum_{j=1}^{k} U_{\varepsilon,x_{\varepsilon,j}} + \omega_\varepsilon$ is a solution of (2.2.11). For this purpose, we only need to prove that $K(\mathbf{x})$ has a critical point in D.

Consider the following problem:

$$
\max_{\mathbf{x}\in D} K(\mathbf{x}).
$$

Thus it is achieved by $\mathbf{x}_\varepsilon \in D$. In order to prove that $\mathbf{x}_\varepsilon$ is a critical point of $K(\mathbf{x})$, we just need to prove that $\mathbf{x}_\varepsilon$ is in the interior of D.

Fix a sufficiently small constant $\beta > 0$ such that $\frac{\beta}{\theta}$ is large, where $\theta > 0$ is the small constant in (2.3.4). We take $\tilde{x}_{\varepsilon,j}$, $j = 1, \cdots, k$, satisfying $|\tilde{x}_{\varepsilon,j} -$

$x_0| \le \varepsilon^\beta$ and $|\tilde{x}_{\varepsilon,j} - \tilde{x}_{\varepsilon,m}| \ge \varepsilon^{1-\beta}$ if $m \ne j$. Then for $\tilde{\mathbf{x}}_\varepsilon = (\tilde{x}_{\varepsilon,1}, \cdots, \tilde{x}_{\varepsilon,k}) \in D$, one has

$$K(\tilde{\mathbf{x}}_\varepsilon) = kV(x_0)^{\frac{p+1}{p-2}-\frac{N}{2}}\varepsilon^N + \varepsilon^N O\big(\varepsilon^{2\beta}\big).$$

Suppose that there exists x_{ε,j_0} such that $x_{\varepsilon,j_0} \in \partial B_\theta(x_0)$. Then by (2.3.7), we have

$$K(\mathbf{x}_\varepsilon) \le V(x_{\varepsilon,j_0})^{\frac{p+1}{p-2}-\frac{N}{2}}\varepsilon^N + \sum_{j\ne j_0} V(x_0)^{\frac{p+1}{p-2}-\frac{N}{2}}\varepsilon^N + o(\varepsilon^N) < K(\tilde{\mathbf{x}}_\varepsilon).$$

But this contradicts the fact that $\mathbf{x}_\varepsilon$ is a maximum point of $K(\mathbf{x})$ in D.

Suppose that there exist x_{ε,j_0} and x_{ε,m_0}, $m_0 \ne j_0$, such that $|x_{\varepsilon,j_0} - x_{\varepsilon,m_0}| = \theta\varepsilon \ln\frac{1}{\varepsilon}$. Then it follows from (2.3.7) that

$$K(\mathbf{x}_\varepsilon) \le kV(x_0)^{\frac{p+1}{p-2}-\frac{N}{2}}\varepsilon^N - B\varepsilon^N e^{-2\sqrt{V(x_0)}\theta\ln\frac{1}{\varepsilon}}|\ln\varepsilon|^{-\frac{N-1}{2}} + \varepsilon^N O(\varepsilon) < K(\tilde{\mathbf{x}}_\varepsilon),$$

for some $B > 0$, which is also a contradiction. Thus, we have proved that $\mathbf{x}_\varepsilon$ is in the interior of D, and so it is a critical point of $K(\mathbf{x})$. □

Remark 2.3.3 It is proved in [53] that if x_0 is a non-degenerate saddle point of $V(x)$, then for any integer $k > 0$, (2.2.11) has a k-peak solution, such that all the peaks cluster at x_0. However, if x_0 is a non-degenerate local maximum point of $V(x)$, then such result is no longer true.

2.4 Existence of Bubbling Solutions for Elliptic Problems with Critical Growth

In this section, we will construct bubbling solutions for the following problem

$$\begin{cases} -\Delta u = u_+^{2^*-1} + \lambda u, \text{ in } \Omega, \\ u \in H_0^1(\Omega), \end{cases} \tag{2.4.1}$$

where $\lambda > 0$ is a small parameter. Note that any nontrivial solution of (2.4.1) is positive. We will construct a solution of the form

$$u_\lambda(x) \approx U_{x_\lambda,\mu_\lambda}(x), \tag{2.4.2}$$

where $\mu_\lambda \to +\infty$ as $\lambda \to 0$, and

$$U_{x,\mu}(y) = \frac{C_0\mu^{\frac{N-2}{2}}}{(1+\mu^2|y-x|^2)^{\frac{N-2}{2}}}, \quad x \in \mathbb{R}^N, \ \mu \in \mathbb{R}_+, \tag{2.4.3}$$

which are all the positive solutions of $-\Delta u = u^{2^*-1}$ in $\mathbb{R}^N$, $u \in D^{1,2}(\mathbb{R}^N)$. It follows from Theorem 2.1.3 that if $x_\lambda \to x_0$ as $\lambda \to 0$, then x_0 must be a critical point of the Robin function $\varphi(x)$.

We will follow the procedure mentioned in subsection 3.2 to construct such solutions for (2.4.1).

The limit problem and the approximate solutions. Since we want to construct a solution of the form (2.4.2), the corresponding limit problem is

$$\begin{cases} -\Delta u = u^{2^*-1}, u > 0, \text{ in } \mathbb{R}^N, \\ u \in D^{1,2}(\mathbb{R}^N), \end{cases} \tag{2.4.4}$$

where $D^{1,2}(\mathbb{R}^N)$ is the completion of the space $C_0^\infty(\mathbb{R}^N)$ under the norm $\|u\| = \left(\int_{\mathbb{R}^N} |\nabla u|^2\right)^{\frac{1}{2}}$. Problem (2.4.4) is invariant under the translation and scaling, and all the solutions of (2.4.4) are given by $U_{x,\mu}$ defined in (2.4.3). We will use $U_{x_\lambda,\mu_\lambda}$ as an approximate solution of (2.4.1). As $U_{x,\mu}$ does not belong to $H_0^1(\Omega)$, we need to make a projection as follows. Let $PU_{x,\mu}$ be the solution of the following problem

$$\begin{cases} -\Delta u = U_{x,\mu}^{2^*-1}, & \text{in } \Omega, \\ u \in H_0^1(\Omega). \end{cases}$$

Let

$$\psi_{x,\mu} = U_{x,\mu} - PU_{x,\mu}.$$

Then we have

$$\begin{cases} -\Delta \psi_{x,\mu} = 0, & \text{in } \Omega, \\ \psi_{x,\mu} = U_{x,\mu}, & \text{on } \partial\Omega. \end{cases}$$

Thus, we have $\psi_{x,\mu} > 0$.

Lemma 2.4.1 *Suppose that $x \in \Omega$ and $\mu > 0$ is large. Then*

$$\psi_{x,\mu}(y) = \frac{C_0 H(y,x)}{\mu^{\frac{N-2}{2}}} + O\left(\frac{1}{\mu^{\frac{N+2}{2}} d^N}\right), \tag{2.4.5}$$

where $d = d(x, \partial\Omega)$, and $H(y,x)$ is the solution of

$$\begin{cases} \Delta u = 0, & \textit{in } \Omega, \\ u = \dfrac{1}{|y-x|^{N-2}}, & \textit{on } \partial\Omega. \end{cases} \tag{2.4.6}$$

Proof. For $y \in \partial\Omega$, one has

$$U_{x,\mu}(y) = \frac{C_0}{\mu^{\frac{N-2}{2}}|y-x|^{N-2}} + O\Big(\frac{1}{\mu^{\frac{N+2}{2}}d^N}\Big).$$

Thus, the function $w = \psi_{x,\mu}(y) - \frac{C_0 H(y,x)}{\mu^{\frac{N-2}{2}}}$ satisfies

$$\begin{cases} \Delta w = 0, & \text{in } \Omega, \\ w = O\Big(\frac{1}{\mu^{\frac{N+2}{2}}d^N}\Big), & \text{on } \partial\Omega. \end{cases}$$

By the maximum principle, we have $|w| = O\Big(\frac{1}{\mu^{\frac{N+2}{2}}d^N}\Big)$. Hence the result follows. □

In the following, we will construct a solution for (2.4.1) of the form

$$u_\lambda = PU_{x_\lambda,\mu_\lambda} + \omega_\lambda,$$

for x_λ near a critical point of $\varphi(x)$ and large $\mu_\lambda > 0$, where $\|\omega_\lambda\|$ is small.

We know that ω_λ satisfies

$$\begin{cases} L_\lambda\omega_\lambda = l_\lambda + R_\lambda(\omega_\lambda), & \text{in } \Omega, \\ \omega_\lambda \in H_0^1(\Omega), \end{cases}$$

where

$$L_\lambda\omega := -\Delta\omega - \lambda\omega - (2^*-1)(PU_{x,\mu})^{2^*-2}\omega,$$

$$l_\lambda = \big(PU_{x,\mu}\big)_+^{2^*-1} - U_{x,\mu}^{2^*-1} + \lambda PU_{x,\mu}, \tag{2.4.7}$$

and

$$R_\lambda(\omega) = \big(PU_{x,\mu}+\omega\big)_+^{2^*-1} - \big(PU_{x,\mu}\big)^{2^*-1} - (2^*-1)(PU_{x,\mu})^{2^*-2}\omega.$$

The energy of the approximate solutions. In the following, we assume that $d = d(x,\partial\Omega) \geq c_0 > 0$. Now we will estimate $I\big(PU_{x,\mu}\big)$, where

$$I(u) = \frac{1}{2}\int_\Omega\big(|\nabla u|^2 - \lambda u^2\big) - \frac{1}{2^*}\int_\Omega u_+^{2^*}.$$

We have

$$
\begin{aligned}
\frac{1}{2}\int_\Omega |\nabla PU_{x,\mu}|^2 &= -\frac{1}{2}\int_\Omega PU_{x,\mu}\Delta(PU_{x,\mu})\\
&= \frac{1}{2}\int_\Omega PU_{x,\mu}U_{x,\mu}^{2^*-1} = \frac{1}{2}\int_\Omega U_{x,\mu}^{2^*} - \frac{1}{2}\int_\Omega \psi_{x,\mu}U_{x,\mu}^{2^*-1}\\
&= \frac{1}{2}\int_{\mathbb{R}^N} U_{0,1}^{2^*} + O\Big(\frac{1}{\mu^N}\Big) - \frac{1}{2}\int_\Omega \frac{C_0 H(y,x)}{\mu^{\frac{N-2}{2}}}U_{x,\mu}^{2^*-1}\\
&\quad + O\Big(\frac{1}{\mu^{\frac{N+2}{2}}}\int_\Omega U_{x,\mu}^{2^*-1}\Big)\\
&= \frac{1}{2}\int_{\mathbb{R}^N} U_{0,1}^{2^*} - \frac{1}{2}\frac{C_0 H(x,x)}{\mu^{N-2}}\int_{\mathbb{R}^N} U_{0,1}^{2^*-1} + O\Big(\frac{\ln\mu}{\mu^N}\Big),
\end{aligned}
$$

since

$$
\begin{aligned}
\int_\Omega \frac{H(y,x)}{\mu^{\frac{N-2}{2}}}U_{x,\mu}^{2^*-1} &= \frac{1}{\mu^{\frac{N-2}{2}}}\int_\Omega \Big(H(x,x) + \langle \nabla H(x,x), y-x\rangle\\
&\quad + O(|y-x|^2)\Big)U_{x,\mu}^{2^*-1}\\
&= \frac{H(x,x)}{\mu^{N-2}}\int_{\mathbb{R}^N} U_{0,1}^{2^*-1} + O\Big(\frac{\ln\mu}{\mu^N}\Big).
\end{aligned}
$$

On the other hand, in view of $PU_{x,\mu} = U_{x,\mu} - \psi_{x,\mu} > 0$, one has $0 \le \frac{\psi_{x,\mu}}{U_{x,\mu}} \le 1$. As a result, if $N \ge 5$, we have that

$$
\begin{aligned}
\int_\Omega (PU_{x,\mu})^{2^*} &= \int_\Omega (U_{x,\mu} - \psi_{x,\mu})^{2^*}\\
&= \int_\Omega U_{x,\mu}^{2^*} - 2^*\int_\Omega U_{x,\mu}^{2^*-1}\psi_{x,\mu} + O\Big(\int_\Omega U_{x,\mu}^{2^*-2}\psi_{x,\mu}^2\Big)\\
&= \int_{\mathbb{R}^N} U_{0,1}^{2^*} - \frac{2^* C_0 H(x,x)}{\mu^{N-2}}\int_{\mathbb{R}^N} U_{0,1}^{2^*-1} + O\Big(\frac{\ln\mu}{\mu^N}\Big),
\end{aligned}
$$

and

$$
\begin{aligned}
\int_\Omega (PU_{x,\mu})^2 &= \int_\Omega (U_{x,\mu} - \psi_{x,\mu})^2\\
&= \int_\Omega U_{x,\mu}^2 - 2\int_\Omega U_{x,\mu}\psi_{x,\mu} + \int_\Omega \psi_{x,\mu}^2\\
&= \frac{1}{\mu^2}\int_{\mathbb{R}^N} U_{0,1}^2 + O\Big(\frac{1}{\mu^{N-2}}\Big).
\end{aligned}
$$

Consequently, for $N \geq 5$, one then has

$$I(PU_{x,\mu}) = \frac{1}{N}\int_{\mathbb{R}^N} U_{0,1}^{2^*} + \frac{1}{2}\frac{C_0\varphi(x)}{\mu^{N-2}}\int_{\mathbb{R}^N} U_{0,1}^{2^*-1} - \frac{\lambda}{\mu^2}\int_{\mathbb{R}^N} U_{0,1}^2 + O\Big(\frac{\ln\mu}{\mu^N} + \frac{\lambda}{\mu^{N-2}}\Big).$$

Thus, the main term is

$$\frac{1}{2}\frac{C_0\varphi(x)}{\mu^{N-2}}\int_{\mathbb{R}^N} U_{0,1}^{2^*-1} - \frac{\lambda}{\mu^2}\int_{\mathbb{R}^N} U_{0,1}^2,$$

which has a critical point (x_0, μ_λ) with $\nabla\varphi(x_0) = 0$ and $\mu_\lambda \sim \lambda^{-1/(N-4)}$ if $N \geq 5$.

To show that the approximate solution $PU_{x,\mu}$ is good, we need to estimate $\|l_\lambda\|$, which is defined in (2.4.7).

Lemma 2.4.2 *Suppose that $N \geq 5$. We have*

$$\|l_\lambda\| \leq \begin{cases} \dfrac{C\lambda}{\mu^2} + \dfrac{C}{\mu^{\frac{N+2}{2}}}, & N \geq 7, \\[2ex] \dfrac{C\lambda(\ln\mu)^{\frac{2}{3}}}{\mu^2} + \dfrac{C(\ln\mu)^{\frac{2}{3}}}{\mu^4}, & N = 6, \\[2ex] \dfrac{C\lambda}{\mu^{\frac{3}{2}}} + \dfrac{C}{\mu^3}, & N = 5. \end{cases}$$

Proof. Since $0 \leq \frac{\psi_{x,\mu}}{U_{x,\mu}} \leq 1$, we have

$$|(PU_{x,\mu})^{2^*-1} - U_{x,\mu}^{2^*-1}| \leq CU_{x,\mu}^{2^*-2}\psi_{x,\mu} \leq \frac{CU_{x,\mu}^{2^*-2}}{\mu^{\frac{N-2}{2}}}.$$

We can check that $\frac{2N(2^*-2)(N-2)}{N+2} < N$ if $N \geq 7$, $\frac{2N(2^*-2)(N-2)}{N+2} = N$ if $N = 6$, and $\frac{2N(2^*-2)(N-2)}{N+2} > N$ if $N \leq 5$. As a result, it follows that

$$\begin{aligned}
\Big|\int_\Omega \big((PU_{x,\mu})^{2^*-1} - U_{x,\mu}^{2^*-1}\big)\eta\Big| &\leq \frac{C}{\mu^{\frac{N-2}{2}}}\int_\Omega U_{x,\mu}^{2^*-2}|\eta| \\
&\leq \frac{C}{\mu^{\frac{N-2}{2}}}\Big(\int_\Omega U_{x,\mu}^{\frac{2N(2^*-2)}{N+2}}\Big)^{\frac{N+2}{2N}}\|\eta\| \\
&\leq \frac{C}{\mu^{\frac{N-2}{2}}}\Big(\int_\Omega\Big(\frac{1}{\mu^{\frac{N-2}{2}}|y-x|^{N-2}}\Big)^{\frac{2N(2^*-2)}{N+2}}\Big)^{\frac{N+2}{2N}}\|\eta\| \leq \frac{C}{\mu^{\frac{N+2}{2}}}\|\eta\|,
\end{aligned}$$

for $N \geq 7$, and that

$$\Big|\int_\Omega \big((PU_{x,\mu})^{2^*-1} - U_{x,\mu}^{2^*-1}\big)\eta\Big|$$

$$\leq \frac{C}{\mu^{N-2}}\Big(\int_{\Omega_\mu} U_{0,1}^{\frac{2N(2^*-2)}{N+2}}\Big)^{\frac{N+2}{2N}}\|\eta\| \leq \begin{cases} \dfrac{C}{\mu^{N-2}}\|\eta\|, & N \leq 5, \\ \dfrac{C(\ln\mu)^{\frac{N+2}{2N}}}{\mu^{N-2}}\|\eta\|, & N = 6, \end{cases}$$

where $\Omega_\mu = \{y : \mu^{-1}y + x \in \Omega\}$.

We also have

$$\Big|\int_\Omega PU_{x,\mu}\eta\Big| \leq C\int_\Omega U_{x,\mu}|\eta| \leq C\Big(\int_\Omega U_{x,\mu}^{\frac{2N}{N+2}}\Big)^{\frac{N+2}{2N}}\|\eta\|$$

$$\leq \begin{cases} \dfrac{C}{\mu^2}\|\eta\|, & N \geq 7, \\ \dfrac{C(\ln\mu)^{\frac{2}{3}}}{\mu^2}\|\eta\|, & N = 6, \\ \dfrac{C}{\mu^{\frac{3}{2}}}\|\eta\|, & N = 5. \end{cases}$$

Combining all the above estimates, we obtain the result. □

Noting that

$$\|l_\lambda\|^2 = o\Big(\frac{1}{\mu^{N-2}} + \frac{\lambda}{\mu^2}\Big),$$

we conclude that the approximate solution $PU_{x,\mu}$ is good.

The non-degeneracy of the solution $U_{0,1}$ and the reduction. By Theorem 6.6.2, the kernel of the linear operator

$$-\Delta\omega - (2^* - 1)U_{0,1}^{2^*-2}\omega, \quad \omega \in D^{1,2}(\mathbb{R}^N),$$

is given by

$$span\Big\{\frac{\partial U_{x,1}}{\partial x_i}\Big|_{x=0}, \ i = 1, \cdots, N, \ \frac{\partial U_{0,\lambda}}{\partial\lambda}\Big|_{\lambda=1}\Big\}.$$

So the kernel is $(N+1)$-dimensional. This shows that the solution $U_{0,1}$ is non-degenerate. We can therefore carry out the reduction procedure, which basically follows exactly the same argument to that in Subsection 3.2.

Define

$$E_\lambda = \Big\{\omega : \omega \in H_0^1(\Omega), \Big\langle\omega, \frac{\partial PU_{x,\mu}}{\partial x_i}\Big\rangle = \Big\langle\omega, \frac{\partial PU_{x,\mu}}{\partial\mu}\Big\rangle = 0, \quad i = 1, \cdots, N\Big\},$$

where $\langle u, v\rangle = \int_\Omega \nabla u \nabla v$.

The projection from $H_0^1(\Omega)$ to E_λ is then defined by

$$Q_\lambda u = u - b_{\lambda,0}\frac{\partial PU_{x,\mu}}{\partial \mu} - \sum_{i=1}^N b_{\lambda,i}\frac{\partial PU_{x,\mu}}{\partial x_i},$$

where the number $b_{\lambda,i}$ is chosen in such a way that $Q_\lambda u \in E_\lambda$. It is easy to check that

$$\|Q_\lambda u\| \le C\|u\|.$$

Proposition 2.4.3 *There exist $\lambda_0 > 0$ and $\rho > 0$ such that for any $\lambda \in (0, \lambda_0]$, $x \in \Omega$ with $d(x, \partial\Omega) \ge c > 0$ and $\mu > 0$ sufficiently large, the operator $Q_\lambda L_\lambda$ is bijective in E_λ. Moreover, one has*

$$\|Q_\lambda L_\lambda \omega\| \ge \rho\|\omega\|, \quad \forall \omega \in E_\lambda. \tag{2.4.8}$$

Proof. The proof of this proposition is very similar to that of Proposition 2.2.3. Therefore, we will just sketch it.

To prove (2.4.8), we argue by contradiction. Suppose that there exist $\lambda_n \to 0$, $x_n \in \Omega$ with $d(x_n, \partial\Omega) \ge c > 0$ and $\mu_n \to +\infty$, $\omega_n \in E_{\lambda_n}$ such that

$$\|Q_{\lambda_n} L_{\lambda_n}\omega_n\| \le \frac{1}{n}\|\omega_n\|. \tag{2.4.9}$$

We may assume $\|\omega_n\| = 1$.

From (2.4.9), we see that

$$\int_\Omega \left(|\nabla\omega_n|^2 - (2^* - 1)(PU_{x_n,\mu_n})^{2^*-2}\omega_n^2\right) = o(1). \tag{2.4.10}$$

Now, defining

$$\tilde\omega_n(y) = \mu_n^{-\frac{N-2}{2}}\omega_n(\mu_n^{-1}y + x_n),$$

one then has

$$\int_{\mathbb{R}^N} |\nabla\tilde\omega_n|^2 \le C.$$

Therefore, we can assume that as $n \to +\infty$,

$$\tilde\omega_n \rightharpoonup \omega \quad \text{weakly in } D^{1,2}(\mathbb{R}^N),$$

and

$$\tilde\omega_n \to \omega \quad \text{strongly in } L^2_{loc}(\mathbb{R}^N).$$

It then follows that $\omega \in D^{1,2}(\mathbb{R}^N)$ satisfies

$$-\Delta\omega - (2^* - 1)U_{0,1}^{2^*-2}\omega = 0.$$

From the non-degeneracy of $U_{0,1}$, we have

$$\omega = c_0 \frac{\partial U_{0,\mu}}{\partial \mu}\Big|_{\mu=1} + \sum_{i=1}^{N} c_i \frac{\partial U_{x,1}}{\partial x_i}\Big|_{x=0}.$$

On the other hand, from $\omega_n \in E_{\lambda_n}$, we can prove easily that

$$\int_{\mathbb{R}^N} \nabla\omega\nabla \frac{\partial U_{x,1}}{\partial x_i}\Big|_{x=0} = \int_{\mathbb{R}^N} \nabla\omega\nabla \frac{\partial U_{0,\mu}}{\partial \mu}\Big|_{\mu=1} = 0, \quad i = 1, \cdots, N.$$

Hence it follows that $\omega = 0$, and consequently, we have

$$\begin{aligned}
\int_\Omega (PU_{x_n,\mu_n})^{2^*-2}\omega_n^2 &\le \int_\Omega U_{x_n,\mu_n}^{2^*-2}\omega_n^2 \\
&= \int_{B_{R\mu_n^{-1}}(x_n)} U_{x_n,\mu_n}^{2^*-2}\omega_n^2 + \int_{\Omega\setminus B_{R\mu_n^{-1}}(x_n)} U_{x_n,\mu_n}^{2^*-2}\omega_n^2 \\
&\le \int_{B_R(0)} U_{0,1}^{2^*-2}\tilde{\omega}_n^2 + \Big(\int_{\Omega\setminus B_{R\mu_n^{-1}}(x_n)} U_{x_n,\mu_n}^{2^*}\Big)^{1-\frac{2}{2^*}} \|\omega_n\|^2 \\
&= o(1),
\end{aligned}$$

which, together with (2.4.10), implies that

$$\int_\Omega |\nabla\omega_n|^2 = o(1).$$

This is a contradiction to $\|\omega_n\| = 1$. □

Now we consider

$$Q_\lambda L_\lambda \omega = Q_\lambda l_\lambda + Q_\lambda R_\lambda(\omega), \quad \omega \in E_\lambda. \tag{2.4.11}$$

It is easy to check that

$$\|R_\lambda(\omega)\| = O\big(\|\omega\|^{\min(2,2^*-1)}\big).$$

Using Proposition 2.4.3 and the contraction mapping theorem, we can prove the following result.

Proposition 2.4.4 *There exists* $\lambda_0 > 0$ *such that for any* $\lambda \in (0, \lambda_0]$, $(x, \mu) \in \Omega \times \mathbb{R}_+$ *with* $d(x, \partial\Omega) \ge c > 0$ *and* $\mu > 0$ *sufficiently large, there is an* $\omega_\lambda \in E_\lambda$ *satisfying* (2.4.11) *and*

$$\|\omega_\lambda\| \le C\|l_\lambda\|.$$

Moreover, the map ω_λ *from* $\{(x, \mu) : d(x, \partial\Omega) \ge c > 0,\ \mu > 0$ *sufficiently large*$\}$ *to* $H_0^1(\Omega)$ *is* C^1.

Proof. As the proof of this proposition is very similar to that of Proposition 2.2.6, we will omit it. □

The finite dimensional problem

We need to choose x and μ, such that $PU_{x,\mu} + \omega_\lambda$ is a solution of (2.4.1). Define

$$K(x, \mu) = I\big(PU_{x,\mu} + \omega_\lambda\big).$$

Similar to Section 2.3, we can prove that if (x, μ) is a critical point of $K(x, \mu)$, then $PU_{x,\mu} + \omega_\lambda$ is a solution of (2.4.1).

Now, expand $K(x, \mu)$ as follows:

$$\begin{aligned}
K(x, \mu) &= I\big(PU_{x,\mu}\big) + O\big(\|l_\lambda\|\|\omega_\lambda\| + \|\omega_\varepsilon\|^2\big)\\
&= I\big(PU_{x,\mu}\big) + O\Big(\frac{\lambda^2}{\mu^2} + \frac{1}{\mu^{N-2+\tau}}\Big)\\
&= \frac{1}{N}\int_{\mathbb{R}^N} U_{0,1}^{2^*} + \frac{1}{2}\frac{C_0\varphi(x)}{\mu^{N-2}}\int_{\mathbb{R}^N} U_{0,1}^{2^*-1} - \frac{\lambda}{\mu^2}\int_{\mathbb{R}^N} U_{0,1}^2\\
&\quad + O\Big(\frac{\lambda^2}{\mu^2} + \frac{1}{\mu^{N-2+\tau}} + \frac{\lambda}{\mu^{N-2}}\Big),
\end{aligned}$$

where $\tau > 0$ is a constant.

We now will discuss the existence of critical point of $K(x, \mu)$. For each fixed $x \in \Omega$, the main term of $K(x, \mu)$ has a critical point $\mu(x)$ that is also a minimum point

$$\mu(x) = L(x)\lambda^{-\frac{1}{N-4}},$$

where

$$L(x) = \left(\frac{C_0(N-2)\varphi(x)\displaystyle\int_{\mathbb{R}^N} U_{0,1}^{2^*-1}}{4\displaystyle\int_{\mathbb{R}^N} U_{0,1}^2}\right)^{\frac{1}{N-4}}. \tag{2.4.12}$$

For any $\mu = t\lambda^{-\frac{1}{N-4}}$ with $t \in [\frac{1}{2}L(x), 2L(x)]$, we have

$$\begin{aligned}
&\frac{1}{2}\frac{C_0\varphi(x)}{\mu^{N-2}}\int_{\mathbb{R}^N} U_{0,1}^{2^*-1} - \frac{\lambda}{\mu^2}\int_{\mathbb{R}^N} U_{0,1}^2\\
&\ge \frac{1}{2}\frac{C_0\varphi(x)}{\mu^{N-2}(x)}\int_{\mathbb{R}^N} U_{0,1}^{2^*-1} - \frac{\lambda}{\mu^2(x)}\int_{\mathbb{R}^N} U_{0,1}^2 + c'(t - L(x))^2\lambda^{\frac{N-2}{N-4}} \qquad (2.4.13)\\
&= -\frac{B_0\lambda^{\frac{N-2}{N-4}}}{\varphi^{\frac{2}{N-4}}(x)} + c'(t - L(x))^2\lambda^{\frac{N-2}{N-4}},
\end{aligned}$$

where $B_0 > 0$ and $c' > 0$ are some constants.

On the other hand, since $\varphi(x) \to +\infty$ as $d(x, \partial\Omega) \to 0$, $\varphi(x)$ always has a minimum point in Ω. Thus, $K(x, \mu)$ has a minimum point (x_λ, μ_λ) satisfying $x_\lambda \to x_0 \in \Omega$ with $\varphi(x_0) = \min_{x\in\Omega} \varphi(x)$, and

$$\mu_\lambda = \left(\left(\frac{C_0(N-2)\varphi(x_0) \int_{\mathbb{R}^N} U_{0,1}^{2^*-1}}{4 \int_{\mathbb{R}^N} U_{0,1}^2} \right)^{\frac{1}{N-4}} + o(1) \right)^{\frac{1}{N-4}} \lambda^{-\frac{1}{N-4}},$$

as $\lambda \to 0$.

Now, we consider the case that the Robin function $\varphi(x)$ has more local minimum points. Suppose that there exists a closed set $S \subset\subset \Omega$ such that

$$\min_{x\in S} \varphi(x) < \min_{x\in\partial S} \varphi(x). \tag{2.4.14}$$

This condition implies that $\varphi(x)$ has a local minimum point in the interior of S. We can now prove the following result:

Theorem 2.4.5 *Suppose that* (2.4.14) *holds. Then there exists* $\lambda_0 > 0$ *such that for any* $\lambda \in (0, \lambda_0]$, (2.4.1) *has a solution of the form*

$$u_\lambda = PU_{x_\lambda,\mu_\lambda} + \omega_\lambda,$$

for some $x_\lambda \in S$ *and* $\mu_\lambda > 0$ *large, and* $\|\omega_\lambda\| = o(1)$. *Moreover, as* $\lambda \to 0$, $x_\lambda \to x_0$ *with* $\varphi(x_0) = \min_{x\in S} \varphi(x)$.

Proof. Write

$$D = \left\{ (x,\mu) : x \in S,\ \mu \in \left[\frac{1}{2} L(x)\lambda^{-\frac{1}{N-4}},\ 2L(x)\lambda^{-\frac{1}{N-4}} \right] \right\},$$

where $L(x)$ is defined as in (2.4.12).

Consider

$$\min_{(x,\mu)\in D} K(x,\mu),$$

which is achieved by $(x_\lambda, \mu_\lambda) \in D$. In order to prove that (x_λ, μ_λ) is a critical point of $K(x, \mu)$, we just need to prove that (x_λ, μ_λ) is in the interior of D.

We take $(x_0, L(x_0)\lambda^{-\frac{1}{N-4}}) \in D$, where $x_0 \in S$ with $\varphi(x_0) = \min_{x\in S} \varphi(x)$. Then it follows that

$$K(x_0, L(x_0)\lambda^{-\frac{1}{N-4}}) = \frac{1}{N} \int_{\mathbb{R}^N} U_{0,1}^{2^*} - \frac{B_0 \lambda^{\frac{N-2}{N-4}}}{\varphi^{\frac{2}{N-4}}(x_0)} + O\left(\lambda^{\frac{N-2}{N-4}+\tau}\right),$$

where $\tau > 0$ is a constant.

Suppose that $x_\lambda \in \partial S$. Then by (2.4.13), we have

$$K(x_\lambda, \mu_\lambda) \geq \frac{1}{N}\int_{\mathbb{R}^N} U_{0,1}^{2^*} - \frac{B_0\lambda^{\frac{N-2}{N-4}}}{\varphi^{\frac{2}{N-4}}(x_\lambda)} + O\big(\lambda^{\frac{N-2}{N-4}+\tau}\big) > K(x_0, L(x_0)\lambda^{-\frac{1}{N-4}}).$$

This is a contradiction, as (x_λ, μ_λ) is a minimum point of $K(x_\lambda, \mu_\lambda)$ in D.

Suppose that $\mu_\lambda = \frac{1}{2}L(x)\lambda^{-\frac{1}{N-4}}$, or $\mu_\lambda = 2L(x)\lambda^{-\frac{1}{N-4}}$. Then by (2.4.13), there exists a constant $c_0 > 0$ such that

$$\begin{aligned} K(x_\lambda, \mu_\lambda) &\geq \frac{1}{N}\int_{\mathbb{R}^N} U_{0,1}^{2^*} - \frac{B_0\lambda^{\frac{N-2}{N-4}}}{\varphi^{\frac{2}{N-4}}(x_\lambda)} + c_0\lambda^{\frac{N-2}{N-4}} + O\big(\lambda^{\frac{N-2}{N-4}+\tau}\big) \\ &> K(x_0, L(x_0)\lambda^{-\frac{1}{N-4}}), \end{aligned}$$

which is also a contradiction. Therefore, we have proved that (x_λ, μ_λ) is in the interior of D, and whence (x_λ, μ_λ) is a critical point of $K(x, \mu)$. □

Here we should note that φ may have a critical point which is not a local minimum point. In this case, we cannot use a minimization procedure to get a critical point for $K(x, \mu)$. Instead, we need to consider the following algebraic equations:

$$\begin{aligned} F_i(x, \mu) := \int_\Omega \Big(&\nabla(PU_{x,\mu} + \omega_\lambda)\nabla\frac{\partial(PU_{x,\mu})}{\partial x_i} - \lambda(PU_{x,\mu} + \omega_\lambda)\frac{\partial(PU_{x,\mu})}{\partial x_i} \\ &- (PU_{x,\mu} + \omega_\lambda)_+^{2^*-1}\frac{\partial(PU_{x,\mu})}{\partial x_i}\Big) \\ = 0, \end{aligned} \tag{2.4.15}$$

for $i = 1, \cdots, N$, and

$$\begin{aligned} F_\mu(x, \mu) := \int_\Omega \Big(&\nabla(PU_{x,\mu} + \omega_\lambda)\nabla\frac{\partial(PU_{x,\mu})}{\partial \mu} - \lambda(PU_{x,\mu} + \omega_\lambda)\frac{\partial(PU_{x,\mu})}{\partial \mu} \\ &- (PU_{x,\mu} + \omega_\lambda)_+^{2^*-1}\frac{\partial(PU_{x,\mu})}{\partial \mu}\Big) \\ = 0. \end{aligned} \tag{2.4.16}$$

In the following discussion, we always assume that $\mu \in [c_0\lambda^{-\frac{1}{N-4}}, c_1\lambda^{-\frac{1}{N-4}}]$ for some large $c_1 > 0$ and small $c_0 > 0$. In order to estimate the left-hand sides of (2.4.15) and (2.4.16), we need the following lemma:

Lemma 2.4.6 *Suppose that $x \in \Omega$ with $d(x, \partial\Omega) \geq c_0 > 0$, and $\mu > 0$ is large; then,*

$$\frac{\partial \psi_{x,\mu}(y)}{\partial \mu} = -\frac{N-2}{2}\frac{C_0 H(y,x)}{\mu^{\frac{N}{2}}} + O\Big(\frac{1}{\mu^{\frac{N+4}{2}}}\Big), \tag{2.4.17}$$

$$\frac{\partial \psi_{x,\mu}(y)}{\partial x_i} = \frac{C_0}{\mu^{\frac{N-2}{2}}}\frac{\partial H(y,x)}{\partial x_i} + O\Big(\frac{1}{\mu^{\frac{N+2}{2}}}\Big), \tag{2.4.18}$$

where $H(y,x)$ is the solution of (2.4.6).

Proof. We have

$$\begin{cases} \Delta \dfrac{\partial \psi_{x,\mu}(y)}{\partial \mu} = 0, & \text{in } \Omega, \\ \dfrac{\partial \psi_{x,\mu}(y)}{\partial \mu} = \dfrac{\partial U_{x,\mu}(y)}{\partial \mu}, & \text{on } \partial\Omega. \end{cases}$$

For $y \in \partial\Omega$, one has

$$\frac{\partial U_{x,\mu}(y)}{\partial \mu} = -\frac{N-2}{2}\frac{C_0}{\mu^{\frac{N}{2}}|y-x|^{N-2}} + O\Big(\frac{1}{\mu^{\frac{N+4}{2}}}\Big),$$

and so (2.4.17) follows.

On the other hand, we have

$$\begin{cases} \Delta \dfrac{\partial \psi_{x,\mu}(y)}{\partial x_i} = 0, & \text{in } \Omega, \\ \dfrac{\partial \psi_{x,\mu}(y)}{\partial x_i} = \dfrac{\partial U_{x,\mu}(y)}{\partial x_i}, & \text{on } \partial\Omega. \end{cases}$$

For $y \in \partial\Omega$, one then has

$$\frac{\partial U_{x,\mu}(y)}{\partial x_i} = \frac{C_0}{\mu^{\frac{N-2}{2}}}\frac{(N-2)(y_i - x_i)}{|y-x|^N} + O\Big(\frac{1}{\mu^{\frac{N+2}{2}}}\Big),$$

which yields (2.4.18). □

By a similar argument to that of the proof of Lemma 2.4.6, we obtain the following estimates:

$$\Big|\frac{\partial (PU_{x,\mu})}{\partial \mu}\Big| \leq C\mu^{-1}U_{x,\mu}, \qquad \Big|\frac{\partial (PU_{x,\mu})}{\partial x_i}\Big| \leq C\mu U_{x,\mu}.$$

We first estimate (2.4.15) and (2.4.16) with $\omega_\lambda = 0$. Note that the left-hand side of (2.4.15) and (2.4.16) are $\frac{\partial I(PU_{x,\mu})}{\partial x_i}$ and $\frac{\partial I(PU_{x,\mu})}{\partial \mu}$, respectively. Therefore, one would expect to differentiate the main term of $I(PU_{x,\mu})$ with respect to x_i or μ to obtain the desired estimates. We will prove this rigorously.

Since $0 < \psi_{x,\mu} < U_{x,\mu}$, we have

$$\begin{aligned}
&\int_\Omega \Big(\nabla PU_{x,\mu}\nabla\frac{\partial(PU_{x,\mu})}{\partial\mu} - (PU_{x,\mu})^{2^*-1}\frac{\partial(PU_{x,\mu})}{\partial\mu}\Big)\\
&= \int_\Omega \big(U_{x,\mu}^{2^*-1} - (PU_{x,\mu})^{2^*-1}\big)\frac{\partial(PU_{x,\mu})}{\partial\mu}\\
&= (2^*-1)\int_\Omega U_{x,\mu}^{2^*-2}\psi_{x,\mu}\frac{\partial U_{x,\mu}}{\partial\mu} + O\Big(\int_\Omega U_{x,\mu}^{2^*-3}\psi_{x,\mu}^2\Big|\frac{\partial U_{x,\mu}}{\partial\mu}\Big|\Big)\\
&= (2^*-1)\int_\Omega U_{x,\mu}^{2^*-2}\frac{\partial U_{x,\mu}}{\partial\mu}\frac{C_0 H(x,x)}{\mu^{\frac{N-2}{2}}} + O\Big(\frac{1}{\mu^{N+1}}\Big)\\
&= \int_{\mathbb{R}^N}\frac{\partial U_{x,\mu}^{2^*-1}}{\partial\mu}\frac{C_0 H(x,x)}{\mu^{\frac{N-2}{2}}} + O\Big(\frac{1}{\mu^{N+1}}\Big)\\
&= -\frac{N-2}{2}\frac{C_0\varphi(x)}{\mu^{N-1}}\int_{\mathbb{R}^N} U_{0,1}^{2^*-1} + O\Big(\frac{1}{\mu^{N+1}}\Big),
\end{aligned}$$

and

$$\begin{aligned}
\lambda\int_\Omega PU_{x,\mu}\frac{\partial(PU_{x,\mu})}{\partial\mu} &= \lambda\int_\Omega U_{x,\mu}\frac{\partial U_{x,\mu}}{\partial\mu} + O\Big(\frac{\lambda}{\mu^{N-1}}\Big)\\
&= \frac{1}{2}\lambda\int_{\mathbb{R}^N}\frac{\partial U_{x,\mu}^2}{\partial\mu} = -\frac{\lambda}{\mu^3}\int_{\mathbb{R}^N} U_{0,1}^2 + O\Big(\frac{\lambda}{\mu^{N-1}}\Big).
\end{aligned}$$

Thus it follows that

$$\begin{aligned}
&\int_\Omega \Big(\nabla PU_{x,\mu}\nabla\frac{\partial(PU_{x,\mu})}{\partial\mu} - \lambda PU_{x,\mu}\frac{\partial(PU_{x,\mu})}{\partial\mu} - (PU_{x,\mu})^{2^*-1}\frac{\partial(PU_{x,\mu})}{\partial\mu}\Big)\\
&= -\frac{N-2}{2}\frac{C_0\varphi(x)}{\mu^{N-1}}\int_{\mathbb{R}^N} U_{0,1}^{2^*-1} + \frac{\lambda}{\mu^3}\int_{\mathbb{R}^N} U_{0,1}^2 + O\Big(\frac{1}{\mu^{N+1}} + \frac{\lambda}{\mu^{N-1}}\Big).
\end{aligned}$$

Similarly, one has

$$\begin{aligned}
&\int_\Omega \Big(\nabla PU_{x,\mu}\nabla\frac{\partial(PU_{x,\mu})}{\partial x_i} - (PU_{x,\mu})^{2^*-1}\frac{\partial(PU_{x,\mu})}{\partial x_i}\Big)\\
&= (2^*-1)\int_\Omega U_{x,\mu}^{2^*-2}\frac{\partial U_{x,\mu}}{\partial x_i}\frac{C_0 H(y,x)}{\mu^{\frac{N-2}{2}}}\,dy + O\Big(\frac{1}{\mu^{N-1}}\Big)\\
&= -(2^*-1)\int_\Omega U_{x,\mu}^{2^*-2}\frac{\partial U_{x,\mu}}{\partial y_i}\frac{C_0 H(y,x)}{\mu^{\frac{N-2}{2}}}\,dy + O\Big(\frac{1}{\mu^{N-1}}\Big)\\
&= \int_\Omega U_{x,\mu}^{2^*-1}\frac{\partial H(y,x)}{\partial y_i}\frac{C_0}{\mu^{\frac{N-2}{2}}}\,dy + O\Big(\frac{1}{\mu^{N-1}}\Big) = \frac{C_0}{2\mu^{N-2}}\frac{\partial\varphi(x)}{\partial x_i}\int_\Omega U_{0,1}^{2^*-1}\\
&\quad + O\Big(\frac{1}{\mu^{N-1}}\Big),
\end{aligned}$$

since $\frac{\partial H(y,x)}{\partial y_i}\Big|_{y=x} = \frac{1}{2}\frac{\partial\varphi(x)}{\partial x_i}$.

From (2.4.5) and (2.4.18),

$$\lambda\int_\Omega PU_{x,\mu}\frac{\partial(PU_{x,\mu})}{\partial x_i}=\lambda\int_\Omega U_{x,\mu}\frac{\partial U_{x,\mu}}{\partial x_i}+O\Big(\frac{\lambda}{\mu^{N-2}}\Big)$$
$$=\lambda\int_{\Omega\setminus B_\delta(x)}U_{x,\mu}\frac{\partial U_{x,\mu}}{\partial x_i}+O\Big(\frac{\lambda}{\mu^{N-2}}\Big)=O\Big(\frac{\lambda}{\mu^{N-2}}\Big).$$

Hence we obtain

$$\int_\Omega\Big(\nabla PU_{x,\mu}\nabla\frac{\partial(PU_{x,\mu})}{\partial x_i}-\lambda PU_{x,\mu}\frac{\partial(PU_{x,\mu})}{\partial x_i}-(PU_{x,\mu})^{2^*-1}\frac{\partial(PU_{x,\mu})}{\partial x_i}\Big)$$
$$=\frac{C_0}{2\mu^{N-2}}\frac{\partial\varphi(x)}{\partial x_i}\int_\Omega U_{0,1}^{2^*-1}+O\Big(\frac{1}{\mu^{N-1}}+\frac{\lambda}{\mu^{N-2}}\Big).$$

To solve (2.4.15) and (2.4.16), we need to prove that the error term ω_λ is negligible in the estimates for the functions $F_i(x,\mu)$ and $F_\mu(x,\mu)$.

In fact, we have

$$\int_\Omega\nabla\omega_\lambda\nabla\frac{\partial(PU_{x,\mu})}{\partial x_i}=0,$$

$$\lambda\int_\Omega\omega_\lambda\frac{\partial(PU_{x,\mu})}{\partial x_i}=O\big(\lambda\|\omega_\lambda\|\big)=O\Big(\frac{\lambda}{\mu^{2+\tau}}\Big),$$

for some small $\tau>0$, and

$$\int_\Omega\Big((PU_{x,\mu}+\omega_\lambda)_+^{2^*-1}-(PU_{x,\mu})^{2^*-1}\Big)\frac{\partial(PU_{x,\mu})}{\partial x_i}$$
$$=(2^*-1)\int_\Omega(PU_{x,\mu})^{2^*-2}\omega_\lambda\frac{\partial(PU_{x,\mu})}{\partial x_i}+O\Big(\frac{1}{\mu^{N-2+\tau}}\Big)$$
$$=(2^*-1)\int_\Omega U_{x,\mu}^{2^*-2}\omega_\lambda\frac{\partial U_{x,\mu}}{\partial x_i}+O\Big(\frac{1}{\mu^{N-2+\tau}}\Big)$$
$$=\int_\Omega\nabla\omega_\lambda\nabla\frac{\partial(PU_{x,\mu})}{\partial x_i}+O\Big(\frac{1}{\mu^{N-2+\tau}}\Big)=O\Big(\frac{1}{\mu^{N-2+\tau}}\Big).$$

We can also obtain similar estimates if we replace $\frac{\partial(PU_{x,\mu})}{\partial x_i}$ by $\frac{\partial(PU_{x,\mu})}{\partial\mu}$. Thus, in conclusion, we have the following estimates for the functions $F_i(x,\mu)$ and $F_\mu(x,\mu)$, defined in (2.4.15) and (2.4.16) as follows:

$$F_i(x,\mu)=\frac{C_0}{2\mu^{N-2}}\frac{\partial\varphi(x)}{\partial x_i}\int_\Omega U_{0,1}^{2^*-1}+O\Big(\frac{1}{\mu^{N-2+\tau}}+\frac{\lambda}{\mu^{N-2}}\Big),\qquad(2.4.19)$$

and

$$F_\mu(x,\mu)=-\frac{N-2}{2}\frac{C_0\varphi(x)}{\mu^{N-1}}\int_{\mathbb{R}^N}U_{0,1}^{2^*-1}+\frac{\lambda}{\mu^3}\int_{\mathbb{R}^N}U_{0,1}^2+O\Big(\frac{1}{\mu^{N-1+\tau}}+\frac{\lambda}{\mu^{N-1}}\Big).\qquad(2.4.20)$$

We are now ready to prove the following result.

Theorem 2.4.7 (Rey [125], 1990) *Suppose that $x_0 \in \Omega$ is an isolated critical point of the Robin function φ satisfying $\deg(\nabla\varphi, B_\delta(x_0), 0) \neq 0$. Then there exists $\lambda_0 > 0$ such that for any $\lambda \in (0, \lambda_0]$, (2.4.1) has a solution of the form*

$$u_\lambda = PU_{x_\lambda,\mu_\lambda} + \omega_\lambda,$$

for some $x_\lambda \in \Omega$ and $\mu_\lambda > 0$ large, and $\|\omega_\lambda\| = o(1)$. Moreover, as $\lambda \to 0$, $x_\lambda \to x_0$.

Proof. Let $\mu = t\lambda^{-\frac{1}{N-4}}$. From (2.4.19) and (2.4.20), we find that $F_i(x, \mu) = 0$ and $F_\mu(x, \mu) = 0$ are equivalent to

$$\nabla\varphi(x) = o(1), \tag{2.4.21}$$

and

$$\begin{aligned} f(x,t) &:= -\frac{N-2}{2}\frac{C_0\varphi(x)}{t^{N-1}}\int_{\mathbb{R}^N} U_{0,1}^{2^*-1} + \frac{1}{t^3}\int_{\mathbb{R}^N} U_{0,1}^2 \\ &= o(1), \end{aligned} \tag{2.4.22}$$

respectively.

Let

$$S = \left\{(x,t) : x \in B_\theta(x_0),\ t \in \left(\frac{1}{2}L(x_0), 2L(x_0)\right)\right\},$$

where $L(x)$ is defined as in (2.4.12). Then we have

$$\deg((\nabla\varphi(x), f(x,t)), S, 0) = \deg(\nabla\varphi(x), B_\theta(x_0), 0) \neq 0.$$

Therefore, (2.4.21) and (2.4.22) have a solution in S. □

Before we close this section, let us point out that the solution u_λ has the following properties.

Proposition 2.4.8 *The following statements on the solution u_λ are valid.*

(i) $\|u_\lambda\|_{L^\infty(\Omega)} \to +\infty$ *as* $\lambda \to 0$.
(ii) There exists a constant $C > 0$, independent of λ, such that

$$|u_\lambda(x)| \le CU_{x_\lambda,\mu_\lambda}(x), \quad \forall x \in \Omega.$$

Proof. Since (i) is obvious, we will only give a proof of (ii). We know that for $\tilde{u}_\lambda(y) = \mu_\lambda^{-\frac{N-2}{2}} u_\lambda(\mu_\lambda^{-1}y + x_\lambda)$, one has

$$-\Delta\tilde{u}_\lambda = \tilde{u}_\lambda^{2^*-1} + \frac{\lambda}{\mu_\lambda^2}\tilde{u}_\lambda, \quad \text{in } \Omega_\lambda,$$

where $\Omega_\lambda = \{y : \mu_\lambda^{-1} y + x_\lambda \in \Omega\}$. Therefore, we just need to show that

$$|\tilde{u}_\lambda(y)| \le \frac{C}{1 + |y|^{N-2}}, \quad \text{if } |y| \text{ is large.}$$

For any $y \in \Omega_\lambda$, one has $|y| \le C\mu_\lambda$. On the other hand, from $\mu_\lambda \sim \lambda^{-\frac{1}{N-4}}$, we find that

$$\frac{\lambda}{\mu_\lambda^2} \le \frac{C}{\mu_\lambda^{N-2}} \le CU_{0,1}(y), \quad y \in \Omega_\lambda. \tag{2.4.23}$$

Using the Kelvin transformation $w_\lambda(y) = |y|^{2-N} \tilde{u}_\lambda\left(\frac{y}{|y|^2}\right)$, it follows from Proposition 6.5.1 that

$$-\Delta w_\lambda = w_\lambda^{2^*-1} + \frac{\lambda}{\mu_\lambda^2 |y|^4} w_\lambda, \quad \text{in } \Omega_\lambda^*,$$

where $\Omega_\lambda^* = \{y : |y|^{-1} \in \Omega_\lambda\}$.

Using (2.4.23), we have that

$$\frac{\lambda}{\mu_\lambda^2 |y|^4} \le |y|^{-4} U_{0,1}(|y|^{-1}) \le C|y|^{N-6}, \quad y \in \Omega_\lambda^*.$$

Set $w_\lambda = 0$ in $\mathbb{R}^N \setminus \Omega_\lambda^*$. Then w_λ satisfies

$$-\Delta w_\lambda \le w_\lambda^{2^*-1} + C|y|^{N-6} w_\lambda, \quad \text{in } \mathbb{R}^N. \tag{2.4.24}$$

On the other hand, it is easy to check that for any small constant $\tau > 0$,

$$\int_{B_\delta(0)} |w_\lambda|^{2^*} \le C \int_{\Omega \setminus B_{\frac{1}{\delta \mu_\lambda}}(x_\lambda)} |u_\lambda|^{2^*} \le C \int_{\mathbb{R}^N \setminus B_{\frac{1}{\delta}}(0)} |U_{0,1}|^{2^*} + C\|\omega_\lambda\|^{2^*} < \tau,$$

uniformly for λ small as $\delta \to 0$. Therefore, for $N \ge 5$, it follows from (2.4.24) that we can use the Moser iteration (Lemma 6.4.6) to prove that

$$|w_\lambda(y)| \le C, \quad y \in B_\delta(0),$$

for a fixed small $\delta > 0$. This, in turn, implies that

$$|\tilde{\omega}_\lambda(y)| \le \frac{C}{1 + |y|^{N-2}}, \quad |y| \ge \frac{1}{\delta}.$$

□

2.5 Further Results and Comments

Local Pohozaev identities were first used by Schoen [130–132] to analyze the blow-up behaviors of bubbling solutions for scalar curvature type elliptic equations. See also [98] and the references therein. In Section 2.1, we basically follow the same procedures as those in the above mentioned papers to locate the concentration point of the peak/bubbling solutions.

The existence of nontrivial solutions for the nonlinear Schrödinger equations or nonlinear field equations in subcritical case has attracted considerable attention in the last four decades. Readers can refer to [14, 16, 22, 44, 72] and the references therein. There are two different methods for finding peak solutions for corresponding singularly perturbed problems. The first one is the finite dimensional reduction arguments discussed in this chapter. Such arguments were first used by Floer and Weinstein [75] to construct peak solutions in one dimensional case. See also [117]. The other one is the variational methods introduced by del Pino and Felmer [59–61, 63], where the nonlinear term is modified and/or a penalty term is added, so as to force the concentration of the solutions to occur at some designated points. The main advantage of the variational methods is that we do not need to fully understand the limit problem. Therefore, such methods can be applied to nonlinear terms $f(u)$, which are more general than u^p. However, it is difficult to use the variational methods to find solutions with clustering peaks. It is also difficult to use them to study elliptic problems with critical growth.

All the basic calculations needed in the discussion of bubbling solutions for elliptic problems with critical growth can be found in [12]. Single bubbling solutions for the Brezis–Nirenberg problem are constructed by Rey [125] under the assumption that the Robin function $\varphi(x)$ has a non-degenerate critical point. The method used to prove Theorem 2.4.5 is adapted from [29], where a simple comparison of the energy leads to the proof of the existence of a critical point for the reduced finite dimensional problem.

Our aim is to explain the essential ideas and the procedure of the reduction arguments in constructing solutions for elliptic problems with a parameter. The nonlinear Schrödinger equation and the Brezis-Nirenberg problem are two typical problems in which we can use these techniques to construct solutions. In the last three decades, such techniques and their invariance have been widely used to study many important problems. For more results, we refer the readers to the monograph [5] and the survey paper [67]. In the following, we briefly discuss some of these results.

The Dirichlet Boundary Problems in Bounded Domains

We now consider the following singularly perturbed problem

$$\begin{cases} -\varepsilon^2 \Delta u + u = u_+^p, & \text{in } \Omega, \\ u \in H_0^1(\Omega), \end{cases} \tag{2.5.1}$$

where $\varepsilon > 0$ is a small parameter, $1 < p < 2^* - 1$, and Ω is a bounded domain in $\mathbb{R}^N$. The functional corresponding to (2.5.1) is then given by

$$I(u) = \frac{1}{2}\int_\Omega \left(\varepsilon^2 |\nabla u|^2 + u^2\right) - \frac{1}{p+1}\int_\Omega u_+^{p+1}, \quad u \in H_0^1(\Omega).$$

In [114], Ni and Wei initiated the study of the least energy solutions for (2.5.1) and proved the following result:

Theorem 2.5.1 *Suppose that u_ε is a least energy solution of* (2.5.1)*, which can be obtained by the mountain pass lemma. Then for $\varepsilon > 0$ small, the solution u_ε has exactly one local maximum point x_ε, and as $\varepsilon \to 0$, $x_\varepsilon \to x_0$, which satisfies $d(x_0) = \max_{x\in\Omega} d(x)$, where $d(x) := d(x, \partial\Omega)$, which is the distance function from x to the boundary $\partial\Omega$.*

If w denotes the solution of (2.2.12), we let $w_{\varepsilon,x}(y) = w\left(\frac{y-x}{\varepsilon}\right)$. The projection $P_\varepsilon w_{\varepsilon,x}$ of $w_{\varepsilon,x}$ to $H_0^1(\Omega)$ is defined as the solution of the following problem

$$\begin{cases} -\varepsilon^2 \Delta u + u = w_{\varepsilon,x}^p, & \text{in } \Omega, \\ u \in H_0^1(\Omega). \end{cases}$$

Then, $\varphi_{\varepsilon,x} = w_{\varepsilon,x} - P_\varepsilon w_{\varepsilon,x}$ satisfies

$$\begin{cases} -\varepsilon^2 \Delta \varphi_{\varepsilon,x} + \varphi_{\varepsilon,x} = 0, & \text{in } \Omega, \\ \varphi_{\varepsilon,x} = w_{\varepsilon,x}, & \text{on } \partial\Omega. \end{cases} \tag{2.5.2}$$

Using a blow-up argument, one can prove that a least energy solution u_ε of (2.5.1) has the following form:

$$u_\varepsilon = P_\varepsilon w_{\varepsilon,x_\varepsilon} + \omega_\varepsilon,$$

where $\|\omega_\varepsilon\|_\varepsilon = o\left(\varepsilon^{\frac{N}{2}}\right)$. To locate the concentration point, we calculate the energy of the main term $P_\varepsilon w_{\varepsilon,x}$ as follows:

$$\begin{aligned} I(P_\varepsilon w_{\varepsilon,x}) &= \frac{1}{2}\int_\Omega w_{\varepsilon,x}^p P_\varepsilon w_{\varepsilon,x} - \frac{1}{p+1}\int_\Omega \left(w_{\varepsilon,x} - \varphi_{\varepsilon,x}\right)_+^{p+1} \\ &= \left(\frac{1}{2} - \frac{1}{p+1}\right)\int_\Omega w_{\varepsilon,x}^{p+1} + \frac{1}{2}\int_\Omega w_{\varepsilon,x}^p \varphi_{\varepsilon,x} + O\left(\int_\Omega w_{\varepsilon,x}^{p-1} \varphi_{\varepsilon,x}^2\right). \end{aligned} \tag{2.5.3}$$

We then have that

$$\Big(\frac{1}{2}-\frac{1}{p+1}\Big)\int_{\Omega} w_{\varepsilon,x}^{p+1}=\Big(\frac{1}{2}-\frac{1}{p+1}\Big)\varepsilon^N\int_{\mathbb{R}^N} w^{p+1}+O\Big(e^{-\frac{(p+1)d(x)}{\varepsilon}}\Big). \tag{2.5.4}$$

To proceed further, it is essential to estimate $\varphi_{\varepsilon,x}$. Ni and Wei [114] used a viscosity solution method to obtain a point-wise estimate for $\varphi_{\varepsilon,x}$ and hence an expansion for $I\big(P_\varepsilon w_{\varepsilon,x}\big)$. Later, the proof of Theorem 2.5.1 is simplified by del Pino and Felmer [62].

On the other hand, we note that to finish the expansion in (2.5.4), it is not necessary to get a very accurate point-wise estimate for $\varphi_{\varepsilon,x}$. Therefore, we can proceed differently. Let $G_\varepsilon(y,x)$ be the Green's function of $-\varepsilon^2\Delta u+u$ in Ω with homogenous Dirichlet boundary condition. Then

$$\varphi_{\varepsilon,x}(y)=\varepsilon^2\int_{\partial\Omega}\frac{\partial G_\varepsilon(z,y)}{\partial\nu}w_{\varepsilon,x}(z)\,dz. \tag{2.5.5}$$

By the exponential decay of $w_{\varepsilon,x}$ and $G_\varepsilon(z,y)$, we obtain

$$0<\varphi_{\varepsilon,x}\le C\varepsilon e^{-\frac{d(x)+d(y)}{\varepsilon}},$$

from which we can derive

$$\begin{aligned}\int_{\Omega} w_{\varepsilon,x}^{p-1}\varphi_{\varepsilon,x}^2&=O\Big(\int_{\Omega} e^{-\frac{(p-1)|z-x|}{\varepsilon}}\varepsilon^2 e^{-\frac{2d(x)+2d(z)}{\varepsilon}}\,dz\Big)\\&=O\Big(\varepsilon^{N+2}e^{-\frac{(2+\sigma)d(x)}{\varepsilon}}\Big),\end{aligned} \tag{2.5.6}$$

where $\sigma>0$ is a constant.

Combining (2.5.3), (2.5.4) and (2.5.6), we obtain

$$I\big(P_\varepsilon w_{\varepsilon,x}\big)=\Big(\frac{1}{2}-\frac{1}{p+1}\Big)\varepsilon^N\int_{\mathbb{R}^N} w^{p+1}+\frac{1}{2}\int_{\Omega} w_{\varepsilon,x}^p\varphi_{\varepsilon,x}+O\Big(\varepsilon^{N+2}e^{-\frac{(2+\sigma)d(x)}{\varepsilon}}\Big). \tag{2.5.7}$$

It remains to estimate the main term $\tau_{\varepsilon,x}=\int_\Omega w_{\varepsilon,x}^p\varphi_{\varepsilon,x}$. We have

$$\begin{aligned}\tau_{\varepsilon,x}&=\int_{\Omega}\big(-\varepsilon^2\Delta w_{\varepsilon,x}+w_{\varepsilon,x}\big)\varphi_{\varepsilon,x}\\&=\varepsilon^2\int_{\partial\Omega}\Big(-\frac{\partial w_{\varepsilon,x}}{\partial\nu}w_{\varepsilon,x}+\frac{\partial\varphi_{\varepsilon,x}}{\partial\nu}w_{\varepsilon,x}\Big).\end{aligned}$$

But from (2.5.2), one has

$$\int_{\partial\Omega}\frac{\partial\varphi_{\varepsilon,x}}{\partial\nu}w_{\varepsilon,x}=\|\varphi_{\varepsilon,x}\|_\varepsilon^2>0,$$

which yields

$$\tau_{\varepsilon,x}\ge-\varepsilon^2\int_{\partial\Omega}\frac{\partial w_{\varepsilon,x}}{\partial\nu}w_{\varepsilon,x}\ge a_0\varepsilon^N e^{-\frac{(2+\theta)d(x)}{\varepsilon}}, \tag{2.5.8}$$

for some constant $a_0 > 0$, where $\theta > 0$ is any fixed small constant. On the other hand, using (2.5.5), we can prove that

$$\tau_{\varepsilon,x} \le a_1 \varepsilon^N e^{-\frac{(2-\theta)d(x)}{\varepsilon}}. \tag{2.5.9}$$

Combining (2.5.8) and (2.5.9), we conclude that the main term $\tau_{\varepsilon,x}$ satisfies

$$a_0 \varepsilon^N e^{-\frac{(2+\theta)d(x)}{\varepsilon}} \le \tau_{\varepsilon,x} \le a_1 \varepsilon^N e^{-\frac{(2-\theta)d(x)}{\varepsilon}}. \tag{2.5.10}$$

From (2.5.10), we conclude that minimizing $\tau_{\varepsilon,x}$ is equivalent to maximizing the distance function $d(x)$. This is the result of Theorem 2.5.1.

Note that $\tau_{\varepsilon,x}$ has a local minimum point near x_0 if x_0 is an isolated local maximum point of the function $d(x)$. By (2.5.7) and (2.5.10), using a minimization procedure similar to that in the proof of Theorem 2.4.5, we can prove the following existence result.

Theorem 2.5.2 *Suppose that x_0 is an isolated local maximum point of the distance function $d(x)$. Then there is an $\varepsilon_0 > 0$ such that for any $\varepsilon \in (0, \varepsilon_0]$, (2.5.1) has a solution $u_\varepsilon = P_\varepsilon w_{\varepsilon,x_\varepsilon} + \omega_\varepsilon$ such that as $\varepsilon \to 0$, $x_\varepsilon \to x_0$ and $\|\omega_\varepsilon\|_\varepsilon = o\big(\varepsilon^{\frac{N}{2}}\big)$.*

When x moves away from the isolated local maximum point x_0 of the function $d(x)$, $\tau_{\varepsilon,x}$ becomes much bigger than τ_{ε,x_0}. This rapid change shows that the rough estimate for $\tau_{\varepsilon,x}$ in (2.5.10) suffices for the proof of Theorem 2.5.2. The estimates here are taken from [50].

In [51], it is shown that if Ω is convex, then for any integer $k \ge 2$, (2.5.1) has no k-peak solutions, while in [50], it is shown that if Ω has nontrivial topology, then for any integer $k \ge 2$, (2.5.1) has a k-peak solution, provided $\varepsilon > 0$ is small. The readers can refer to [29, 144] for other results on the effect of the domain geometry on the existence of multi-peak solutions for (2.5.1).

The Neumann Boundary Problems in Bounded Domains

In their pioneering paper [113], Ni and Takagi studied the following Neumann problem:

$$\begin{cases} -\varepsilon^2 \Delta u + u = u_+^p, & \text{in } \Omega, \\ \dfrac{\partial u}{\partial \nu} = 0, & \text{on } \partial\Omega, \end{cases} \tag{2.5.11}$$

where $\varepsilon > 0$ is a small parameter, $1 < p < 2^* - 1$, Ω is a bounded domain in $\mathbb{R}^N$, and ν is the unit outward normal to the boundary $\partial\Omega$. They studied the asymptotic behavior of the least energy solutions to (2.5.11) and proved the following result.

Theorem 2.5.3 *Suppose that u_ε is a least energy solution of* (2.5.11), *which can be obtained by the mountain pass lemma. Then for $\varepsilon > 0$ small, u_ε has exactly one local maximum point x_ε and $x_\varepsilon \in \partial\Omega$. Moreover, as $\varepsilon \to 0$, $x_\varepsilon \to x_0 \in \partial\Omega$, and x_0 satisfies $H(x_0) = \max_{x\in\partial\Omega} H(x)$, where $H(x)$ is the mean curvature of the boundary $\partial\Omega$ at x.*

If w denotes the solution of (2.2.12), we let $w_{\varepsilon,x}(y) = w\big(\frac{y-x}{\varepsilon}\big)$. Similar to the Dirichlet problem, the projection $P_\varepsilon w_{\varepsilon,x}$ of $w_{\varepsilon,x}$ to $H_0^1(\Omega)$ is defined as the solution of the following problem:

$$\begin{cases} -\varepsilon^2\Delta u + u = w_{\varepsilon,x}^p, & \text{in } \Omega, \\ \dfrac{\partial u}{\partial \nu} = 0, & \text{on } \partial\Omega. \end{cases}$$

Then, $\varphi_{\varepsilon,x} = w_{\varepsilon,x} - P_\varepsilon w_{\varepsilon,x}$ satisfies

$$\begin{cases} -\varepsilon^2\Delta\varphi_{\varepsilon,x} + \varphi_{\varepsilon,x} = 0, & \text{in } \Omega, \\ \dfrac{\partial \varphi_{\varepsilon,x}}{\partial \nu} = \dfrac{\partial w_{\varepsilon,x}}{\partial \nu}, & \text{on } \partial\Omega. \end{cases} \tag{2.5.12}$$

By a blow-up argument, one can see that a least energy solution u_ε of (2.5.1) has the following form:

$$u_\varepsilon = P_\varepsilon w_{\varepsilon,x_\varepsilon} + \omega_\varepsilon,$$

for some $x_\varepsilon \in \partial\Omega$, where $\|\omega_\varepsilon\|_\varepsilon = o\big(\varepsilon^{\frac{N}{2}}\big)$.

The functional corresponding to (2.5.11) is given by

$$I(u) = \frac{1}{2}\int_\Omega \big(\varepsilon^2|\nabla u|^2 + u^2\big) - \frac{1}{p+1}\int_\Omega u_+^{p+1}, \quad u \in H^1(\Omega).$$

Let x be a point on $\partial\Omega$ and write $P_\varepsilon w_{\varepsilon,x}$ for an approximate solution for (2.5.11). We then have the following energy expansion for the approximate solution $P_\varepsilon w_{\varepsilon,x}$:

$$I\big(P_\varepsilon w_{\varepsilon,x}\big) = \Big(\frac{1}{2} - \frac{1}{p+1}\Big)\varepsilon^N \int_{\mathbb{R}^N_+} w^{p+1} - BH(x)\varepsilon^{N+1} + o(\varepsilon^{N+1}), \tag{2.5.13}$$

where $B > 0$ is a constant. To minimize the energy of $P_\varepsilon w_{\varepsilon,x}$, we just need to maximize the mean curvature function $H(x)$. This is the result of Theorem 2.5.3.

Suppose that $H(x)$ has an isolated local maximum or local minimum point x_0. After carrying out the reduction argument, we can use (2.5.13) and a minimization procedure if x_0 is a local maximum point of $H(x)$, or a maximization procedure if x_0 is a local minimum point, to prove the following result:

Theorem 2.5.4 *Suppose that $x_0 \in \partial\Omega$ is either an isolated local maximum point or an isolated local minimum point of the mean curvature function $H(x)$. Then there is an $\varepsilon_0 > 0$ such that for any $\varepsilon \in (0, \varepsilon_0]$,* (2.5.11) *has a solution $u_\varepsilon = P_\varepsilon w_{\varepsilon,x_\varepsilon} + \omega_\varepsilon$ such that $x_\varepsilon \in \partial\Omega$, and as $\varepsilon \to 0$, $x_\varepsilon \to x_0$ and $\|\omega_\varepsilon\|_\varepsilon = o\big(\varepsilon^{\frac{N}{2}}\big)$.*

Similar to Theorem 2.3.2, from (2.5.13), we can also construct solutions with k-peaks clustering at an isolated local minimum point of the mean curvature function $H(x)$. See [49, 85]. Results on the existence of peak solution concentrating at a non-degenerate critical point of $H(x)$ can be found in [145].

Now we consider the peak solution concentrating at an interior point in Ω. Similar to the Dirichlet problem, we can calculate $I\big(P_\varepsilon w_{\varepsilon,x}\big)$ as follows:

$$\begin{aligned} I\big(P_\varepsilon w_{\varepsilon,x}\big) &= \frac{1}{2}\int_\Omega w_{\varepsilon,x}^p P_\varepsilon w_{\varepsilon,x} - \frac{1}{p+1}\int_\Omega \big(w_{\varepsilon,x} - \varphi_{\varepsilon,x}\big)_+^{p+1} \\ &= \Big(\frac{1}{2} - \frac{1}{p+1}\Big)\varepsilon^N \int_{\mathbb{R}^N} w^{p+1} + \frac{1}{2}\int_\Omega w_{\varepsilon,x}^p \varphi_{\varepsilon,x} \\ &\quad + O\Big(\int_\Omega w_{\varepsilon,x}^{p-1}\varphi_{\varepsilon,x}^2\Big) + O\Big(e^{-\frac{(p+1)d(x)}{\varepsilon}}\Big). \end{aligned}$$

To estimate $\varphi_{\varepsilon,x}$, we have

$$\varphi_{\varepsilon,x}(y) = \varepsilon^2 \int_{\partial\Omega} G_\varepsilon(z, y) w_{\varepsilon,x}(z)\, dz, \tag{2.5.14}$$

where $G_\varepsilon(y, x)$ is the Green's function of $-\varepsilon^2\Delta u + u$ in Ω with homogenous Neumann boundary condition. Then,

$$|\varphi_{\varepsilon,x}| \le C\varepsilon e^{-\frac{d(x)+d(y)}{\varepsilon}},$$

from which we can derive

$$\int_\Omega w_{\varepsilon,x}^{p-1}\varphi_{\varepsilon,x}^2 = O\Big(\varepsilon^{N+2} e^{-\frac{(2+\sigma)d(x)}{\varepsilon}}\Big).$$

To estimate the main term $\tau_{\varepsilon,x} = \int_\Omega w_{\varepsilon,x}^p \varphi_{\varepsilon,x}$, we have

$$\begin{aligned} \tau_{\varepsilon,x} &= \int_\Omega \big(-\varepsilon^2\Delta w_{\varepsilon,x} + w_{\varepsilon,x}\big)\varphi_{\varepsilon,x} \\ &= \varepsilon^2 \int_{\partial\Omega}\Big(-\frac{\partial w_{\varepsilon,x}}{\partial\nu}\varphi_{\varepsilon,x} + \frac{\partial w_{\varepsilon,x}}{\partial\nu} w_{\varepsilon,x}\Big). \end{aligned}$$

But from (2.5.12), one has

$$\int_{\partial\Omega} \frac{\partial w_{\varepsilon,x}}{\partial\nu}\varphi_{\varepsilon,x} = \|\varphi_{\varepsilon,x}\|_\varepsilon^2 > 0,$$

which gives

$$\tau_{\varepsilon,x} \le \varepsilon^2 \int_{\partial\Omega} \frac{\partial w_{\varepsilon,x}}{\partial\nu} w_{\varepsilon,x} \le -a_0 \varepsilon^N e^{-\frac{(2+\theta)d(x)}{\varepsilon}}, \tag{2.5.15}$$

for some constant $a_0 > 0$, where $\theta > 0$ is any fixed small constant. On the other hand, using (2.5.14), we can prove that

$$\tau_{\varepsilon,x} \ge -a_1 \varepsilon^N e^{-\frac{(2-\theta)d(x)}{\varepsilon}}. \tag{2.5.16}$$

Combining (2.5.15) and (2.5.16), we conclude that the main term $\tau_{\varepsilon,x}$ satisfies

$$-a_1 \varepsilon^N e^{-\frac{(2-\theta)d(x)}{\varepsilon}} \le \tau_{\varepsilon,x} \le -a_0 \varepsilon^N e^{-\frac{(2+\theta)d(x)}{\varepsilon}}. \tag{2.5.17}$$

From (2.5.17), we see that unlike the Dirichlet problem, for the Neumann problem, the energy of $P_\varepsilon w_{\varepsilon,x}$ decreases if x moves away from a local maximum point of $d(x)$. Therefore, we can use a maximization procedure to prove the following result.

Theorem 2.5.5 *Suppose that x_0 is an isolated local maximum point of the distance function $d(x)$. Then there is an $\varepsilon_0 > 0$ such that for any $\varepsilon \in (0, \varepsilon_0]$,* (2.5.11) *has a solution $u_\varepsilon = P_\varepsilon w_{\varepsilon,x_\varepsilon} + \omega_\varepsilon$ with the property that as $\varepsilon \to 0$, $x_\varepsilon \to x_0$ and $\|\omega_\varepsilon\|_\varepsilon = o(\varepsilon^{\frac{N}{2}})$.*

Note also that the interaction of two peaks always becomes weaker if one peak moves away from the other. Similar to Theorem 2.3.2, we have the following result:

Theorem 2.5.6 *For any integer $k \ge 0$, there is an $\varepsilon_0 > 0$ such that for any $\varepsilon \in (0, \varepsilon_0]$,* (2.5.11) *has a solution of the form*

$$u_\varepsilon = \sum_{j=1}^{k} P_\varepsilon w_{\varepsilon,x_{\varepsilon,j}} + \omega_\varepsilon,$$

where as $\varepsilon \to 0$, $x_{\varepsilon,j} \to x_j$ for some $x_j \in \Omega$ and $x_i \ne x_j$ if $i \ne j$, and $\|\omega_\varepsilon\|_\varepsilon = o(\varepsilon^{\frac{N}{2}})$.

The above estimates are taken from [152]. Theorem 2.5.6 is proved in [83, 152]. This theorem shows that the Neumann problem (2.5.11) always has multi-peak solutions. Moreover, one can also construct solution u_ε for (2.5.11), such that u_ε has k_1 boundary peaks and k_2 interior peaks. See [52, 84]. Finally, Lin, Ni and Wei [101] proved that for $\varepsilon > 0$ small, (2.5.11) has a solution with k interior peaks, where k is an integer, which can depend on ε.

Theorem 2.5.3 is generalized by Ni-Pan-Takagi [112] to the critical case $p = 2^* - 1$. It is convenient to rewrite (2.5.11) to the following problem in the critical case:

$$\begin{cases} -\Delta u + \lambda u = u_+^{2^*-1}, & \text{in } \Omega, \\ \dfrac{\partial u}{\partial \nu} = 0, & \text{on } \partial\Omega, \end{cases} \tag{2.5.18}$$

where $\lambda > 0$ is a large parameter. For problem (2.5.18), we can use $U_{x,\mu}$ defined in (2.4.3) as an approximate solution, where $x \in \bar{\Omega}$ and $\mu > 0$ is a large constant. Let

$$J(u) = \frac{1}{2}\int_{\Omega}\left(|\nabla u|^2 + \lambda u^2\right) - \frac{1}{2^*}\int_{\Omega}|u|^{2^*}.$$

If $x \in \partial\Omega$, we can prove

$$J\left(U_{x,\mu}\right) = A - \frac{BH(x)}{\mu} + \frac{B_1\lambda}{\mu^2} + o\left(\frac{1}{\mu} + \frac{\lambda}{\mu^2}\right), \tag{2.5.19}$$

where A, B and B_1 are some positive constants. The main term in (2.5.19) is given by

$$f(x,\mu) := -\frac{BH(x)}{\mu} + \frac{B_1\lambda}{\mu^2}.$$

If we minimize $f(x,\mu)$ on $\partial\Omega \times (0,+\infty)$, then the minimum point (x_0,μ_0) satisfies $H(x_0) = \max_{x\in\partial\Omega} H(x)$ and $\mu_0 = \frac{BH(x_0)}{2B_1\lambda}$. Moreover, if x_0 is an isolated local maximum point (or an isolated local minimum point) of $H(x)$ with $H(x_0) > 0$, then we can construct a single bubbling solution concentrating at x_0. In [146], similar to Theorem 2.3.2, if x_0 is an isolated local minimum point of $H(x)$ with $H(x_0) > 0$, Wei–Yan proved that for any integer $k > 0$, (2.5.18) has a k-bubbling solution clustering at x_0, provided $\lambda > 0$ is large enough.

For other results on the singularly perturbed Neumann problems with critical growth, the readers can refer to [2–4, 82, 126–129, 141–143] and the references therein.

Elliptic Problems with Critical Growth

We now go back to the Brezis-Nirenberg problem, and we want to construct a multi-bubbling solution of the form

$$u_\lambda = \sum_{j=1}^{k} PU_{x_{\lambda,j},\mu_{\lambda,j}} + \omega_\lambda,$$

where $U_{x,\mu}$ is defined by (2.4.3). To find the equations for $(x_{\lambda,j}, \mu_{\lambda,j})$, we calculate

$$I\Big(\sum_{j=1}^{k} PU_{x_{\lambda,j},\mu_{\lambda,j}}\Big) = kA + B_1 \sum_{j=1}^{k} \frac{\varphi(x_{\lambda,j})}{\mu_{\lambda,j}^{N-2}} - B_2 \sum_{j=1}^{k} \frac{\lambda}{\mu_{\lambda,j}^{2}}$$
$$- B_3 \sum_{j\neq i}^{k} \frac{G(x_{\lambda,j}, x_{\lambda,i})}{\mu_{\lambda,j}^{\frac{N-2}{2}} \mu_{\lambda,i}^{\frac{N-2}{2}}} + h.o.t.,$$

where A, and B_l, $l = 1, 2, 3$, are some positive constants, and $G(y, x)$ is the Green's function of $-\Delta$ in Ω with homogenous Dirichlet boundary condition. (See [12, 15]). Therefore, we see that the main problem is to study the existence of critical points with $x_{\lambda,j} \in \Omega$, $\mu_{\lambda,j} > 0$ for the following function:

$$B_1 \sum_{j=1}^{k} \frac{\varphi(x_{\lambda,j})}{\mu_{\lambda,j}^{N-2}} - B_2 \sum_{j=1}^{k} \frac{\lambda}{\mu_{\lambda,j}^{2}} - B_3 \sum_{j\neq i}^{k} \frac{G(x_{\lambda,j}, x_{\lambda,i})}{\mu_{\lambda,j}^{\frac{N-2}{2}} \mu_{\lambda,i}^{\frac{N-2}{2}}}.$$

This is a very delicate problem. In [81], Grossi and Takahashi proved that if Ω is convex, this function, as a function of $x = (x_1, \cdots, x_k)$ for any fixed $\mu = (\mu_1, \cdots, \mu_k)$, $(\mu_j > 0,\ j = 1, \cdots, k)$, has no critical point. On the other hand, if Ω is a dumb-bell-shaped domain with long and thin handles, this function does have a stable critical point. It is still an open problem when such a function has a stable critical point.

The Brezis-Nirenberg problem is a perturbation of the following critical problem:

$$\begin{cases} -\Delta u = u_+^{2^*-1}, & \text{in } \Omega, \\ u \in H_0^1(\Omega). \end{cases} \tag{2.5.20}$$

There are several ways to perturb (2.5.20). We list some of them. The first one is

$$\begin{cases} -\Delta u = u_+^{2^*-1-\varepsilon}, & \text{in } \Omega, \\ u \in H_0^1(\Omega). \end{cases} \tag{2.5.21}$$

The second one is

$$\begin{cases} -\Delta u = u_+^{2^*-1+\varepsilon}, & \text{in } \Omega, \\ u \in H_0^1(\Omega), \end{cases} \tag{2.5.22}$$

and the last one is

$$\begin{cases} -\Delta u = u_+^{2^*-1}, & \text{in } \Omega \setminus \cup_{j=1}^{k} B_{\varepsilon_j}(p_j), \\ u \in H_0^1(\Omega \setminus \cup_{j=1}^{k} B_{\varepsilon_j}(p_j)). \end{cases} \tag{2.5.23}$$

Here, $\varepsilon_j > 0$ is a small parameter with $p_j \in \Omega$, $j = 1, \cdots, k$.

Problem (2.5.21) is almost the same as the Brezis-Nirenberg problem. Problem (2.5.22) is more delicate. By the Pohozaev identity, (2.5.22) has no nontrivial solution if Ω is star-shaped. On the other hand, if Ω has a suitably small hole, del Pino, Felmer and Musso [64] proved that (2.5.22) has a solution with two bubbles, provided $\varepsilon > 0$ is small.

For the multiplicity of positive solutions for (2.5.23), the readers can refer to [80, 96].

The Lazer-McKenna Conjecture

In the early 1980s, Lazer and McKenna [93, 94] made a conjecture for the following problem:

$$\begin{cases} -\Delta u = f(u) - s\varphi_1, & \text{in } \Omega, \\ u = 0, & \text{on } \partial\Omega, \end{cases} \tag{2.5.24}$$

where Ω is a bounded domain in $\mathbb{R}^N$, $\frac{f(t)}{t} \to +\infty$ as $t \to +\infty$ and $\lim_{t\to-\infty} \frac{f(t)}{t} < \lambda_1$, λ_1 and $\varphi_1 > 0$ are the first eigenvalue and the the first eigenfunction of $-\Delta$ with homogenous Dirichlet boundary condition, respectively. A typical case is $f(t) = t_+^p + \lambda t$ for some $p \in (1, 2^* - 1]$ and $\lambda < \lambda_1$. The conjecture is, as $s \to +\infty$, the number of the solutions for (2.5.24) is unbounded. In [54, 55, 57, 97, 147], positive results for this conjecture are given by constructing solutions with many peaks clustering at a point, which is either a maximum point of φ_1, or some specific point on the boundary of Ω.

The Non-smooth Elliptic Problems

In all the problems discussed so far, we can carry out the reduction procedure in a standard Sobolev space. Such techniques cannot be used anymore if we consider the following non-smooth elliptic problem:

$$\begin{cases} -\varepsilon^2 \Delta u = (u-1)_+^p, & \text{in } \Omega, \\ u \in H_0^1(\Omega), \end{cases} \tag{2.5.25}$$

where $p \in [0, 1]$, and Ω is a bounded domain in $\mathbb{R}^2$. The cases $p = 1$ and $p = 0$ correspond to the plasma problem and the vortex patch problems in the fluid dynamics, respectively. For (2.5.25), the coefficient for corresponding linear operator is not smooth. The invertibility of such operator can only be studied in $W^{1,p}(\Omega)$ for some $p > 2$. The readers can refer to [39, 40] for more details.

Schrödinger Equations with Critical Frequency or Decay Potentials

In the nonlinear Schrödinger equation (2.2.11), $V(x) \geq 0$ is called critical frequency if $V(x)$ satisfies $\{x \in \mathbb{R}^N : V(x) = 0\} \neq \emptyset$. For this case, Byeon and

Wang have proved in [25, 26] that the least energy solution must concentrate at the zero-point set of $V(x)$ and the asymptotic behavior of the least energy solution was also analyzed if ε is sufficiently small. In [32, 33, 36], the solutions concentrating at zero points or stable critical points of $V(x)$ were studied. Another interesting case for (2.2.11) is that $V(x)$ decays at infinity and, especially, is compactly supported. In [6], Ambrosetti and Malchiodi conjectured the existence of solutions for (2.2.11) if $V(x)$ decays faster than $|x|^{-2}$ at infinity. This conjecture was partly answered in [11, 27, 37, 69] by combining the singularly perturbed methods with a penalized argument.

Sign-Changing Solutions

When we use the critical point theories to study the existence of nontrivial solutions, except for the mountain pass solution, the other solutions are most likely sign-changing. The methods introduced in this chapter can also be used to construct sign-changing solutions. A simple case is that we can prove the following result for (2.2.11) without any difficulty.

Theorem 2.5.7 *Suppose that $V(y)$ has k different critical points p_j, $j = 1, \cdots, k$, which satisfies $deg(\nabla V, B_\delta(p_j), 0) \neq 0$. Then there exists $\varepsilon_0 > 0$ such that for any $\varepsilon \in (0, \varepsilon_0]$, (2.2.11) has a solution of the form*

$$u_\varepsilon = \sum_{j=1}^{k} (-1)^{\sigma(j)} U_{\varepsilon, x_{\varepsilon,j}} + \omega_\varepsilon,$$

for some $x_{\varepsilon,j} \in B_\delta(p_j)$, and $\|\omega_\varepsilon\|_\varepsilon = O(\varepsilon^{\frac{N}{2}+1})$, where $\sigma(j)$ is either 1 or 0.

For results on the construction of sign-changing solutions, the readers can refer to [17, 58, 66] and the references therein.

Solutions Concentrating at Higher Dimensional Manifolds

A problem proposed by Ni in [111] is whether or not singularly perturbed elliptic problems admit solutions concentrating at higher dimensional sets or manifolds. Let us briefly discuss the nonlinear Schrödinger equation (2.2.11) with radial potential $V(|y|)$.

If we want to construct a radial solution $u_\varepsilon(|y|)$, which concentrates at the sphere $|y| = r_0 > 0$, then after a simple blow-up analysis, we see that the limit problem is the following ordinary differential equation

$$-w'' + V(r_0)w = w^p, \quad \text{in } (-\infty, +\infty),$$

with $w(0) = \max_{t \in \mathbb{R}} w(t)$. Let w_0 be the solution of

$$-w'' + w = w^p, \quad \text{in } (-\infty, +\infty),$$

with $w_0(0) = \max_{t \in \mathbb{R}} w_0(t)$.

We will use

$$U_{\varepsilon,r_\varepsilon}(r) = (V(r_\varepsilon))^{\frac{1}{p-1}} w_0\Big(\frac{\sqrt{V(r_\varepsilon)}}{\varepsilon}(r - r_\varepsilon)\Big)$$

as an approximate solution, where $r_\varepsilon \to r_0$ as $\varepsilon \to 0$. We now calculate

$$\begin{aligned} I(U_{\varepsilon,r_\varepsilon}) &= \frac{\omega_{N-1}}{2}\int_{-\infty}^{+\infty} r^{N-1}\big(\varepsilon^2|U'_{\varepsilon,r_\varepsilon}|^2 + V(r)U^2_{\varepsilon,r_\varepsilon}\big) \\ &\quad - \frac{\omega_{N-1}}{p+1}\int_{-\infty}^{+\infty} r^{N-1}U^{p+1}_{\varepsilon,r_\varepsilon} \\ &= \varepsilon\omega_{N-1}\Big(\frac{1}{2} - \frac{1}{p+1}\Big)\Big(\int_{-\infty}^{+\infty} w_0^{p+1}\Big) r_\varepsilon^{N-1}(V(r_\varepsilon))^{\frac{p+1}{p-1}-\frac{1}{2}} + O(\varepsilon^2), \end{aligned} \tag{2.5.26}$$

where ω_{N-1} is the area of the unit sphere $\mathbb{S}^{N-1}$. From (2.5.26), Ambrosetti, Malchiodi and Ni [7] prove that if $r_0 > 0$ is an isolated local maximum point, or an isolated local minimum point of $r^{N-1}(V(r))^{\frac{p+1}{p-1}-\frac{1}{2}}$, then (2.2.11) has a radial solution u_ε, which concentrates at the sphere $|y| = r_0 > 0$.

Let us point out that if $V(x)$ is not radially symmetric, the construction of solutions for (2.5.26), which concentrate at higher dimensional manifolds, is much more complicated. Here, we need to deal with many small eigenvalues of the linear operators for the approximate solutions. See [65] for results on the non-radial case.

The readers can refer to [7, 8, 18, 19, 27, 56, 95, 107, 108] and the references therein for other results on solutions concentrating at higher dimensional manifolds.

3
Local Uniqueness of Solutions

In Chapter 2, we construct peak solutions for the following nonlinear Schrödinger problem:

$$\begin{cases} -\varepsilon^2 \Delta u + V(y)u = u_+^p, \ u > 0, & \text{in } \mathbb{R}^N, \\ u \in H^1(\mathbb{R}^N), \end{cases} \tag{3.0.1}$$

where $\varepsilon > 0$ is a small parameter, $V(y)$ satisfies $0 < V_0 \le V(y) \le V_1$, and $u_+ = u$ if $u \ge 0$, $u_+ = 0$ if $u < 0$. These solutions concentrate at some isolated critical points p_j of $V(y)$ that satisfy $\deg(\nabla V, B_\delta(p_j), 0) \neq 0$. Note that an isolated critical point p_j of $V(y)$ satisfying $\deg(\nabla V, B_\delta(p_j), 0) \neq 0$ can be very degenerate, and the degeneracy at each direction x_i can be very different. An isolated local maximum point or local minimum point p of $V(y)$ satisfies $\deg(\nabla V, B_\delta(p_j), 0) \neq 0$. Here one should note that the property $\deg(\nabla V, B_\delta(p_j), 0) \neq 0$ guarantees that the critical point p_j of $V(y)$ is stable. In other words, any small perturbation of $V(y)$ in C^1 still has a critical point near p_j.

In this chapter, we will consider the local uniqueness problems. We will prove that if two sequences of solutions have the same concentration set and each point in the concentration set is non-degenerate, then the solutions must coincide whenever the parameter $\varepsilon > 0$ is small enough.

Consider the following Brezis–Nirenberg problem:

$$\begin{cases} -\Delta u = u_+^{2^*-1} + \lambda u, \text{ in } \Omega, \\ u \in H_0^1(\Omega). \end{cases} \tag{3.0.2}$$

It was proven in Chapter 2 that if the Robin function $\varphi(x)$ has a critical point x_0 satisfying $\deg(\nabla\varphi, B_\delta(x_0), 0) \neq 0$, then for small enough $\lambda > 0$, (3.0.2) has a solution u_λ which concentrates at x_0 as $\lambda \to 0$. In this chapter, we will consider the local uniqueness problems for (3.0.2) for small $\lambda > 0$, with the additional assumption that x_0 is a non-degenerate critical point of $\varphi(x)$.

3.1 Local Uniqueness of Peak Solutions for Nonlinear Schrödinger Equations

In this section, we will study the local uniqueness of peak solutions for (3.0.1). We assume that (3.0.1) has a solution u_ε of the form

$$u_\varepsilon = \sum_{j=1}^{k} U_{\varepsilon,x_{\varepsilon,j}} + \omega_\varepsilon, \tag{3.1.1}$$

where as $\varepsilon \to 0$, $x_{\varepsilon,j} \to p_j$, $p_j \neq p_m$ for every $m \neq j$, and $\|\omega_\varepsilon\|_\varepsilon = o\big(\varepsilon^{\frac{N}{2}}\big)$.

Firstly, we prove the following lemma.

Lemma 3.1.1 *If ω_ε in (3.1.1) satisfies $\|\omega_\varepsilon\|_\varepsilon = o\big(\varepsilon^{\frac{N}{2}}\big)$, then there are constants $C > 0$ large and $\tau > 0$ small, such that*

$$|\omega_\varepsilon(x)| \le C \sum_{j=1}^{k} e^{-\frac{\tau|x-x_{\varepsilon,j}|}{\varepsilon}}, \quad \forall\, x \in \mathbb{R}^N,$$

and

$$|\nabla\omega_\varepsilon(x)| \le Ce^{-\frac{\tau}{\varepsilon}}, \quad \forall\, x \in \partial B_\delta(x_{\varepsilon,j}),\ j = 1, \cdots, k.$$

Proof. For $j \in \{1, \cdots, k\}$, let $\tilde{\omega}_{\varepsilon,j} = \omega_\varepsilon(\varepsilon y + x_{\varepsilon,j})$. Then we have that $\|\tilde{\omega}_{\varepsilon,j}\|_{H^1(\mathbb{R}^N)} = o(1)$ and that

$$\begin{aligned} &- \Delta\tilde{\omega}_{\varepsilon,j} + V(\varepsilon y + x_{\varepsilon,j})\tilde{\omega}_{\varepsilon,j} \\ &= \Big(\sum_{i=1}^{k} U_{\varepsilon,x_{\varepsilon,i}}(\varepsilon y + x_{\varepsilon,j}) + \tilde{\omega}_{\varepsilon,j}\Big)_+^p - \Big(\sum_{i=1}^{k} U_{\varepsilon,x_{\varepsilon,i}}(\varepsilon y + x_{\varepsilon,j})\Big)^p + f_\varepsilon(y), \end{aligned}$$

where

$$\begin{aligned} f_\varepsilon(y) =& \Big(\sum_{i=1}^{k} U_{\varepsilon,x_{\varepsilon,i}}(\varepsilon y + x_{\varepsilon,j})\Big)^p - \sum_{i=1}^{k} U^p_{\varepsilon,x_{\varepsilon,i}}(\varepsilon y + x_{\varepsilon,j}) \\ &- \sum_{i=1}^{k} \big(V(\varepsilon y + x_{\varepsilon,j}) - V(x_{\varepsilon,i})\big)U_{\varepsilon,x_{\varepsilon,i}}(\varepsilon y + x_{\varepsilon,j}). \end{aligned}$$

Since

$$\|f_\varepsilon(y)\|_{L^\infty(\mathbb{R}^N)} = o(1),$$

by applying the Moser iteration, one deduces that

$$\|\omega_\varepsilon\|_{L^\infty(\mathbb{R}^N)} = \|\tilde{\omega}_{\varepsilon,j}\|_{L^\infty(\mathbb{R}^N)} \le C\|\tilde{\omega}_{\varepsilon,j}\|_{H^1(\mathbb{R}^N)} + \|f_\varepsilon(y)\|_{L^\infty(\mathbb{R}^N)} = o(1),$$

where $C > 0$ is independent of ε.

Proceeding as we prove Lemma 2.1.2 in Chapter 2, we can verify the results. □

From Theorem 2.1.1, p_j is a critical point of $V(y)$. For simplicity, we assume that each p_j is isolated and is non-degenerate in the sense that the Hessian matrix $\left(\frac{\partial^2 V(y)}{\partial y_i \partial y_j}\right)$ of V at p_j is invertible.

In the sequel, we will frequently employ the following local Pohozaev identity (Theorem 6.2.1):

$$\begin{aligned}
&-\varepsilon^2 \int_{\partial B_\delta(x_{\varepsilon,j})} \frac{\partial u_\varepsilon}{\partial \nu}\frac{\partial u_\varepsilon}{\partial y_i} + \frac{1}{2}\varepsilon^2 \int_{\partial B_\delta(x_{\varepsilon,j})} \nu_i |\nabla u_\varepsilon|^2 \\
&= \frac{1}{p+1}\int_{\partial B_\delta(x_{\varepsilon,j})} u_\varepsilon^{p+1}\nu_i - \frac{1}{2}\int_{\partial B_\delta(x_{\varepsilon,j})} V(y)u_\varepsilon^2 \nu_i \\
&\quad + \int_{B_\delta(x_{\varepsilon,j})} \frac{\partial V}{\partial y_i} u_\varepsilon^2, \quad i = 1, \cdots, N.
\end{aligned} \tag{3.1.2}$$

In order to obtain a local uniqueness result, we need to obtain an estimate for $|x_{\varepsilon,j} - p_j|$.

Lemma 3.1.2 *Let $u_\varepsilon = \sum_{j=1}^k U_{\varepsilon,x_{\varepsilon,j}} + \omega_\varepsilon$ be a solution of (3.0.1). Then,*

$$\nabla V(x_{\varepsilon,j}) = o(\varepsilon).$$

Proof. Employing Lemma 3.1.1, we see that there exists a $\theta > 0$, such that

$$|u_\varepsilon(x)| + |\nabla u_\varepsilon(x)| \le Ce^{-\frac{\theta}{\varepsilon}}, \quad \forall\, x \in \partial B_\delta(x_{\varepsilon,j}),\ j = 1, \cdots, k.$$

Hence, using the local Pohozaev identity (3.1.2), we obtain

$$\int_{B_\delta(x_{\varepsilon,j})} \frac{\partial V}{\partial y_i} u_\varepsilon^2 = \left(e^{-\tau/\varepsilon}\right), \tag{3.1.3}$$

where $\tau > 0$ is a small constant.

On the other hand, by Lemma 3.1.1, we see that

$$\begin{aligned}
&\int_{B_\delta(x_{\varepsilon,j})}\Big(\frac{\partial V(y)}{\partial y_i}-\frac{\partial V(x_{\varepsilon,j})}{\partial y_i}\Big)u_\varepsilon^2\\
=&\int_{B_\delta(x_{\varepsilon,j})}\big\langle\nabla^2V(x_{\varepsilon,j}),y-x_{\varepsilon,j}\big\rangle u_\varepsilon^2+O\Big(\int_{B_\delta(x_{\varepsilon,j})}|y-x_{\varepsilon,j}|^2u_\varepsilon^2\Big)\\
=&\int_{B_\delta(x_{\varepsilon,j})}\big\langle\nabla^2V(x_{\varepsilon,j}),y-x_{\varepsilon,j}\big\rangle\big(U^2_{\varepsilon,x_{\varepsilon,j}}+2U_{\varepsilon,x_{\varepsilon,j}}\omega_\varepsilon+\omega_\varepsilon^2\big)+O\big(\varepsilon^{N+2}\big)\\
&+O\Big(\int_{B_\delta(x_{\varepsilon,j})}|y-x_{\varepsilon,j}|\Big(\sum_{i\neq j}U^2_{\varepsilon,x_{\varepsilon,i}}+2\sum_{i\neq j}U_{\varepsilon,x_{\varepsilon,i}}U_{\varepsilon,x_{\varepsilon,j}}+\sum_{i\neq j}U_{\varepsilon,x_{\varepsilon,i}}\omega_\varepsilon\Big)\Big)\\
=&\int_{B_\delta(x_{\varepsilon,j})}\big\langle\nabla^2V(x_{\varepsilon,j}),y_i-x_{\varepsilon,j}\big\rangle\big(U^2_{\varepsilon,x_{\varepsilon,j}}+2U_{\varepsilon,x_{\varepsilon,j}}\omega_\varepsilon+\omega_\varepsilon^2\big)+O\big(\varepsilon^{N+2}\big).
\end{aligned}\tag{3.1.4}$$

Now, by symmetry, we have

$$\int_{B_\delta(x_{\varepsilon,j})}\big\langle\nabla^2V(x_{\varepsilon,j}),y-x_{\varepsilon,j}\big\rangle U^2_{\varepsilon,x_{\varepsilon,j}}=0.\tag{3.1.5}$$

Applying the Hölder inequality, we obtain

$$\begin{aligned}
&\Big|\int_{B_\delta(x_{\varepsilon,j})}\big\langle\nabla^2V(x_{\varepsilon,j}),y-x_{\varepsilon,j}\big\rangle U_{\varepsilon,x_{\varepsilon,j}}\omega_\varepsilon\Big|\\
\le&C\Big(\int_{B_\delta(x_{\varepsilon,j})}|y-x_{\varepsilon,j}|^2U^2_{\varepsilon,x_{\varepsilon,j}}\Big)^{\frac12}\|\omega_\varepsilon\|_\varepsilon=o(\varepsilon^{N+1}),
\end{aligned}\tag{3.1.6}$$

and similarly, by Lemma 3.1.1, one has

$$\begin{aligned}
\Big|\int_{B_\delta(x_{\varepsilon,j})}\big\langle\nabla^2V(x_{\varepsilon,j}),y-x_{\varepsilon,j}\big\rangle\omega_\varepsilon^2\Big|\le C\Big(\int_{B_\delta(x_{\varepsilon,j})}|y-x_{\varepsilon,j}|^2\omega_\varepsilon^2\Big)^{\frac12}\|\omega_\varepsilon\|_\varepsilon\\
=o(\varepsilon^{N+1}).
\end{aligned}\tag{3.1.7}$$

Inserting (3.1.5), (3.1.6) and (3.1.7) into (3.1.4), we obtain

$$\int_{B_\delta(x_{\varepsilon,j})}\Big(\frac{\partial V(y)}{\partial y_i}-\frac{\partial V(x_{\varepsilon,j})}{\partial y_i}\Big)u_\varepsilon^2=o(\varepsilon^{N+1}),$$

which, together with (3.1.3), yields

$$\frac{\partial V(x_{\varepsilon,j})}{\partial y_i}\int_{B_\delta(x_{\varepsilon,j})}u_\varepsilon^2=o(\varepsilon^{N+1}),$$

and the result follows, since

$$\int_{B_\delta(x_{\varepsilon,j})} u_\varepsilon^2 = \varepsilon^N\big(B + o(1)\big)$$

for some $B > 0$. □

Suppose that for $\varepsilon > 0$ small, (3.0.1) has solutions of the form

$$u_\varepsilon^{(i)} = \sum_{j=1}^{k} U_{\varepsilon, x_{\varepsilon,j}^{(i)}} + \omega_\varepsilon^{(i)}, \quad i = 1, 2, \tag{3.1.8}$$

where as $\varepsilon \to 0$,

$$x_{\varepsilon,j}^{(i)} \to p_j, \quad j = 1, \cdots, k, \quad \|\omega_\varepsilon^{(i)}\|_\varepsilon = o(\varepsilon^{\frac{N}{2}}). \tag{3.1.9}$$

We will prove the following local uniqueness result.

Theorem 3.1.3 (Cao-Noussair-Yan [34], 1998; Grossi [79], 2002) *Suppose that* (3.0.1) *has solutions of the form* (3.1.8), *satisfying* (3.1.9). *If* p_j *is an isolated non-degenerate critical point of* $V(y)$, $j = 1, \cdots, k$, *then there exists* $\varepsilon_0 > 0$, *such that for any* $\varepsilon \in (0, \varepsilon_0]$, $u_\varepsilon^{(1)} = u_\varepsilon^{(2)}$.

To prove Theorem 3.1.3, we argue by contradiction. Suppose that $u_\varepsilon^{(1)} \neq u_\varepsilon^{(2)}$. We define

$$\xi_\varepsilon = \frac{u_\varepsilon^{(1)} - u_\varepsilon^{(2)}}{\|u_\varepsilon^{(1)} - u_\varepsilon^{(2)}\|_{L^\infty(\mathbb{R}^N)}}.$$

Then, ξ_ε satisfies

$$-\varepsilon^2\Delta\xi_\varepsilon + V(y)\xi_\varepsilon = \frac{(u_\varepsilon^{(1)})^p - (u_\varepsilon^{(2)})^p}{\|u_\varepsilon^{(1)} - u_\varepsilon^{(2)}\|_{L^\infty(\mathbb{R}^N)}}.$$

Write

$$\begin{aligned}(u_\varepsilon^{(1)})^p - (u_\varepsilon^{(2)})^p &= \int_0^1 \frac{d}{dt}(tu_\varepsilon^{(1)} + (1-t)u_\varepsilon^{(2)})^p\,dt \\ &= p\int_0^1 (tu_\varepsilon^{(1)} + (1-t)u_\varepsilon^{(2)})^{p-1}\,dt(u_\varepsilon^{(1)} - u_\varepsilon^{(2)}) \\ &=: C_\varepsilon(y)(u_\varepsilon^1 - u_\varepsilon^2).\end{aligned}$$

We then have

$$-\varepsilon^2\Delta\xi_\varepsilon + V(y)\xi_\varepsilon - C_\varepsilon(y)\xi_\varepsilon = 0.$$

Since $\|\xi_\varepsilon\|_{L^\infty(\mathbb{R}^N)} = 1$, we will obtain a contradiction by showing $\|\xi_\varepsilon\|_{L^\infty(\mathbb{R}^N)} \to 0$ as $\varepsilon \to 0$. This consists of two parts: (i) the estimate of ξ_ε in $B_{R\varepsilon}(x_{\varepsilon,j}^{(1)})$, $j = 1, \cdots, k$, and (ii) the estimate of ξ_ε in $\mathbb{R}^N \setminus \cup_{j=1}^k B_{R\varepsilon}(x_{\varepsilon,j}^{(1)})$.

From now on, we always assume that each critical point p_j is non-degenerate, though we can obtain a local uniqueness result under weaker conditions. By Lemma 3.1.2, and taking the non-degeneracy of p_j into account, it follows from

$$\nabla V(x_{\varepsilon,j}^{(h)}) = o(\varepsilon), \ \ h = 1, 2$$

that we have

$$|x_{\varepsilon,j}^{(h)} - p_j| = o(\varepsilon), \quad h = 1, 2,$$

which, in turn, implies that

$$|x_{\varepsilon,j}^{(1)} - x_{\varepsilon,j}^{(2)}| = o(\varepsilon). \tag{3.1.10}$$

Estimate (3.1.10) tells us that the corresponding peaks in the two solutions are very close in terms of ε.

Now we estimate ξ_ε in $\mathbb{R}^N \setminus \cup_{j=1}^k B_{R\varepsilon}(x_{\varepsilon,j}^{(1)})$. For this, we have the following lemma.

Lemma 3.1.4 *There exist constants $C > 0$ and $\theta > 0$, such that*

$$|\xi_\varepsilon(y)| \le C \sum_{j=1}^k e^{-\theta \frac{|y - x_{\varepsilon,j}^{(1)}|}{\varepsilon}}, \quad \forall\, y \in \mathbb{R}^N \setminus \cup_{j=1}^k B_{R\varepsilon}(x_{\varepsilon,j}^{(1)}), \tag{3.1.11}$$

and

$$|\nabla \xi_\varepsilon| \le C e^{-\frac{\theta}{\varepsilon}}, \quad \forall\, x \in \partial B_\delta(x_{\varepsilon,j}^{(1)}), \ j = 1, \cdots, k. \tag{3.1.12}$$

Proof. Given any small $\tau > 0$, it follows from Lemma 3.1.1 and (3.1.10) that there exists a large $R > 0$ such that

$$|C_\varepsilon(y)| \le p(u_\varepsilon^{(1)} + u_\varepsilon^{(2)})^{p-1} \le \tau, \quad \forall\, y \in \mathbb{R}^N \setminus \cup_{j=1}^k B_{R\varepsilon}(x_{\varepsilon,j}^{(1)}),$$

since

$$|y - x_{\varepsilon,j}^{(2)}| \ge |y - x_{\varepsilon,j}^{(1)}| - |x_{\varepsilon,j}^{(1)} - x_{\varepsilon,j}^{(2)}| \ge (R + o(1))\varepsilon.$$

Consequently, for every $y \in \mathbb{R}^N \setminus \cup_{j=1}^k B_{R\varepsilon}(x_{\varepsilon,j}^{(1)})$, we have

$$V(y) - C_\varepsilon(y) \ge \min_{y \in \mathbb{R}^N} V(y) - \tau > 2\theta,$$

for some small $\theta > 0$. Using

$$-\varepsilon^2 \Delta \xi_\varepsilon + (V(y) - C_\varepsilon(y))\xi_\varepsilon = 0,$$

along with the comparison theorem (Theorem 6.4.3), we can prove (3.1.11).

The estimate (3.1.12) can be verified by using the L^p estimate for the elliptic equations (Theorem 6.4.5). □

Now we estimate ξ_ε in $B_{R\varepsilon}(x^{(1)}_{\varepsilon,j})$, $j = 1, \cdots, k$. We aim to prove $\xi_\varepsilon = o(1)$ in $B_{R\varepsilon}(x^{(1)}_{\varepsilon,j})$, $j = 1, \cdots, k$. For this purpose, we will use the blow-up argument. Let

$$\tilde{\xi}_{\varepsilon,j}(y) = \xi_\varepsilon(\varepsilon y + x^{(1)}_{\varepsilon,j}), \quad j = 1, \cdots, k.$$

It is easy to see that

$$\begin{aligned} &u^{(1)}_\varepsilon(\varepsilon y + x^{(1)}_{\varepsilon,j}) \\ &= U_{\varepsilon, x^{(1)}_{\varepsilon,j}}(\varepsilon y + x^{(1)}_{\varepsilon,j}) + \sum_{m \neq j}^{k} U_{\varepsilon, x^{(1)}_{\varepsilon,m}}(\varepsilon y + x^{(1)}_{\varepsilon,j}) + \omega^{(1)}_\varepsilon(\varepsilon y + x^{(1)}_{\varepsilon,j}) \\ &\to U_j, \end{aligned} \tag{3.1.13}$$

uniformly in $B_R(0)$ for any $R > 0$, where

$$U_j(y) = \big(V(p_j)\big)^{\frac{1}{p-1}} w\Big(\sqrt{V(p_j)}y\Big),$$

which is a solution of

$$-\Delta u + V(p_j)u = u^p, \ u > 0, \quad \text{in } \mathbb{R}^N,$$

and $u \in H^1(\mathbb{R}^N)$.

On the other hand, in view of (3.1.10), one has

$$\begin{aligned} u^{(2)}_\varepsilon(\varepsilon y + x^{(1)}_{\varepsilon,j}) &= U_{\varepsilon, x^{(2)}_{\varepsilon,j}}(\varepsilon y + x^{(1)}_{\varepsilon,j}) + o(1) \\ &= U_{\varepsilon, x^{(2)}_{\varepsilon,j}}(\varepsilon y + x^{(2)}_{\varepsilon,j} + o(\varepsilon)) + o(1) \\ &= U_j + o(1), \ \text{ uniformly in } B_R(0) \text{ for any } R > 0. \end{aligned} \tag{3.1.14}$$

Combining (3.1.13) and (3.1.14), we obtain

$$C_\varepsilon(\varepsilon y + x^{(1)}_{\varepsilon,j}) = pU_j^{p-1} + o(1), \quad \text{uniformly in } B_R(0) \text{ for any } R > 0.$$

Now $|\tilde{\xi}_\varepsilon| \leq 1$. By using the L^p estimate (Theorem 6.4.5), the Sobolev embedding theorem (Theorem 6.3.6) and the Schauder estimate (Theorem 6.4.4), we conclude that $\tilde{\xi}_\varepsilon \to \xi_j$ in $C^2(B_R(0))$ for any $R > 0$, and ξ_j satisfies

$$-\Delta \xi_j + V(p_j)\xi_j - pU_j^{p-1}\xi_j = 0, \quad \text{in } \mathbb{R}^N.$$

Thus

$$\xi_j = \sum_{i=1}^{N} b_{ji} \frac{\partial U_j}{\partial y_i}, \tag{3.1.15}$$

for some constants b_{ji}, $i = 1, \cdots, N$. If we can prove $b_{ji} = 0$ for every $i = 1, \cdots, N$ and $j = 1, \cdots, k$, then $\xi_\varepsilon = o(1)$ in $B_{R\varepsilon}(x_{\varepsilon,j}^{(1)})$ for $j = 1, \cdots, k$. To achieve this goal, we will use the Pohozaev identities (3.1.2). Applying (3.1.2) to $u_\varepsilon^{(i)}$, we obtain

$$\begin{aligned}
&\frac{\varepsilon^2}{2}\int_{\partial B_\tau(x_{\varepsilon,j}^{(1)})} \frac{\partial\big(u_\varepsilon^{(1)}+u_\varepsilon^{(2)}\big)}{\partial\nu}\frac{\partial \xi_\varepsilon}{\partial y_i} + \frac{\varepsilon^2}{2}\int_{\partial B_\tau(x_{\varepsilon,j}^{(1)})} \frac{\partial \xi_\varepsilon}{\partial\nu}\frac{\partial\big(u_\varepsilon^{(1)}+u_\varepsilon^{(2)}\big)}{\partial y_i}\\
&-\frac{1}{2}\varepsilon^2\int_{\partial B_\tau(x_{\varepsilon,j}^{(1)})} \nu_i\big\langle\nabla\big(u_\varepsilon^{(1)}+u_\varepsilon^{(2)}\big),\nabla\xi_\varepsilon\big\rangle\\
=&\frac{1}{p+1}\int_{\partial B_\tau(x_{\varepsilon,j}^{(1)})} D_\varepsilon(x)\xi_\varepsilon\nu_i - \frac{1}{2}\int_{\partial B_\tau(x_{\varepsilon,j}^{(1)})} V(y)\big(u_\varepsilon^{(1)}+u_\varepsilon^{(2)}\big)\xi_\varepsilon\nu_i\\
&+\int_{B_\tau(x_{\varepsilon,j}^{(1)})}\frac{\partial V}{\partial y_i}\big(u_\varepsilon^{(1)}+u_\varepsilon^{(2)}\big)\xi_\varepsilon,\ i=1,\cdots,N,
\end{aligned} \tag{3.1.16}$$

where

$$D_\varepsilon(x) = \frac{(u_\varepsilon^{(1)})^{p+1}-(u_\varepsilon^{(2)})^{p+1}}{u_\varepsilon^{(1)}-u_\varepsilon^{(2)}} = (p+1)\int_0^1 (tu_\varepsilon^{(1)}+(1-t)u_\varepsilon^{(2)})^p\,dt.$$

Proof of Theorem 3.1.3. It follows by Lemma 3.1.4 that (3.1.16) can be reduced to

$$\int_{B_\tau(x_{\varepsilon,j}^{(1)})}\frac{\partial V}{\partial y_i}\big(u_\varepsilon^{(1)}+u_\varepsilon^{(2)}\big)\xi_\varepsilon = O\big(e^{-\tau_1/\varepsilon}\big),\ i=1,\cdots,N,$$

where $\tau_1 > 0$ is a small constant. As a result,

$$\int_{B_{\tau\varepsilon^{-1}}(0)}\frac{\partial V(\varepsilon y + x_{\varepsilon,j}^{(1)})}{\partial y_i}\big(u_\varepsilon^{(1)}(\varepsilon y+x_{\varepsilon,j}^{(1)})+u_\varepsilon^{(2)}(\varepsilon y+x_{\varepsilon,j}^{(1)})\big)\tilde{\xi}_\varepsilon = O\big(\varepsilon^{-N}e^{-\tau_1/\varepsilon}\big). \tag{3.1.17}$$

It follows from the fact that $|x_{\varepsilon,j}^{(1)} - p_j| = o(\varepsilon)$ that

$$\frac{\partial V(x_{\varepsilon,j}^{(1)})}{\partial y_i} = \frac{\partial V(p_j)}{\partial p_{j,i}} + O(|x_{\varepsilon,j}^{(1)} - p_j|) = o(\varepsilon),$$

which implies

$$\int_{B_{\tau\varepsilon^{-1}}(0)}\frac{\partial V(x_{\varepsilon,j}^{(1)})}{\partial y_i}\big(u_\varepsilon^{(1)}(\varepsilon y+x_{\varepsilon,j}^{(1)})+u_\varepsilon^{(2)}(\varepsilon y+x_{\varepsilon,j}^{(1)})\big)\tilde{\xi}_\varepsilon = o(\varepsilon).$$

Hence (3.1.17) gives

$$\begin{aligned}
&\int_{B_{\tau\varepsilon^{-1}}(0)}\Big(\frac{\partial V(\varepsilon y + x_{\varepsilon,j}^{(1)})}{\partial y_i} - \frac{\partial V(x_{\varepsilon,j}^{(1)})}{\partial y_i}\Big)\big(u_\varepsilon^{(1)}(\varepsilon y+x_{\varepsilon,j}^{(1)})+u_\varepsilon^{(2)}(\varepsilon y+x_{\varepsilon,j}^{(1)})\big)\tilde{\xi}_\varepsilon\\
&= o(\varepsilon),
\end{aligned}$$

which also implies that

$$\varepsilon \sum_{h=1}^{N} \int_{B_{\tau\varepsilon^{-1}}(0)} \frac{\partial^2 V(x_{\varepsilon,j}^{(1)})}{\partial y_i \partial y_h} y_h \big(u_\varepsilon^{(1)}(\varepsilon y + x_{\varepsilon,j}^{(1)}) + u_\varepsilon^{(2)}(\varepsilon y + x_{\varepsilon,j}^{(1)})\big)\tilde{\xi}_\varepsilon = o(\varepsilon). \tag{3.1.18}$$

Letting $\varepsilon \to 0$ in (3.1.18), it follows from (3.1.13), (3.1.14), and (3.1.15) that

$$\sum_{h=1}^{N} \int_{\mathbb{R}^N} \frac{\partial^2 V(y)}{\partial y_i \partial y_h}\Big|_{y=p_j} y_h U_j \sum_{l=1}^{N} b_{jl} \frac{\partial U_j}{\partial y_l} = 0.$$

Hence

$$\sum_{h=1}^{N} \frac{\partial^2 V(y)}{\partial y_i \partial y_h}\Big|_{y=p_j} b_{jh} \int_{\mathbb{R}^N} y_h U_j \frac{\partial U_j}{\partial y_h} = 0, \ i = 1, \cdots, N. \tag{3.1.19}$$

But

$$\int_{\mathbb{R}^N} y_h U_j \frac{\partial U_j}{\partial x_h} = \frac{1}{2} \int_{\mathbb{R}^N} y_h \frac{\partial U_j^2}{\partial y_h} = -\frac{1}{2} \int_{\mathbb{R}^N} U_j^2 < 0,$$

and we obtain from (3.1.19) that

$$\left(\frac{\partial^2 V(y)}{\partial y_i \partial y_h}\Big|_{y=p_j} \right)_{N\times N} (b_j)^T = (0, \cdots, 0)^T, \ j = 1, \cdots, k,$$

where $b_j = (b_{j1}, \cdots, b_{jN})$. By the non-degeneracy of p_j, we conclude that $b_{jh} = 0, h = 1, \cdots, N, j = 1, \cdots, k$.

In conclusion, we have proven that $\xi_\varepsilon = o(1)$ in $B_{R\varepsilon}(x_{\varepsilon,j}^{(1)})$, $j = 1, \cdots, k$, which, together with Lemma 3.1.4, gives $\|\xi_\varepsilon\|_{L^\infty(\mathbb{R}^N)} = o(1)$. This is a contradiction of $\|\xi_\varepsilon\|_{L^\infty(\mathbb{R}^N)} = 1$.

□

Remark 3.1.5 Here we should take note that the non-degeneracy of the critical point p_j of $V(y)$ is not essential in our proof of the local uniqueness result. What we really need is that in the expansion of $V(y)$ at p_j, the order in each direction is the same. For example, if $V(y)$ has the expansion (after suitable rotation)

$$V(y) = V(p_j) + \sum_{i=1}^{N} a_{ji}|y_i - p_{ji}|^\alpha + O\big(|y - p_j|^{\alpha+1}\big), \quad y \in B_\theta(p_j), \tag{3.1.20}$$

and

$$\frac{\partial V(y)}{\partial y_i} = \alpha a_{ji}|y_i - p_{ji}|^{\alpha-2}(y_i - p_{ji}) + O\big(|y - p_j|^\alpha\big), \quad y \in B_\theta(p_j), \tag{3.1.21}$$

for some $\alpha \geq 2$ and $a_{ji} \neq 0$, we can then use the same argument to obtain the local uniqueness result. Conditions (3.1.20) and (3.1.21) are used in [31].

3.2 Local Uniqueness of Bubbling Solutions for Equations with Critical Growth

In this section, we will discuss the local uniqueness result for (3.0.2). Suppose that (3.0.2) has a solution u_λ of the form

$$u_\lambda = U_{x_\lambda,\mu_\lambda} + \omega_\lambda, \tag{3.2.1}$$

which satisfies the property that as $\lambda \to 0$,

$$x_\lambda \to x_0 \in \Omega, \quad \mu_\lambda \to +\infty, \quad \|\omega_\lambda\| \to 0. \tag{3.2.2}$$

We will prove the following result.

Theorem 3.2.1 (Glangetas [78], 1993) *Assume that* $N \geq 6$. *Suppose that* (3.0.2) *has solutions of the form* (3.2.1), *satisfying* (3.2.2). *If* x_0 *is an isolated non-degenerate critical point of the Robin function* $\varphi(x)$, *then there exists* $\lambda_0 > 0$, *such that for any* $\lambda \in (0, \lambda_0]$, u_λ *is unique.*

By Theorem 6.2.1, any solution u of (3.0.2) satisfies following Pohozaev identities:

$$\begin{aligned}
&\int_{\partial B_\delta(x_\lambda)} \Big((\nabla u \cdot \nu)((x - x_\lambda) \cdot \nabla u) - \frac{|\nabla u|^2}{2}(x - x_\lambda) \cdot \nu \Big) \\
&+\frac{N-2}{2} \int_{\partial B_\delta(x_\lambda)} u \frac{\partial u}{\partial \nu} + \int_{\partial B_\delta(x_\lambda)} \Big(\frac{1}{2^*}|u|^{2^*} + \frac{\lambda}{2}u^2 \Big)((x - x_\lambda) \cdot \nu) \\
&= \lambda \int_{B_\delta(x_\lambda)} u^2,
\end{aligned} \tag{3.2.3}$$

and for $i = 1, \cdots, N$,

$$\int_{\partial B_\delta(x_\lambda)} \frac{\partial u}{\partial x_i} \frac{\partial u}{\partial \nu} - \frac{1}{2} \int_{\partial B_\delta(x_\lambda)} |\nabla u|^2 \nu_i + \int_{\partial B_\delta(x_\lambda)} \Big(\frac{1}{2^*}|u|^{2^*} + \frac{\lambda}{2}u^2 \Big) \nu_i = 0, \tag{3.2.4}$$

where ν denotes the unit outward normal to the boundary $\partial B_\delta(x_\lambda)$.

We recall that in the Pohozaev identities (3.1.2), all the surface integrals are negligible since the solution u_ε is exponentially small on $\partial B_\delta(x_\lambda)$. On the other hand, since the Pohozaev identities (3.2.4) do not involve any volume integral, we expect that those surface integrals in (3.2.4) play an essential role in proving the local uniqueness of bubbling solutions for problem (3.0.2). Therefore, we see that to prove Theorem 3.2.1, a crucial step is to obtain an adequate expansion for the solution u_λ on $\partial B_\delta(x_\lambda)$, which in turn requires us to obtain a better estimate for the error term ω_λ in (3.2.2), as well as an accurate estimate for both the location x_λ and the height μ_λ. For this purpose, we need to change the location x_λ and the height μ_λ a bit.

Consider

$$\inf\Big\{\|u_\lambda - PU_{x,\mu}\|^2 : \ \mu \in [c_0\mu_\lambda, c_1\mu_\lambda], \ |x - x_\lambda| \le \frac{1}{c_0^2\mu_\lambda}\Big\}, \tag{3.2.5}$$

where $c_1 > 0$ is a large constant and $c_0 > 0$ is a small constant. Then, (3.2.5) is achieved by some $(\bar{x}_\lambda, \bar{\mu}_\lambda)$. Since

$$\|u_\lambda - PU_{\bar{x}_\lambda,\bar{\mu}_\lambda}\| \le \|u_\lambda - PU_{x_\lambda,\mu_\lambda}\| = \|\omega_\lambda\| \to 0,$$

we find that

$$\begin{aligned}
o(1) =& \|u_\lambda - PU_{\bar{x}_\lambda,\bar{\mu}_\lambda}\|^2 = \|PU_{\bar{x}_\lambda,\bar{\mu}_\lambda} - PU_{x_\lambda,\mu_\lambda}\|^2 + O\Big(\|u_\lambda - PU_{x_\lambda,\mu_\lambda}\|\Big) \\
=& \|PU_{\bar{x}_\lambda,\bar{\mu}_\lambda}\|^2 + \|PU_{x_\lambda,\mu_\lambda}\|^2 + O\Big(\int_\Omega U_{\bar{x}_\lambda,\bar{\mu}_\lambda}^{2^*-1} U_{x_\lambda,\mu_\lambda}\Big) + o(1) \\
=& \|PU_{\bar{x}_\lambda,\bar{\mu}_\lambda}\|^2 + \|PU_{x_\lambda,\mu_\lambda}\|^2 + O\Big(\frac{1}{\frac{\bar{\mu}_\lambda}{\mu_\lambda} + \frac{\mu_\lambda}{\bar{\mu}_\lambda} + \bar{\mu}_\lambda\mu_\lambda|\bar{x}_\lambda - x_\lambda|^2}\Big) + o(1) \\
=& 2\int_{\mathbb{R}^N} U_{0,1}^{2^*} + O\Big(\frac{1}{\frac{\bar{\mu}_\lambda}{\mu_\lambda} + \frac{\mu_\lambda}{\bar{\mu}_\lambda} + \bar{\mu}_\lambda\mu_\lambda|\bar{x}_\lambda - x_\lambda|^2}\Big) + o(1).
\end{aligned}$$

Hence we conclude $\bar{\mu}_\lambda \in (c_0\mu_\lambda, c_1\mu_\lambda)$ and $|\bar{x}_\lambda - x_\lambda| < \frac{1}{c_0^2\mu_\lambda}$. For such $\bar{\mu}_\lambda$ and $\bar{x}_\lambda$, the function $\omega_{\bar{x}_\lambda,\bar{\mu}_\lambda} = u_\lambda - PU_{\bar{x}_\lambda,\bar{\mu}_\lambda}$ satisfies

$$\omega_{\bar{x}_\lambda,\bar{\mu}_\lambda} \in E_{\bar{x}_\lambda,\bar{\mu}_\lambda},$$

and $\|\omega_{\bar{x}_\lambda,\bar{\mu}_\lambda}\| \le \|u_\lambda - PU_{x_\lambda,\mu_\lambda}\| \to 0$, where

$$E_{x,\mu} \in \Big\{\omega : \omega \in H_0^1(\Omega), \int_\Omega \nabla\omega\nabla\frac{\partial PU_{x,\mu}}{\partial x_j} = \int_\Omega \nabla\omega\frac{\partial PU_{x,\mu}}{\partial \mu} = 0, \quad j = 1, \cdots, N\Big\}.$$

To simplify the notation, we still use (x_λ, μ_λ) to denote $(\bar{x}_\lambda, \bar{\mu}_\lambda)$.

With this change of the location x_λ and the height μ_λ, $\omega_{x_\lambda,\mu_\lambda}$ has the property that $\omega_{x_\lambda,\mu_\lambda} \in E_{x_\lambda,\mu_\lambda}$. Therefore, we can use similar arguments to those in Propositions 2.4.3 and 2.4.4, and in Lemma 2.4.2, to deduce the following estimate for $\|\omega_{x_\lambda,\mu_\lambda}\|$:

$$\|\omega_{x_\lambda,\mu_\lambda}\| \le \begin{cases} \dfrac{C\lambda}{\mu_\lambda^2} + \dfrac{C}{\mu_\lambda^{\frac{N+2}{2}}}, & N \ge 7, \\[2ex] \dfrac{C\lambda(\ln\mu_\lambda)^{\frac{2}{3}}}{\mu_\lambda^2} + \dfrac{C(\ln\mu_\lambda)^{\frac{2}{3}}}{\mu_\lambda^4}, & N = 6, \\[2ex] \dfrac{C\lambda}{\mu_\lambda^{\frac{3}{2}}} + \dfrac{C}{\mu_\lambda^3}, & N = 5. \end{cases} \tag{3.2.6}$$

Moreover, it follows from Proposition 2.4.8 that

$$|u_\lambda(x)| \le CU_{x_\lambda,\mu_\lambda}(x), \quad \forall x \in \Omega. \tag{3.2.7}$$

We now turn to the estimate of u_λ in $\Omega \setminus B_\delta(x_\lambda)$. Recall that Lemma 2.1.5 in Chapter 2 gives

$$u_\lambda(x) = A_\lambda \mu_\lambda^{-\frac{N-2}{2}} G(x, x_\lambda) + O\Big(\frac{\lambda}{\mu_\lambda^{\frac{N-2}{2}}} + \frac{1}{\mu_\lambda^{\frac{N}{2}}}\Big), \quad x \in \Omega \setminus B_\delta(x_\lambda), \tag{3.2.8}$$

and

$$\nabla u_\lambda(x) = A_\lambda \mu_\lambda^{-\frac{N-2}{2}} \nabla_x G(x, x_\lambda) + O\Big(\frac{\lambda}{\mu_\lambda^{\frac{N-2}{2}}} + \frac{1}{\mu_\lambda^{\frac{N}{2}}}\Big), \quad x \in \Omega \setminus B_\delta(x_\lambda), \tag{3.2.9}$$

where $A_\lambda = \int_{B_{\frac{1}{2}d\mu_\lambda}(0)} \tilde{u}_\lambda^{2^*-1}(y)$, and $\tilde{u}_\lambda(y) = \mu_\lambda^{-\frac{N-2}{2}} u_\lambda(\mu_\lambda^{-1}y + x_\lambda)$.

On the other hand, using (3.2.4) and (3.2.9), we can locate the bubble from the following relation:

$$\nabla\varphi(x_\lambda) = O\Big(\lambda + \frac{1}{\mu_\lambda}\Big). \tag{3.2.10}$$

See the proof of (2.1.20) in Chapter 2.

Combining (3.2.8), (3.2.9), and (3.2.3), we can also determine the height of the bubble as follows. We have

$$\begin{aligned}\frac{A_\lambda^2}{\mu_\lambda^{N-2}}\int_{\partial B_\delta(x_\lambda)} & \Big((\nabla G(x,x_\lambda)\cdot \nu)((x-x_\lambda)\cdot\nabla G(x,x_\lambda)) \\ & -\frac{|\nabla G(x,x_\lambda)|^2}{2}(x-x_\lambda)\cdot\nu\Big) \\ & +\frac{N-2}{2}\frac{A_\lambda^2}{\mu_\lambda^{N-2}}\int_{\partial B_\delta(x_\lambda)} G(x,x_\lambda)\frac{\partial G(x,x_\lambda)}{\partial \nu} \\ & =\lambda\int_{B_\delta(x_\lambda)} u_\lambda^2+O\Big(\frac{\lambda}{\mu_\lambda^{N-2}}+\frac{1}{\mu_\lambda^{N-1}}\Big).\end{aligned}$$

From Proposition 6.2.4, we obtain

$$\frac{A_\lambda^2(N-2)H(x_\lambda,x_\lambda)}{2\mu_\lambda^{N-2}}=\lambda\int_{B_\delta(x_\lambda)} u_\lambda^2+O\Big(\frac{\lambda}{\mu_\lambda^{N-2}}+\frac{1}{\mu_\lambda^{N-1}}\Big). \tag{3.2.11}$$

Moreover, we have

$$\int_{B_\delta(x_\lambda)} u_\lambda^2=\frac{1}{\mu_\lambda^2}\Big(\int_{\mathbb{R}^N} U_{0,1}^2+o(1)\Big)$$

and

$$A_\lambda=\int_{\mathbb{R}^N} U_{0,1}^{2^*-1}+o(1).$$

So we obtain

$$\begin{aligned}\frac{(N-2)H(x_\lambda,x_\lambda)\Big(\int_{\mathbb{R}^N} U_{0,1}^{2^*-1}+o(1)\Big)^2}{2\mu_\lambda^{N-2}}=&\ \frac{\lambda\Big(\int_{\mathbb{R}^N} U_{0,1}^2+o(1)\Big)}{\mu_\lambda^2}\\ &+O\Big(\frac{1}{\mu_\lambda^N}+\frac{\lambda}{\mu_\lambda^{N-2}}\Big),\end{aligned}$$

which gives

$$\lambda=\mu_\lambda^{4-N}\big(B_0+o(1)\big), \tag{3.2.12}$$

where $B_0>0$ is some constant.

To prove the local uniqueness result, we need to improve the estimate in (3.2.10) and (3.2.12). The following lemma improves Lemma 2.1.5 in Chapter 2.

Lemma 3.2.2 *Assume that* $N \geq 6$. *Suppose that* u_λ *is the solution of* (3.0.2) *satisfying* (3.2.1) *and* (3.2.2). *Then, for* $x \in \Omega \setminus B_\delta(x_\lambda)$, *we have*

$$u_\lambda(x) = A_\lambda \mu_\lambda^{-\frac{N-2}{2}} G(x, x_\lambda) + O\Big(\frac{\ln \mu_\lambda}{\mu_\lambda^{\frac{N+2}{2}}}\Big) \tag{3.2.13}$$

and

$$\nabla u_\lambda(x) = A_\lambda \mu_\lambda^{-\frac{N-2}{2}} \nabla_x G(x, x_\lambda) + O\Big(\frac{\ln \mu_\lambda}{\mu_\lambda^{\frac{N+2}{2}}}\Big), \tag{3.2.14}$$

where $A_\lambda = \int_{B_{\frac{1}{2}\delta}(x_\lambda)} \tilde{u}_\lambda^{2^*-1}$.

Proof. We note that

$$u_\lambda(x) = \int_\Omega G(x, y)\Big(u_\lambda^{2^*-1}(y) + \lambda u_\lambda(y)\Big)\, dy.$$

Then by (3.2.7) and (3.2.12), one has that

$$\begin{aligned}
\lambda \int_\Omega G(x, y) u_\lambda(y)\, dy &\leq C\lambda \int_\Omega G(x, y) U_{x_\lambda, \mu_\lambda}(y)\, dy \\
&\leq C\lambda \int_\Omega \frac{1}{|y - x|^{N-2}} \frac{1}{\mu_\lambda^{\frac{N-2}{2}} |y - x_\lambda|^{N-2}}\, dy \\
&= C\lambda \int_{B_{\frac{1}{2}\delta}(x_\lambda)} \frac{1}{|y - x|^{N-2}} \frac{1}{\mu_\lambda^{\frac{N-2}{2}} |y - x_\lambda|^{N-2}}\, dy \\
&\quad + C\lambda \int_{\Omega \setminus B_{\frac{1}{2}\delta}(x_\lambda)} \frac{1}{|y - x|^{N-2}} \frac{1}{\mu_\lambda^{\frac{N-2}{2}} |y - x_\lambda|^{N-2}}\, dy \\
&\leq \frac{C\lambda}{\mu_\lambda^{\frac{N-2}{2}}} \leq \frac{C}{\mu_\lambda^{\frac{N-2}{2} + N - 4}}.
\end{aligned}$$

On the other hand,

$$\begin{aligned}
&\int_{\Omega \setminus B_{\frac{1}{2}\delta}(x_\lambda)} G(x, y) u_\lambda^{2^*-1}(y)\, dy \\
&\leq \frac{C}{\mu_\lambda^{\frac{N+2}{2}}} \int_{\Omega \setminus B_{\frac{1}{2}\delta}(x_\lambda)} \frac{1}{|y - x|^{N-2}} \frac{1}{|y - x_\lambda|^{N+2}}\, dy \leq \frac{C}{\mu_\lambda^{\frac{N+2}{2}}}.
\end{aligned}$$

Moreover,

$$\begin{aligned}
&\int_{B_{\frac{1}{2}d}(x_\lambda)} G(x,y)u_\lambda^{2^*-1}(y)\,dy\\
&= G(x,x_\lambda)\int_{B_{\frac{1}{2}\delta}(x_\lambda)} u_\lambda^{2^*-1} + \int_{B_{\frac{1}{2}\delta}(x_\lambda)} \big\langle \nabla G(x,x_\lambda), y-x_\lambda\big\rangle u_\lambda^{2^*-1}(y)\,dy\\
&\quad +O\Big(\int_{B_{\frac{1}{2}\delta}(x_\lambda)} |y-x_\lambda|^2 u_\lambda^{2^*-1}\Big).
\end{aligned}$$

It is now easy to verify that

$$\begin{aligned}
\int_{B_{\frac{1}{2}\delta}(x_\lambda)} |y-x_\lambda|^2 u_\lambda^{2^*-1} &\le C\int_{B_{\frac{1}{2}\delta}(x_\lambda)} |y-x_\lambda|^2 U_{x_\lambda,\mu_\lambda}^{2^*-1}\\
&= \frac{C}{\mu_\lambda^{\frac{N+2}{2}}}\int_{B_{\frac{1}{2}\delta\mu_\lambda}(0)} |y|^2 U_{0,1}^{2^*-1} = O\Big(\frac{\ln\mu_\lambda}{\mu_\lambda^{\frac{N+2}{2}}}\Big),
\end{aligned}$$

and by the symmetry, one has

$$\begin{aligned}
&\int_{B_{\frac{1}{2}\delta}(x_\lambda)} \big\langle \nabla G(x,x_\lambda), y-x_\lambda\big\rangle u_\lambda^{2^*-1}(y)\,dy\\
&= \int_{B_{\frac{1}{2}\delta}(x_\lambda)} \big\langle \nabla G(x,x_\lambda), y-x_\lambda\big\rangle U_{x_\lambda,\mu_\lambda}^{2^*-1}(y)\,dy\\
&+ O\Big(\int_{B_{\frac{1}{2}\delta}(x_\lambda)} |y-x_\lambda| U_{x_\lambda,\mu_\lambda}^{2^*-2}|\omega_{x_\lambda,\mu_\lambda}|\Big)\\
&= O\Big(\int_{B_{\frac{1}{2}\delta}(x_\lambda)} |y-x_\lambda| U_{x_\lambda,\mu_\lambda}^{2^*-2}|\omega_{x_\lambda,\mu_\lambda}|\Big) = O\Big(\int_{B_{\frac{1}{2}\delta}(x_\lambda)} \frac{1}{\mu_\lambda^2|y-x_\lambda|^3}|\omega_{x_\lambda,\mu_\lambda}|\Big)\\
&= O\Big(\frac{1}{\mu_\lambda^2}\Big)\|\omega_{x_\lambda,\mu_\lambda}\| = O\Big(\frac{1}{\mu_\lambda^{\frac{N+2}{2}}}\Big).
\end{aligned}$$

So, by (3.2.6), we have proved (3.2.13).

Similarly, from

$$\nabla u_\lambda(x) = \int_\Omega \nabla_x G(x,y)\Big(u_\lambda^{2^*-1}(y) + \lambda u_\lambda(y)\Big)\,dy,$$

we can prove (3.2.14).

□

Using Lemma 3.2.2, we can now improve the estimates in (3.2.10) and (3.2.12) as follows:

Lemma 3.2.3 *Assume that $N \ge 6$. Suppose that u_λ is the solution of* (3.0.2) *satisfying* (3.2.1) *and* (3.2.2). *Then we have*

$$\nabla\varphi(x_\lambda) = O\Big(\frac{\ln\mu_\lambda}{\mu_\lambda^2}\Big) \tag{3.2.15}$$

and

$$\lambda = \mu_\lambda^{4-N}\Big(B_0 + O\Big(\frac{1}{\mu_\lambda^2}\Big)\Big), \tag{3.2.16}$$

where $B_0 > 0$ *is the constant in* (3.2.12).

Proof. Firstly, by using Lemma 3.2.2 and the Pohozaev identity (3.2.4), we can obtain (3.2.15).

On the other hand, we have

$$\begin{aligned}\int_{B_\delta(x_\lambda)} u_\lambda^2 &= \int_{B_\delta(x_\lambda)} (PU_{x_\lambda,\mu_\lambda})^2 + O\Big(\int_{B_\delta(x_\lambda)} U_{x_\lambda,\mu_\lambda}|\omega_\lambda| + \|\omega_\lambda\|^2\Big)\\ &= \frac{1}{\mu_\lambda^2}\Big(\int_{\mathbb{R}^N} U_{0,1}^2 + O\Big(\frac{1}{\mu_\lambda^{N-4}}\Big)\Big) + O\Big(\frac{1}{\mu_\lambda}\|\omega_\lambda\| + \|\omega_\lambda\|^2\Big)\\ &= \frac{1}{\mu_\lambda^2}\Big(\int_{\mathbb{R}^N} U_{0,1}^2 + O\Big(\frac{1}{\mu_\lambda^2}\Big)\Big)\end{aligned} \tag{3.2.17}$$

and

$$\begin{aligned}\int_{B_\delta(x_\lambda)} |u_\lambda|^{2^*-1} &= \int_{B_\delta(x_\lambda)} (PU_{x_\lambda,\mu_\lambda})^{2^*-1} + O\Big(\int_{B_\delta(x_\lambda)} |U_{x_\lambda,\mu_\lambda}|^{2^*-2}|\omega_\lambda|\Big)\\ &= \frac{1}{\mu_\lambda^{\frac{N-2}{2}}}\Big(\int_{\mathbb{R}^N} U_{0,1}^{2^*-1} + O\Big(\frac{1}{\mu_\lambda^2}\Big)\Big)\\ &\quad + O\Big(\Big(\int_{B_\delta(x_\lambda)} |U_{x_\lambda,\mu_\lambda}|^{\frac{2N(2^*-2)}{N+2}}\Big)^{\frac{N+2}{2N}}\|\omega_\lambda\|\Big)\\ &= \frac{1}{\mu_\lambda^{\frac{N-2}{2}}}\Big(\int_{\mathbb{R}^N} U_{0,1}^{2^*-1} + O\Big(\frac{1}{\mu_\lambda^2}\Big)\Big).\end{aligned} \tag{3.2.18}$$

Inserting (3.2.17) and (3.2.18) into (3.2.11), we obtain

$$\begin{aligned}&\frac{(N-2)H(x_\lambda,x_\lambda)\Big(\int_{\mathbb{R}^N} U_{0,1}^{2^*-1} + O\Big(\frac{1}{\mu_\lambda^2}\Big)\Big)^2}{2\mu_\lambda^{N-2}}\\ &= \frac{\lambda\Big(\int_{\mathbb{R}^N} U_{0,1}^2 + O\Big(\frac{1}{\mu_\lambda^2}\Big)\Big)}{\mu_\lambda^2} + O\Big(\frac{1}{\mu_\lambda^N} + \frac{\lambda}{\mu_\lambda^{N-2}}\Big),\end{aligned} \tag{3.2.19}$$

which gives

$$\lambda = \mu_\lambda^{4-N}\Big(B_0 + O\Big(\frac{1}{\mu_\lambda^2}\Big)\Big).$$

□

By Lemma 3.2.3, if the critical point x_0 of φ is non-degenerate, we have the following estimate:

$$|x_\lambda - x_0| = O\Big(\frac{\ln \mu_\lambda}{\mu_\lambda^2}\Big), \quad N \geq 6. \tag{3.2.20}$$

We now turn to prove Theorem 3.2.1. Here, we argue by contradiction. Suppose that (3.0.2) has two different solutions $u_\lambda^{(1)}$ and $u_\lambda^{(2)}$. We will use $x_\lambda^{(k)}$ and $\mu_\lambda^{(k)}$ to denote the center and the height of the bubbles appearing in $u_\lambda^{(k)}$, respectively.

Let

$$\xi_\lambda = \frac{u_\lambda^{(1)} - u_\lambda^{(2)}}{\|u_\lambda^{(1)} - u_\lambda^{(2)}\|_{L^\infty(\Omega)}}.$$

Then, ξ_λ satisfies $\|\xi_\lambda\|_{L^\infty(\Omega)} = 1$ and

$$-\Delta \xi_\lambda = f(y, u_\lambda^{(1)}, u_\lambda^{(2)}), \tag{3.2.21}$$

where

$$f(y, u_\lambda^{(1)}, u_\lambda^{(2)}) = \frac{1}{\|u_\lambda^{(1)} - u_\lambda^{(2)}\|_{L^\infty(\Omega)}}\big((u_\lambda^{(1)})^{2^*-1} - (u_\lambda^{(2)})^{2^*-1}\big) + \lambda\xi_\lambda.$$

Write

$$f(y, u_\lambda^{(1)}, u_\lambda^{(2)}) = c_\lambda(y)\xi_\lambda(y) + \lambda\xi_\lambda,$$

where

$$c_\lambda(y) = (2^* - 1)\int_0^1 \big(tu_\lambda^{(1)}(y) + (1-t)u_\lambda^{(2)}(y)\big)^{2^*-2}\,dt.$$

Since

$$\big|U_{x_\lambda^{(1)},\mu_\lambda^{(1)}} - U_{x_\lambda^{(2)},\mu_\lambda^{(2)}}\big| \leq CU_{x_\lambda^{(1)},\mu_\lambda^{(1)}}\Big(\mu_\lambda^{(1)}|x_\lambda^{(1)} - x_\lambda^{(2)}| + \frac{|\mu_\lambda^{(1)} - \mu_\lambda^{(2)}|}{\mu_\lambda^{(1)}}\Big),$$

we obtain from (3.2.20) and (3.2.16) that

$$\big|U_{x_\lambda^{(1)},\mu_\lambda^{(1)}} - U_{x_\lambda^{(2)},\mu_\lambda^{(2)}}\big| \leq \frac{CU_{x_\lambda^{(1)},\mu_\lambda^{(1)}} \ln \mu_\lambda^{(1)}}{\mu_\lambda^{(1)}},$$

which in turns implies

$$\begin{aligned} c_\lambda(y) &= (2^* - 1)U^{2^*-2}_{x_\lambda^{(1)},\mu_\lambda^{(1)}} \\ &\quad + O\Big(\Big(U_{x_\lambda^{(1)},\mu_\lambda^{(1)}}\frac{\ln \mu_\lambda^{(1)}}{\mu_\lambda^{(1)}} + |\omega_\lambda^{(1)}| + |\omega_\lambda^{(2)}|\Big)^{2^*-2}\Big). \end{aligned} \tag{3.2.22}$$

We will obtain a contradiction by proving $\|\xi_\lambda\|_{L^\infty(\Omega)} \to 0$ as $\lambda \to 0$. We first prove the following estimate for ξ_λ.

Lemma 3.2.4 *One has the following estimate:*

$$|\xi_\lambda(x)| \le \frac{C}{(1+(\mu_\lambda^{(1)})^2|x-x_\lambda^{(1)}|^2)^{\frac{N-2}{2}}}. \tag{3.2.23}$$

Proof. Let $G_\lambda(y,x)$ be the Green's function of $-\Delta u - \lambda u$ in Ω with homogenous Dirichlet boundary condition. It then follows that

$$0 < G_\lambda(y,x) \le \frac{C}{|y-x|^{N-2}}.$$

We have

$$\xi_\lambda(x) = \int_\Omega G_\lambda(y,x)c_\lambda(y)\xi_\lambda(y)\,dy.$$

From Proposition 2.4.8, we have $u_\lambda^{(k)} \le CU_{x_\lambda^{(k)},\mu_\lambda^{(k)}}$. In view of (3.2.20) and (3.2.16), we then obtain

$$|c_\lambda(y)| \le \frac{C(\mu_\lambda^{(1)})^2}{(1+(\mu_\lambda^{(1)})^2|y-x_\lambda^{(1)}|^2)^2}.$$

Since $|\xi_\lambda| \le 1$, one has

$$\begin{aligned}|\xi_\lambda(x)| \le & C\int_\Omega \frac{1}{|y-x|^{N-2}}\frac{(\mu_\lambda^{(1)})^2}{(1+(\mu_\lambda^{(1)})^2|y-x_\lambda^{(1)}|^2)^2}dy\\ \le & C\int_{\mathbb{R}^N} \frac{1}{|y-\mu_\lambda^{(1)}x|^{N-2}}\frac{1}{(1+|y-\mu_\lambda^{(1)}x_\lambda^{(1)}|^2)^2}dy\\ \le & \frac{C}{1+(\mu_\lambda^{(1)})^2|x-x_\lambda^{(1)}|^2}.\end{aligned} \tag{3.2.24}$$

From (3.2.24), we have

$$|c_\lambda(y)\xi_\lambda(y)| \le \frac{C(\mu_\lambda^{(1)})^2}{(1+(\mu_\lambda^{(1)})^2|y-x_\lambda^{(1)}|^2)^2}|\xi_\lambda| \le \frac{C(\mu_\lambda^{(1)})^2}{(1+(\mu_\lambda^{(1)})^2|y-x_\lambda^{(1)}|^2)^3}.$$

As a result,

$$\begin{aligned}|\xi_\lambda(x)| \le & C\int_{\mathbb{R}^N} \frac{1}{|y-x|^{N-2}}\frac{(\mu_\lambda^{(1)})^2}{(1+(\mu_\lambda^{(1)})^2|y-x_\lambda^{(1)}|^2)^3}dy\\ \le & \frac{C}{(1+(\mu_\lambda^{(1)})^2|x-x_\lambda^{(1)}|^2)^2}.\end{aligned}$$

We can continue this process to obtain

$$|\xi_\lambda(x)| \le \frac{C}{(1+(\mu_\lambda^{(1)})^2|x-x_\lambda^{(1)}|^2)^{\frac{N-2}{2}}}.$$

□

From (3.2.23), we find that if $R>0$ is large, then $|\xi_\lambda(x)| \le \frac{1}{2}$ for $x \in \Omega \setminus B_{R(\mu_\lambda^{(1)})^{-1}}(x_\lambda^{(1)})$. To obtain a contradiction, we need to prove that $|\xi_\lambda(x)| = o(1)$ for $x \in B_{R(\mu_\lambda^{(1)})^{-1}}(x_\lambda^{(1)})$.

Let

$$\tilde{\xi}_\lambda(y) = \xi_\lambda\Big(\frac{1}{\mu_\lambda^{(1)}}y + x_\lambda^{(1)}\Big).$$

We then have the following result:

Lemma 3.2.5 *We have that*

$$\tilde{\xi}_\lambda(y) \to \sum_{k=0}^{N} b_k \psi_k(y),$$

in $C^1(B_R(0))$ for any $R>0$, where b_k, $k=0,\cdots,N$, are some constants, and

$$\psi_0 = \frac{\partial U_{0,\lambda}}{\partial \lambda}\Big|_{\lambda=1}, \quad \psi_j = \frac{\partial U_{0,1}}{\partial y_j}, \quad j=1,\cdots,N.$$

Proof. In view of $|\tilde{\xi}_\lambda| \le 1$, we may assume that $\tilde{\xi}_\lambda \to \xi$ in $C^1(B_R(0))$ for any $R>0$. Then from (3.2.21) and (3.2.22), we find that ξ satisfies

$$-\Delta\xi = (2^*-1)U_{0,1}^{2^*-2}\xi, \quad \text{in } \mathbb{R}^N.$$

This gives

$$\xi = \sum_{k=0}^{N} b_k \psi_k.$$

□

To finish the proof, we only need to prove that $b_0 = b_1 = \cdots = b_N = 0$. This will be achieved by using the Pohozaev identities. Using (3.2.3) and (3.2.4), we obtain

$$\begin{aligned} &-\int_{\partial B_\delta(x_\lambda^{(1)})} \frac{\partial \xi_\lambda}{\partial \nu}\frac{\partial u_\lambda^{(1)}}{\partial x_i} - \int_{\partial B_\delta(x_\lambda^{(1)})} \frac{\partial u_\lambda^{(2)}}{\partial \nu}\frac{\partial \xi_\lambda}{\partial x_i} \\ &+\frac{1}{2}\int_{\partial B_\delta(x_\lambda^{(1)})} \langle \nabla(u_\lambda^{(1)}+u_\lambda^{(2)}), \nabla\xi_\lambda\rangle \nu_i = \int_{\partial B_\delta(x_\lambda^{(1)})} C_\lambda(x)\xi_\lambda \nu_i \end{aligned} \tag{3.2.25}$$

and

$$
\begin{aligned}
&\int_{\partial B_\delta(x_\lambda^{(1)})} \frac{\partial \xi_\lambda}{\partial \nu} \langle x - x_\lambda, \nabla u_\lambda^{(1)} \rangle + \int_{\partial B_\delta(x_\lambda^{(1)})} \frac{\partial u_\lambda^{(2)}}{\partial \nu} \langle x - x_\lambda, \nabla \xi_\lambda \rangle \\
&- \frac{1}{2} \int_{\partial B_\delta(x_\lambda^{(1)})} \langle \nabla (u_\lambda^{(1)} + u_\lambda^{(2)}), \nabla \xi_\lambda \rangle \langle x - x_\lambda, \nu \rangle \\
&+ \frac{N-2}{2} \int_{\partial B_\delta(x_\lambda^{(1)})} \frac{\partial \xi_\lambda}{\partial \nu} u_\lambda^{(1)} + \frac{N-2}{2} \int_{\partial B_\delta(x_\lambda^{(1)})} \frac{\partial u_\lambda^{(2)}}{\partial \nu} \xi_\lambda \\
= &- \int_{\partial B_\delta(x_\lambda^{(1)})} C_\lambda(x) \xi_\lambda \langle x - x_\lambda^{(1)}, \nu \rangle + \lambda \int_{B_\delta(x_\lambda^{(1)})} (u_\lambda^{(1)} + u_\lambda^{(2)}) \xi_\lambda,
\end{aligned}
\tag{3.2.26}
$$

where

$$
C_\lambda(x) = \int_0^1 \left(t u_\lambda^{(1)} + (1-t) u_\lambda^{(2)} \right)^{2^*-1} dt + \frac{\lambda}{2} \left(u_\lambda^{(1)} + u_\lambda^{(2)} \right).
$$

Similar to (3.2.22), we can deduce

$$
\begin{aligned}
C_\lambda(x) = U^{2^*-1}_{x_\lambda^{(1)}, \mu_\lambda^{(1)}} &+ O\Big(\Big(U_{x_\lambda^{(1)}, \mu_\lambda^{(1)}} \frac{\ln \mu_\lambda^{(1)}}{\mu_\lambda^{(1)}} + |\omega_\lambda^{(1)}| + |\omega_\lambda^{(2)}| \Big)^{2^*-1} \Big) \\
&+ \frac{\lambda}{2} \left(u_\lambda^{(1)} + u_\lambda^{(2)} \right).
\end{aligned}
$$

To use (3.2.25) and (3.2.26), we need the following estimates for ξ_λ.

Lemma 3.2.6 *We have the following estimate:*

$$
\begin{aligned}
\xi_\lambda(x) = &B_\lambda G(x_\lambda^{(1)}, x) + \frac{1}{(\mu_\lambda^{(1)})^{N-1}} \sum_{h=1}^{N} B_{\lambda,h} \partial_h G(x_\lambda^{(1)}, x) \\
&+ O\Big(\frac{\ln \mu_\lambda^{(1)}}{(\mu_\lambda^{(1)})^N} \Big), \quad \text{in } C^1\big(\Omega \setminus B_{2\theta}(x_\lambda^{(1)})\big),
\end{aligned}
\tag{3.2.27}
$$

where $\theta > 0$ is any fixed small constant, $\partial_h G(y, x) = \frac{\partial G(y,x)}{\partial y_h}$,

$$
B_\lambda = \int_{B_\theta(x_\lambda^{(1)})} f_\lambda^*,
$$

$$
B_{\lambda,h} = \frac{1}{(\mu_{j,\lambda}^{(1)})^2} \int_{B_{\theta \mu_\lambda^{(1)}}(0)} y_h f_\lambda^* \Big(\frac{1}{\mu_\lambda^{(1)}} y + x_\lambda^{(1)} \Big)
$$

and $f_\lambda^*(y) = f(y, u_\lambda^{(1)}(y), u_\lambda^{(2)}(y))$.

Proof. For $y \in \Omega \setminus B_\theta(x_\lambda^{(1)})$, it follows from Lemma 3.2.4 that we have

$$\begin{aligned}\xi_\lambda(x) &= \int_\Omega G(y,x) f_\lambda^*(y)\,dy = \int_{B_\theta(x_\lambda^{(1)})} G(y,x) f_\lambda^*(y)\,dy \\ &\quad + \int_{\Omega\setminus B_\theta(x_\lambda)} G(y,x) f_\lambda^*(y)\,dy \\ &= B_\lambda G(x_\lambda^{(1)}, x) + \int_{B_\theta(x_\lambda^{(1)})} \Big(G(y,x) - G(x_\lambda^{(1)}, x)\Big) f_\lambda^*(y)\,dy \\ &\quad + O\Big(\frac{1}{(\mu_\lambda^{(1)})^N} + \frac{\lambda}{(\mu_\lambda^{(1)})^{N-2}}\Big).\end{aligned}$$

On the other hand, using Lemma 3.2.4, we have that

$$\begin{aligned}&\int_{B_\theta(x_\lambda^{(1)})} \big(G(y,x) - G(x_\lambda^{(1)}, x)\big) f_\lambda^*(y)\,dy \\ &= \int_{B_\theta(x_\lambda^{(1)})} \big\langle \partial G(x_\lambda^{(1)}, x),\ y - x_\lambda^{(1)}\big\rangle f_\lambda^*(y)\,dy \\ &\quad + O\Big(\frac{1}{|x - x_\lambda^{(1)}|^N} \int_{B_\theta(x_\lambda^{(1)})} |y - x_\lambda^{(1)}|^2 |f_\lambda^*|\Big) \\ &= \frac{1}{(\mu_\lambda^{(1)})^{N-1}} \int_{B_{\theta\mu_\lambda^{(1)}}(0)} \big\langle \partial G(x_\lambda^{(1)}, x),\ y\big\rangle \frac{1}{(\mu_\lambda^{(1)})^2} f_\lambda^*\Big(\frac{1}{\mu_\lambda^{(1)}} y + x_\lambda^{(1)}\Big)\,dy \\ &\quad + \frac{1}{|x - x_\lambda^{(1)}|^N} O\Big(\int_{B_\theta(x_\lambda^{(1)})} \frac{(\mu_\lambda^{(1)})^2 |y - x_\lambda^{(1)}|^2}{(1 + \mu_\lambda^{(1)}|y - x_\lambda^{(1)}|)^{N+2}} \\ &\quad + \int_{B_\theta(x_\lambda^{(1)})} \frac{\lambda |y - x_\lambda^{(1)}|^2}{(1 + \mu_\lambda^{(1)}|y - x_\lambda^{(1)}|)^{N-2}}\Big) \\ &= \frac{1}{(\mu_\lambda^{(1)})^{N-1}} \int_{B_{\theta\mu_\lambda^{(1)}}(0)} \big\langle \partial G(x_\lambda^{(1)}, x),\ y\big\rangle \frac{1}{(\mu_\lambda^{(1)})^2} f_\lambda^*\Big(\frac{1}{\mu_\lambda^{(1)}} y + x_\lambda^{(1)}\Big)\,dy \\ &\quad + O\Big(\frac{\ln \mu_\lambda^{(1)}}{(\mu_\lambda^{(1)})^N |x - x_\lambda^{(1)}|^N} + \frac{\lambda}{(\mu_\lambda^{(1)})^{N-2} |x - x_\lambda^{(1)}|^N}\Big).\end{aligned}$$

Noting that $\lambda \le C(\mu_\lambda^{(1)})^{4-N} \le C(\mu_\lambda^{(1)})^{-2}$ if $N \ge 6$, we obtain

$$\xi_\lambda = B_\lambda G(x_\lambda^{(1)}, x) + \frac{1}{(\mu_\lambda^{(1)})^{N-1}} \sum_{h=1}^N B_{\lambda,h,} \partial_h G(x_\lambda^{(1)}, x) + O\Big(\frac{\ln \mu_\lambda^{(1)}}{(\mu_\lambda^{(1)})^N}\Big), \tag{3.2.28}$$

in $C\big(\Omega \setminus B_{2\theta}(x_\lambda^{(1)})\big)$.

Similarly, we can prove that (3.2.28) holds in $C^1\big(\Omega \setminus B_{2\theta}(x_\lambda^{(1)})\big)$. □

Proof of Theorem 3.2.1. It is easy to check that

$$|B_\lambda| \le \frac{C}{(\mu_\lambda^{(1)})^{N-2}}.$$

Define

$$Q(\xi, u, \delta) = -\int_{\partial B_\delta(x_\lambda^{(1)})} \frac{\partial \xi}{\partial \nu}\frac{\partial u}{\partial x_i} - \int_{\partial B_\delta(x_\lambda^{(1)})} \frac{\partial u}{\partial \nu}\frac{\partial \xi}{\partial x_i}$$
$$+ \int_{\partial B_\delta(x_\lambda^{(1)})} \langle \nabla u, \nabla \xi\rangle \nu_i.$$

Inserting (3.2.14), (3.2.27) into (3.2.25), we are led to

$$Q\Big(B_\lambda G(x_\lambda^{(1)}, x) + \sum_{h=1}^N \frac{B_{\lambda,h}\partial_h G(x_\lambda^{(1)}, x)}{(\mu_\lambda^{(1)})^{N-1}}, \frac{A_\lambda G(x, x_\lambda^{(1)})}{(\mu_\lambda^{(1)})^{\frac{N-2}{2}}}, \delta\Big)$$
$$= \frac{1}{(\mu_\lambda^{(1)})^{N-2}} O\Big(\frac{\ln \mu_\lambda^{(1)}}{(\mu_\lambda^{(1)})^{\frac{N+2}{2}}} + \frac{\lambda}{(\mu_\lambda^{(1)})^{\frac{N-2}{2}}}\Big). \tag{3.2.29}$$

Since $A_\lambda \ge a_0 > 0$, it follows from (3.2.29) that

$$Q\Big(B_\lambda G(x_\lambda^{(1)}, x) + \frac{1}{(\mu_\lambda^{(1)})^{N-1}} \sum_{h=1}^N B_{\lambda,h}\partial_h G(x_\lambda^{(1)}, x), G(x, x_\lambda^{(1)}), \delta\Big)$$
$$= \frac{1}{(\mu_\lambda^{(1)})^{N-2}} O\Big(\frac{\ln \mu_\lambda^{(1)}}{(\mu_\lambda^{(1)})^2} + \lambda\Big). \tag{3.2.30}$$

But from Proposition 6.2.3, we find

$$Q(B_\lambda G(x_\lambda^{(1)}, x), G(x, x_\lambda^{(1)}), \delta) = B_\lambda Q(G(x_\lambda^{(1)}, x), G(x, x_\lambda^{(1)}), \delta)$$
$$= B_\lambda \frac{\partial \varphi(x_\lambda^{(1)})}{\partial x_i} = O\Big(\frac{\ln \mu_\lambda^{(1)}}{(\mu_\lambda^{(1)})^N}\Big).$$

Hence we obtain from (3.2.30) that

$$\sum_{h=1}^N B_{\lambda,h} Q(\partial_h G(x_\lambda^{(1)}, x), G(x, x_\lambda^{(1)}), \delta) = O\Big(\frac{\ln \mu_\lambda^{(1)}}{\mu_\lambda^{(1)}}\Big),$$

which, together with Proposition 6.2.5 and the non-degeneracy of the critical point x_0, gives

$$B_{\lambda,h} = O\Big(\frac{\ln \mu_\lambda^{(1)}}{\mu_\lambda^{(1)}}\Big).$$

On the other hand, we have that as $\lambda \to 0$,

$$\begin{aligned} B_{\lambda,h} \to & \int_{\mathbb{R}^N} y_h U_{0,1}^{2^*-2}\Big(\sum_{k=0}^{N} b_k \psi_k(y)\Big) = b_h \int_{\mathbb{R}^N} y_h U_{0,1}^{2^*-2} \psi_h(y) \\ & = -\frac{b_h}{2^*-1} \int_{\mathbb{R}^N} U_{0,1}^{2^*-1}. \end{aligned}$$

Thus $b_h = 0$, $h = 1, \cdots, N$.

Now we insert (3.2.14) and (3.2.27) into (3.2.26) to obtain

$$\begin{aligned} & 2A_\lambda B_\lambda \int_{\partial B_\delta(x_\lambda^{(1)})} \frac{\partial G(x_\lambda^{(1)}, x)}{\partial \nu} \langle x - x_\lambda^{(1)}, \nabla G(x_\lambda^{(1)}, x)\rangle \\ & - A_\lambda B_\lambda \int_{\partial B_\delta(x_\lambda^{(1)})} \langle \nabla G(x_\lambda^{(1)}, x), \nabla G(x_\lambda^{(1)}, x)\rangle \langle x - x_\lambda^{(1)}, \nu\rangle \\ & + (N-2) A_\lambda B_\lambda \int_{\partial B_\delta(x_\lambda^{(1)})} \frac{\partial G(x_\lambda^{(1)}, x)}{\partial \nu} G(x_\lambda^{(1)}, x) \\ & = \lambda (\mu_\lambda^{(1)})^{\frac{N-2}{2}} \int_{B_\delta(x_\lambda^{(1)})} (u_\lambda^{(1)} + u_\lambda^{(2)}) \xi_\lambda + \frac{1}{(\mu_\lambda^{(1)})^{N-2}} O\Big(\frac{\ln \mu_\lambda^{(1)}}{(\mu_\lambda^{(1)})^2} + \lambda\Big), \end{aligned}$$

where $A_\lambda = \int_{\mathbb{R}^N} U_{0,1}^{2^*-1} + o(1)$.

On the other hand, it follows from Proposition 6.2.4 that

$$\begin{aligned} & 2 \int_{\partial B_\delta(x_\lambda^{(1)})} \frac{\partial G(x_\lambda^{(1)}, x)}{\partial \nu} \langle x - x_\lambda^{(1)}, \nabla G(x_\lambda^{(1)}, x)\rangle \\ & - \int_{\partial B_\delta(x_\lambda^{(1)})} \langle \nabla G(x_\lambda^{(1)}, x), \nabla G(x_\lambda^{(1)}, x)\rangle \langle x - x_\lambda^{(1)}, \nu\rangle \\ & + (N-2) \int_{\partial B_\delta(x_\lambda^{(1)})} \frac{\partial G(x_\lambda^{(1)}, x)}{\partial \nu} G(x_\lambda^{(1)}, x) \\ & = (N-2) H(x_\lambda^{(1)}, x_\lambda^{(1)}). \end{aligned}$$

Also, we have

$$B_\lambda = \frac{2^*-1}{(\mu_\lambda^{(1)})^{N-2}} \Big(\int_{\mathbb{R}^N} U_{0,1}^{2^*-2} b_0 \psi_0 + o(1)\Big),$$

and

$$\lambda(\mu_\lambda^{(1)})^{\frac{N-2}{2}} \int_{B_\delta(x_\lambda^{(1)})} (u_\lambda^{(1)} + u_\lambda^{(2)})\xi_\lambda = \frac{2\lambda}{(\mu_\lambda^{(1)})^2}\Big(b_0 \int_{\mathbb{R}^N} U_{0,1}\psi_0 + o(1)\Big).$$

Thus we obtain

$$\frac{(N-2)(2^*-1)H(x_\lambda^{(1)}, x_\lambda^{(1)})}{(\mu_\lambda^{(1)})^{N-2}} \int_{\mathbb{R}^N} U_{0,1}^{2^*-1}\Big(\int_{\mathbb{R}^N} U_{0,1}^{2^*-2} b_0\psi_0 + o(1)\Big)$$
$$= \frac{2\lambda}{(\mu_\lambda^{(1)})^2}\Big(b_0 \int_{\mathbb{R}^N} U_{0,1}\psi_0 + o(1)\Big).$$

For any $\mu > 0$, one has

$$\int_{\mathbb{R}^N} U_{0,\mu}^2 = \frac{1}{\mu^2}\int_{\mathbb{R}^N} U_{0,1}^2, \quad \int_{\mathbb{R}^N} U_{0,\mu}^{2^*-1} = \frac{1}{\mu^{\frac{N-2}{2}}}\int_{\mathbb{R}^N} U_{0,1}^{2^*-1}.$$

Differentiating with respect to μ and letting $\mu = 1$, we obtain

$$\int_{\mathbb{R}^N} U_{0,1}\psi_0 = -\int_{\mathbb{R}^N} U_{0,1}^2, \quad (2^*-1)\int_{\mathbb{R}^N} U_{0,1}^{2^*-2}\psi_0 = -\frac{N-2}{2}\int_{\mathbb{R}^N} U_{0,1}^{2^*-1}.$$

Thus

$$\frac{(N-2)^2}{2}\frac{H(x_\lambda^{(1)}, x_\lambda^{(1)})}{(\mu_\lambda^{(1)})^{N-2}}\Big(\int_{\mathbb{R}^N} U_{0,1}^{2^*-1}\Big)^2 b_0$$
$$= \frac{2\lambda}{(\mu_\lambda^{(1)})^2}\int_{\mathbb{R}^N} U_{0,1}^2 b_0 + o(1).$$

But since $\mu_\lambda^{(1)}$ and λ satisfy (3.2.19), this yields $b_0 = 0$. Therefore, we have proved that $\xi_\lambda \to 0$ uniformly in $B_R(0)$ for any $R > 0$. Hence it follows that $\|\xi_\lambda\|_{L^\infty(\Omega)} \to 0$ as $\lambda \to 0$. This is a contradiction to $\|\xi_\lambda\|_{L^\infty(\Omega)} = 1$.

□

Remark 3.2.7 If $N = 5$, then $\lambda = \mu_\lambda^{-1}\big(B_0 + O(\mu_\lambda^{-2})\big)$. Hence for any $x \in \Omega \setminus B_\delta(x_\lambda)$, we have

$$u_\lambda(x) = A_\lambda \mu_\lambda^{-\frac{3}{2}} G(x, x_\lambda) + O\Big(\frac{\lambda}{\mu_\lambda^{\frac{3}{2}}} + \frac{\ln \mu_\lambda}{\mu_\lambda^{\frac{7}{2}}}\Big) \tag{3.2.31}$$

and

$$\nabla u_\lambda(x) = A_\lambda \mu_\lambda^{-\frac{3}{2}} \nabla_x G(x, x_\lambda) + O\Big(\frac{\lambda}{\mu_\lambda^{\frac{3}{2}}} + \frac{\ln \mu_\lambda}{\mu_\lambda^{\frac{7}{2}}}\Big). \tag{3.2.32}$$

Therefore, it follows from (3.2.31) and (3.2.32) that if x_0 is a non-degenerate critical point of $\varphi(x)$, then

$$|x_\lambda - x_0| = O\Big(\lambda + \frac{\ln \mu_\lambda}{\mu_\lambda^2}\Big) = O\Big(\frac{1}{\mu_\lambda}\Big). \tag{3.2.33}$$

Unfortunately, (3.2.33) does not guarantee that (3.2.22) is true.

3.3 Further Results and Comments

Local uniqueness for the bubbling solution of (3.0.2) was first studied in [78], where the main idea to obtain a local uniqueness result was to prove the uniqueness of solution for the reduced finite dimensional problem by counting the local degree. The same idea is also used to study the local uniqueness of peak or bubbling solutions for other nonlinear elliptic problems (see, for example, [34, 115, 122]). One of the disadvantages in the degree counting method is that one has to estimate the second order derivatives of the error term. Even for (3.0.2), the estimates for the second order derivatives of the error term are lengthy and technical. For some nonlinear problems, the solutions may have infinitely many bubbles. In this case, the equations to determine the locations and the height of bubbles are infinitely dimensional. So the degree is not defined for such equations. In other problems, the solution may not be C^2 everywhere. In this case, the estimate for the second order derivatives of the error term is technically complicated.

The use of the Pohozaev identities to obtain a local uniqueness result for subcritical problems first appears in [79]. Note that all the terms in the Pohozaev identities involve first-order derivatives only. Thus this method totally avoids the the estimate for the second-order derivatives of the error term.

Let us now discuss two problems to illustrate the effectiveness of the local Pohozaev identities on the local uniqueness problem for some bubbling solutions.

The first problem we will discuss is the following prescribed scalar curvature problem

$$-\Delta u = K(x)u^{2^*-1}, \quad u > 0, \quad x \in \mathbb{R}^N, \tag{3.3.1}$$

where $K(x)$ satisfies

$$K(x) = K(0) - a|x|^m + O(|x|^{m+1}), \ \ \nabla K(x) = -am|x|^{m-2}x + O(|x|^m),$$
$$\forall\, x \in B_\delta(0), \tag{3.3.2}$$

with $K(0) > 0$, $a > 0$, $m \in (N-2, N)$. Without loss of generality, one may assume that $K(0) = 1$.

In [153], Yan proved that (3.3.1) has a solution with two bubbles separating far from each other under the condition that $K(x)$ has a sequence of unbounded local maximum points. Li-Wei-Xu [100] extended this result and proved the following theorem:

Theorem 3.3.1 *Assume that $N \geq 5$. Suppose that $K(x)$ is periodic in x_1 with period 1 and satisfies (3.3.2). Then there is a large integer $L_0 > 0$ such that for any integer $L \geq L_0$, (3.3.1) has a solution u_L, which satisfies*

$$u_L = \sum_{j=-\infty}^{+\infty} U_{x_{j,L},\mu_{j,L}} + \omega_L, \tag{3.3.3}$$

where $U_{x,\mu}(y)$ is the positive solution of $-\Delta u = u^{2^*-1}$ in $\mathbb{R}^N$ with $U_{x,\mu}(x) = \max_{y\in\mathbb{R}^N} U_{x,\mu}(y)$,

$$x_{j,L} = (jL, 0, \cdots, 0) + o_L(1), \quad \mu_{j,L} = L^{\frac{N-2}{m-N+2}}\big(B_0 + o_L(1)\big) \quad \text{for some } B_0 > 0, \tag{3.3.4}$$

and

$$|\omega_L(x)| = o_L(1) \sum_{j=-\infty}^{+\infty} \frac{\mu_{j,L}^{\frac{N-2}{2}}}{(1+\mu_{j,L}|x-x_{j,L}|)^{\frac{N-2}{2}+\tau}} \quad \text{for some constant } \tau > 1. \tag{3.3.5}$$

The solutions in (3.3.3) have infinitely many bubbles. The degree counting method can not be used to discuss the local uniqueness of such solutions. In [68], by using the local Pohozaev identities, the following result was proved:

Theorem 3.3.2 *Assume that $N \geq 5$. Suppose that $K(x)$ is periodic in x_1 with period 1 and satisfies (3.3.2). If $u_L^{(1)}$ and $u_L^{(2)}$ are two sequences of solutions of (3.3.1) satisfying (3.3.3)–(3.3.5), then $u_L^{(1)} = u_L^{(2)}$ provided $L > 0$ is large enough.*

The results in Theorems 3.3.1 and 3.3.2 were generalized to the following poly-harmonic problem in [89]:

$$(-\Delta)^m u = K(y) u^{\frac{N+2m}{N-2m}}, \quad u > 0 \quad \text{in } \mathbb{R}^N,$$

where $N > 2m+2$, $m \in \mathbb{N}_+$.

Another problem to consider is the following vortex patch problem in planar ideal fluid flow:

$$\begin{cases} -\Delta u = \lambda 1_{\{u>\kappa\}} & \text{in } \Omega, \\ u \in H_0^1(\Omega), \end{cases} \tag{3.3.6}$$

where Ω is a bounded domain in $\mathbb{R}^2$, $\kappa > 0$ is a constant, $\lambda > 0$ is a large parameter, and 1_S is the characteristics function of the set S. The solution u_λ is not C^2 on the free boundary $\{x : u_\lambda = \kappa\}$.

In [30], the following uniqueness result was proved:

Theorem 3.3.3 *Suppose that $x_0 \in \Omega$ is an isolated non-degenerate critical point of the Robin function in Ω. Then, for large $\lambda > 0$,* (3.3.6) *has a unique solution u_λ that satisfies*

$$\{x : u_\lambda > \kappa\} \to \{x_0\},$$

as $\lambda \to +\infty$.

Local uniqueness of solutions with infinitely many bubbles involving fractional Laplacian operators was studied in [88].

Theorem 3.2.1 is due to Glangetas [78]. The proof of this theorem given here is adapted from Cao-Guo-Peng-Yan [30]. In [34, 79], only single peak solutions were studied. Here, we proceed in a similar spirit to that in [79] to prove Theorem 3.1.3.

4

Construction of Infinitely Many Solutions

In this chapter, we will study the multiplicity of solutions for the following Schrödinger equations:

$$\begin{cases} -\Delta u + V(y)u = u^p, \quad u > 0, \quad \text{in } \mathbb{R}^N, \\ u \in H^1(\mathbb{R}^N), \end{cases} \tag{4.0.1}$$

where $p \in \left(1, \frac{N+2}{N-2}\right)$ if $N \geq 3$; $p \in (1, \infty)$ if $N = 2$, and $V \in C(\mathbb{R}^N)$ satisfies $\lim_{|y|\to+\infty} V(y) = V_\infty > 0$.

In Chapter 1, it is shown that if $V(y) \leq V_\infty$ for $y \in \mathbb{R}^N$, (4.0.1) has a least energy solution, while if $V(y) \geq V_\infty$ for $y \in \mathbb{R}^N$, and $V(y) \not\equiv V_\infty$, (4.0.1) does not have a least energy solution. In Chapter 2, various peak solutions are constructed for the following singularly perturbed problems:

$$\begin{cases} -\varepsilon^2\Delta u + V(y)u = u^p, \quad u > 0, \quad \text{in } \mathbb{R}^N, \\ u \in H^1(\mathbb{R}^N), \end{cases} \tag{4.0.2}$$

where $\varepsilon > 0$ is a small parameter. These solutions concentrate at some critical points of $V(y)$. In particular, if $V(y)$ has an isolated local maximum point y_0, then the number of solutions of (4.0.2) increases as ε becomes smaller. But for (4.0.1), we do not have any multiplicity result that can be proved by variational methods.

In this chapter, we will study the multiplicity of solutions for the following problems:

$$\begin{cases} -\Delta u + V(|y|)u = |u|^{p-1}u, \quad \text{in } \mathbb{R}^N, \\ u \in H^1(\mathbb{R}^N), \end{cases} \tag{4.0.3}$$

where $1 < p < \frac{N+2}{N-2}$ if $N \geq 3$, $1 < p < +\infty$ if $N = 2$. The functional corresponding to (4.0.3) is given by

$$I(u) = \frac{1}{2}\int_{\mathbb{R}^N} \big(|\nabla u|^2 + V(|y|)u^2 - \frac{1}{p+1}\int_{\mathbb{R}^N} |u|^{p+1}.$$

Let

$$H^1_{rad}(\mathbb{R}^N) = \big\{u : u \in H^1(\mathbb{R}^N), \quad u \text{ is radially symmetric}\big\},$$

and let

$$L^q_{rad}(\mathbb{R}^N) = \big\{u : u \in L^q(\mathbb{R}^N), \ u \text{ is radially symmetric}\big\}.$$

Since the imbedding of $H^1_{rad}(\mathbb{R}^N)$ to $L^q_{rad}(\mathbb{R}^N)$ is compact for $q \in (2, 2^*)$, we can use the symmetric mountain pass lemma to prove that (4.0.3) possesses infinitely many radially symmetric solutions. But these solutions may not be always positive. Note that if $V(r)$ is non-decreasing, any positive solution of (4.0.3) is radial by [76].

4.1 The Main Result

We will always assume that V is positive and

$$\lim_{r\to+\infty} V(r) = V_\infty > 0.$$

We now try to answer two questions on the existence of positive solutions for (4.0.3).

(i) Does (4.0.3) possess many positive solutions?
(ii) Is any positive solution of (4.0.3) radially symmetric?

By [76], to obtain a non-radial positive solution for (4.0.3), $V(r)$ cannot be non-decreasing in r. In this chapter, we assume that $V(r)$ satisfies the following condition:

(V) There are constants $a > 0$, $m > 1$, $\theta > 0$ and $V_\infty > 0$, such that

$$V(r) = V_\infty + \frac{a}{r^m} + O\Big(\frac{1}{r^{m+\theta}}\Big) \tag{4.1.1}$$

as $r \to +\infty$.

Condition (4.1.1) implies that $V(r)$ is decreasing near infinity. This is a local assumption. Without loss of generality, we may assume that $V_\infty = 1$.

We will prove the following multiplicity result of positive solutions for (4.0.3).

Theorem 4.1.1 (Wei–Yan [148], 2010) *If $V(r)$ satisfies (V), then problem (4.0.3) has infinitely many non-radial positive solutions.*

It seems difficult to prove Theorem 4.1.1 by using variational methods. In the following, we will use the techniques in the singularly perturbed elliptic problems to construct non-radial solutions for (4.0.3) directly. Such techniques work for problems with a parameter such as (4.0.2). If we want to apply those techniques for problem (4.0.3), a parameter should be created.

Let us recall the global compactness result for a (P.-S.) sequence $u_n \geq 0$ of the functional $I(u)$. There exist $u \in H^1(\mathbb{R}^N)$, a non-negative integer k, $x_n^{(j)} \in \mathbb{R}^N$ and $|x_n^{(j)}| \to \infty$, $j = 1, \cdots, k$, such that

$$u_n(x) = u(x) + \sum_{j=1}^{k} U(x - x_n^{(j)}) + \omega_n, \tag{4.1.2}$$

$$|x_n^{(j)} - x_n^{(i)}| \to \infty, \quad j \neq i, \tag{4.1.3}$$

and $\|\omega_n\| \to 0$, as $n \to +\infty$, where $U(x) = U(|x|)$ is the solution of the following problem:

$$\begin{cases} -\Delta u + u = u^p, \quad u > 0, \quad \text{in } \mathbb{R}^N, \\ u \in H^1(\mathbb{R}^N). \end{cases}$$

The decomposition (4.1.2) shows that the loss of compactness of a non-negative (P.-S.) sequence is due to the 'bump' $U(x - x_n^{(j)})$, which lies near infinity. Moreover, the bumps are separated from one another in view of (4.1.3).

In the following, we will use this loss of the compactness to build up solutions. More precisely, we want to construct solutions u_k for (4.0.3) such that

$$u_k \approx \sum_{j=1}^{k} U(x - x_k^{(j)}), \tag{4.1.4}$$

for some large integer k, with $x_k^{(j)} \in \mathbb{R}^N$ satisfying $|x_k^{(j)}| \to +\infty$ and $|x_k^{(j)} - x_k^{(i)}| \to \infty$, $j \neq i$ as $k \to +\infty$. Here, the large integer k, which is the number of the bumps in the approximate solutions, is used as the parameter in the construction of solutions for (4.0.3).

Let us outline the main idea in the proof of Theorem 4.1.1. We will construct solutions with large number of bumps near the infinity. The crucial step is to determine the location $x_k^{(j)}$ of the bump in (4.1.4).

Since $V(y)$ is radially symmetric, we will construct positive solutions for (4.0.3), which are non-radial, but still keep some of the symmetry of the problem. For this purpose, we choose the point x_j as follow: For any large integer $k > 0$, let

$$x_j = \Big(r\cos\frac{2(j-1)\pi}{k}, r\sin\frac{2(j-1)\pi}{k}, 0\Big), \quad j = 1, \cdots, k,$$

where 0 is the zero vector in $\mathbb{R}^{N-2}$ and $r \in [r_0 k \ln k, r_1 k \ln k]$ for some $r_1 > r_0 > 0$.

Denote $U_{x_j}(y) = U(y - x_j)$. Let

$$W_r(y) = \sum_{j=1}^{k} U_{x_j}(y).$$

Then $W_r(y)$ is even in y_h for $h = 2, \cdots, N$, and

$$U(r\cos\theta, r\sin\theta, y'') = U\Big(r\cos\Big(\theta + \frac{2\pi j}{k}\Big), r\sin\Big(\theta + \frac{2\pi j}{k}\Big), y''\Big),$$

where we set $y = (y', y'')$, $y' \in \mathbb{R}^2$, $y'' \in \mathbb{R}^{N-2}$. We define

$$\begin{aligned} H_s = \Big\{u :\ & u \in H^1(\mathbb{R}^N), u \text{ is even in } y_h, h = 2, \cdots, N, \\ & u(r\cos\theta, r\sin\theta, y'') \\ & = u\Big(r\cos\Big(\theta + \frac{2\pi j}{k}\Big), r\sin\Big(\theta + \frac{2\pi j}{k}\Big), y''\Big)\Big\}. \end{aligned}$$

We will prove Theorem 4.1.1 by establishing the following result.

Theorem 4.1.2 *Suppose that $V(r)$ satisfies (V). Then there is an integer $k_0, > 0$ such that for any integer $k \geq k_0$,* (4.0.3) *has a solution u_k of the form*

$$u_k = W_{r_k}(y) + \omega_k,$$

where $\omega_k \in H_s$, $r_k \in [r_0 k \ln k, r_1 k \ln k]$ and as $k \to +\infty$,

$$\int_{\mathbb{R}^N} (|\nabla \omega_k|^2 + \omega_k^2) \to 0.$$

By Theorem 4.1.2, (4.0.1) has solutions with a large number of bumps near infinity. Thus the energy of these solutions can be very large.

4.2 Energy Expansion for the Approximate Solutions

We will follow the procedure outlined in Chapter 2 to construct solutions with many bumps for (4.0.3). In this section, we will give the energy expansion for the approximate solutions. Recall that

$$x_j = \left(r\cos\frac{2(j-1)\pi}{k}, r\sin\frac{2(j-1)\pi}{k}, 0\right), \quad j = 1, \cdots, k.$$

We then define

$$\Omega_j = \left\{y = (y', y'') \in \mathbb{R}^2 \times \mathbb{R}^{N-2} : \left\langle \frac{(y', 0)}{|y'|}, \frac{x_j}{|x_j|}\right\rangle \ge \cos\frac{\pi}{k}\right\}, \quad j = 1, \cdots, k.$$

First, we have the following basic estimate.

Lemma 4.2.1 *For $\eta \in (0, 1]$, there is a constant $C > 0$ depending on η, such that*

$$\sum_{j=2}^{k} U_{x_j}(y) \le Ce^{-\eta|x_1|\frac{\pi}{k}} e^{-(1-\eta)|y-x_1|}, \quad \forall\, y \in \Omega_1.$$

Proof. For $y \in \Omega_1$, one has $|y - x_j| \ge |y - x_1|$. We claim that $|y - x_j| \ge \frac{1}{2}|x_j - x_1|$ if $y \in \Omega_1$. In fact, if $|y - x_1| \ge \frac{1}{2}|x_j - x_1|$, then $|y - x_j| \ge |y - x_1| \ge \frac{1}{2}|x_j - x_1|$. If $|y - x_1| \le \frac{1}{2}|x_j - x_1|$, then we have

$$|y - x_j| \ge |x_j - x_1| - |y - x_1| \ge \frac{1}{2}|x_j - x_1|.$$

It then follows that for any $y \in \Omega_1$, one has

$$\begin{aligned} U_{x_j}(y) &\le Ce^{-\eta|y-x_j|} e^{-(1-\eta)|y-x_j|} \\ &\le Ce^{-\frac{1}{2}\eta|x_j-x_1|} e^{-(1-\eta)|y-x_1|}. \end{aligned} \tag{4.2.1}$$

Thus, we conclude that

$$\begin{aligned} \sum_{j=2}^{k} U_{x_j}(y) &\le Ce^{-(1-\eta)|y-x_1|} \sum_{j=2}^{k} e^{-\frac{1}{2}\eta|x_j-x_1|} \\ &\le Ce^{-(1-\eta)|y-x_1|} \sum_{j=2}^{k} e^{-\eta|x_1|\sin\frac{(j-1)\pi}{k}} \\ &\le C_1 e^{-(1-\eta)|y-x_1|} \sum_{j=1}^{[\frac{k}{2}]} e^{-\eta|x_1|\sin\frac{j\pi}{k}}. \end{aligned}$$

Let $\tau > 0$ be a fixed small constant. Then

$$\sum_{j \ge k\pi\tau} e^{-\eta|x_1|\sin\frac{j\pi}{k}} \le ke^{-\eta|x_1|\sin\tau} \le C_\eta e^{-\eta|x_1|\frac{\pi}{k}}.$$

On the other hand, in view of

$$\sin\frac{j\pi}{k} - \sin\frac{\pi}{k} = 2\sin\frac{(j-1)\pi}{2k}\cos\frac{(j+1)\pi}{2k} \ge \frac{c_0(j-1)\pi}{2k},$$

for some small $c_0 > 0$ if $j \le k\pi\tau$, we see

$$\sum_{j \le k\pi\tau} e^{-\eta|x_1|\sin\frac{j\pi}{k}} \le e^{-\eta|x_1|\sin\frac{\pi}{k}} \sum_{j \le k\pi\tau} e^{-\eta|x_1|\frac{c_0(j-1)\pi}{2k}} \le C_\eta e^{-\eta|x_1|\frac{\pi}{k}}.$$

Hence we have proved this lemma. □

We now denote $r = |x_1|$. In the following, we always assume that

$$r \in S_k := \Big[\Big(\frac{m}{2\pi} - \beta\Big)k\ln k, \Big(\frac{m}{2\pi} + \beta\Big)k\ln k\Big], \tag{4.2.2}$$

where m is the constant in the expansion for V and $\beta > 0$ is a fixed small constant.

Proposition 4.2.2 *We have*

$$I(U_{x_1}) = A + \frac{B_1}{r^m} + O\big(\frac{1}{k^{m+\theta}}\big),$$

where

$$A = \Big(\frac{1}{2} - \frac{1}{p+1}\Big)\int_{\mathbb{R}^N} U^{p+1}, \quad B_1 = \frac{a}{2}\int_{\mathbb{R}^N} U^2.$$

Proof. We have

$$\begin{aligned} I(U_{x_1}) &= \frac{1}{2}\int_{\mathbb{R}^N} \big(|\nabla U|^2 + U^2\big) - \frac{1}{p+1}\int_{\mathbb{R}^N} U^{p+1} \\ &\quad + \frac{1}{2}\int_{\mathbb{R}^N} \big(V(|y|) - 1\big)U_{x_1}^2 \\ &= A + \frac{1}{2}\int_{\mathbb{R}^N} \big(V(|y - x_1|) - 1\big)U^2. \end{aligned} \tag{4.2.3}$$

On the other hand, for any small $\tau > 0$, using (V), we obtain

$$\begin{aligned} &\int_{\mathbb{R}^N} \big(V(|y - x_1|) - 1\big)U^2 \\ &= \int_{B_{\frac{1}{2}r}(0)} \big(V(|y - x_1|) - 1\big)U^2 + O\Big(\int_{\mathbb{R}^N \setminus B_{\frac{1}{2}r}(0)} U^2\Big) \\ &= \int_{B_{\frac{1}{2}r}(0)} \Big(\frac{a}{|y - x_1|^m} + O\Big(\frac{1}{|y - x_1|^{m+\theta}}\Big)\Big)U^2 + O\big(e^{-(1-\tau)r}\big). \end{aligned} \tag{4.2.4}$$

But for any $\alpha > 0$, one has

$$\frac{1}{|y - x_1|^\alpha} = \frac{1}{|x_1|^\alpha}\Big(1 + O\Big(\frac{|y|}{|x_1|}\Big)\Big), \quad y \in B_{\frac{1}{2}|x_1|}(0).$$

Thus it follows that

$$\int_{B_{\frac{1}{2}r}(0)} \frac{1}{|y-x_1|^\alpha} U^2 = \frac{1}{|x_1|^\alpha} \int_{\mathbb{R}^N} U^2 + O\Big(\frac{1}{|x_1|^{\alpha+1}}\Big). \tag{4.2.5}$$

Inserting (4.2.5) into (4.2.4), we obtain

$$\int_{\mathbb{R}^N} (V(|y-x_1|)-1)U^2 = \frac{B_1}{r^m} + O\Big(\frac{1}{k^{m+\theta}\ln^{m+\theta} k}\Big). \tag{4.2.6}$$

Consequently, the result follows from (4.2.3) and (4.2.6).

□

Proposition 4.2.3 *There is a small constant $\sigma > 0$, such that*

$$I(W_r) = k\Big(A + \frac{B_1}{r^m} - B_2(1+o(1))e^{-\frac{2\pi r}{k}}\Big(\frac{2\pi r}{k}\Big)^{-\frac{N-1}{2}} + O\Big(\frac{1}{k^{m+\sigma}}\Big)\Big), \tag{4.2.7}$$

where A and B_1 are the constants in Proposition 4.2.2, and $B_2 > 0$ is a positive constant.

Proof. Using the symmetry, one has the following calculations:

$$\begin{aligned}\int_{\mathbb{R}^N} (|\nabla W_r|^2 + W_r^2) &= \sum_{j=1}^k \sum_{i=1}^k \int_{\mathbb{R}^N} U_{x_j}^p U_{x_i} \\ &= k\int_{\mathbb{R}^N} U^{p+1} + 2k\sum_{i=2}^k \int_{\mathbb{R}^N} U_{x_1}^p U_{x_i}.\end{aligned} \tag{4.2.8}$$

Recall that

$$\Omega_j = \Big\{y = (y', y'') \in \mathbb{R}^2 \times \mathbb{R}^{N-2} : \Big\langle \frac{(y',0)}{|y'|}, \frac{x_j}{|x_j|}\Big\rangle \ge \cos\frac{\pi}{k}\Big\}, \quad j = 1, \cdots, k.$$

By a similar argument to that in the proof of (4.2.6), we deduce

$$\begin{aligned}&\int_{\mathbb{R}^N} (V(|y|)-1)W_r^2 = k\int_{\Omega_1} (V(|y|)-1)W_r^2 \\ &= k\int_{\Omega_1} (V(|y|)-1)\Big(U_{x_1} + O\big(e^{-\frac{1}{2}|x_1|\frac{\pi}{k}}e^{-\frac{1}{2}|y-x_1|}\big)\Big)^2 \\ &= k\int_{\Omega_1} (V(|y|)-1)U_{x_1}^2 \\ &+ kO\Big(\int_{\Omega_1} |V(|y|)-1|\Big(U_{x_1}e^{-\frac{1}{2}|x_1|\frac{\pi}{k}}e^{-\frac{1}{2}|y-x_1|} + e^{-|x_1|\frac{\pi}{k}}e^{-|y-x_1|}\Big)\Big) \\ &= k\Big(\frac{B_1}{|x_1|^m} + O\Big(\frac{1}{k^{m+\theta}}\Big)\Big).\end{aligned} \tag{4.2.9}$$

Suppose that $p \le 3$. Then, for any $y \in \Omega_1$, one has

$$W_r^{p+1} = U_{x_1}^{p+1} + (p+1)U_{x_1}^p \sum_{j=2}^k U_{x_j} + O\Big(U_{x_1}^{\frac{p+1}{2}} \Big(\sum_{j=2}^k U_{x_j}\Big)^{\frac{p+1}{2}}\Big).$$

Using Lemma 4.2.1, we have

$$\begin{aligned} U_{x_1}^{\frac{p+1}{2}} \Big(\sum_{j=2}^k U_{x_j}\Big)^{\frac{p+1}{2}} &= U_{x_1}^{\frac{p+1}{2}} \sum_{j=2}^k U_{x_j} \Big(\sum_{j=2}^k U_{x_j}\Big)^{\frac{p-1}{2}} \\ &\le Ce^{-\eta \frac{p-1}{2} \frac{|x_1|\pi}{k}} U_{x_1}^{(1-\eta)p} \sum_{j=2}^k U_{x_j}. \end{aligned}$$

But for any $r \in S_k$, one has the following estimate:

$$\begin{aligned} \int_{\mathbb{R}^N} U_{x_1}^{(1-\eta)p} \sum_{j=2}^k U_{x_j} &\le C \sum_{j=2}^k e^{-|x_j - x_1|} \\ &\le Ce^{-\frac{2\pi}{k} r} \le \frac{C}{k^{m-\beta}}. \end{aligned}$$

Hence it follows that for $p \in (1, 3]$ and small enough $\beta > 0$, we have

$$\begin{aligned} \int_{\mathbb{R}^N} W_r^{p+1} &= k \int_{\Omega_1} W_r^{p+1} = k \int_{\Omega_1} \Big(U_{x_1}^{p+1} + (p+1) \sum_{i=2}^k U_{x_1}^p U_{x_i}\Big) \\ &\quad + O\Big(\sum_{i=2}^k e^{-\eta \frac{p-1}{2} \frac{|x_1|\pi}{k}} e^{-|x_i - x_1|}\Big) \\ &= k\Big(\int_{\mathbb{R}^N} U_{x_1}^{p+1} + (p+1) \sum_{i=2}^k \int_{\mathbb{R}^N} U_{x_1}^p U_{x_i} + O\Big(\frac{1}{k^{m+\sigma}}\Big)\Big). \end{aligned} \tag{4.2.10}$$

Now, suppose that $p > 3$. Then, for any $y \in \Omega_1$, we have

$$W_r^{p+1} = U_{x_1}^{p+1} + (p+1)U_{x_1}^p \sum_{j=2}^k U_{x_j} + O\Big(U_{x_1}^{p-1} \Big(\sum_{j=2}^k U_{x_j}\Big)^2\Big).$$

Since $p - 1 > 2$, by a similar argument to that in the proof of (4.2.10), we can obtain the following estimate for $p > 3$:

$$\begin{aligned}\int_{\mathbb{R}^N} W_r^{p+1} &= k\int_{\Omega_1} W_r^{p+1}\\ &= k\Big(\int_{\mathbb{R}^N} U_{x_1}^{p+1} + (p+1)\sum_{i=2}^{k}\int_{\mathbb{R}^N} U_{x_1}^p U_{x_i} \qquad (4.2.11)\\ &\quad + O\Big(\frac{1}{k^{m+\sigma}}\Big)\Big).\end{aligned}$$

Combining (4.2.8), (4.2.9), (4.2.10) and (4.2.11), we are led to

$$I(W_r) = k\Big(A + \frac{B_1}{r^m} - \frac{1}{2}\sum_{i=2}^{k}\int_{\mathbb{R}^N} U_{x_1}^p U_{x_i} + O\Big(\frac{1}{k^{m+\sigma}}\Big)\Big).$$

But by Lemma 6.1.4, we have

$$\int_{\mathbb{R}^N} U_{x_1}^p U_{x_i} = \frac{a_0(1+o(1))e^{-|x_1-x_i|}}{|x_1-x_i|^{\frac{N-1}{2}}}.$$

Consequently, there is a constant $B_2 > 0$ such that

$$\begin{aligned}&\sum_{i=2}^{k}\int_{\mathbb{R}^N} U_{x_1}^p U_{x_i}\\ &= \frac{a_0(1+o(1))e^{-|x_1-x_2|}}{|x_1-x_2|^{\frac{N-1}{2}}}\sum_{i=2}^{k} e^{-(|x_i-x_1|-|x_2-x_1|)}\frac{|x_1-x_2|^{\frac{N-1}{2}}}{|x_i-x_1|^{\frac{N-1}{2}}}\\ &= B_2(1+o(1))e^{-\frac{2\pi|x_1|}{k}}\Big(\frac{2\pi|x_1|}{k}\Big)^{-\frac{N-1}{2}}.\end{aligned}$$

Hence the result follows. □

Now, the main term in expansion (4.2.7) for $I(W_r)$ is given by

$$\frac{B_1}{r^m} - B_2 e^{-\frac{2\pi r}{k}}\Big(\frac{2\pi r}{k}\Big)^{-\frac{N-1}{2}}.$$

This function has a maximum point $r_k \in S_k$. Therefore, we expect that for large k, (4.0.3) has a solution $u_k \approx W_{r_k}$, $r_k \in S_k$. To achieve this goal, we need to show that the approximate solution W_r is good enough. We will now estimate l_k, which is defined as follows:

$$\langle l_k, \phi\rangle = \sum_{i=1}^{k}\int_{\mathbb{R}^N}(V(|y|)-1)U_{x_i}\phi + \int_{\mathbb{R}^N}\Big(W^p - \sum_{i=1}^{k} U_{x_i}^p\Big)\phi.$$

Lemma 4.2.4 *There is a small $\sigma > 0$ such that*

$$\|l_k\| \le \frac{C}{k^{\frac{m-1}{2}+\sigma}}.$$

Proof. By the symmetry of the problem, one has the following calculations:

$$\begin{aligned}\sum_{i=1}^{k}\int_{\mathbb{R}^N}(V(|y|)-1)U_{x_i}\phi &= k\int_{\mathbb{R}^N}(V(|y|)-1)U_{x_1}\phi\\ &= k\int_{\mathbb{R}^N}(V(|y-x_1|)-1)U\phi(y-x_1)\\ &\le k\Big(\int_{\mathbb{R}^N}(V(|y-x_1|)-1)^2U^2\Big)^{\frac{1}{2}}\Big(\int_{\mathbb{R}^N}\phi^2\Big)^{\frac{1}{2}}\\ &\le \frac{Ck}{r^m}\|\phi\| \le \frac{C}{k^{\frac{m-1}{2}+\sigma}}\|\phi\|,\end{aligned} \tag{4.2.12}$$

since $m > 1$.

On the other hand, for any $y \in \Omega_1$, we have

$$U_{x_i}^p \le CU_{x_1}^{p-1}U_{x_i}.$$

Thus it follows from (4.2.1) that whenever $\eta \in (0, 1)$, one has

$$\begin{aligned}&\Big|\int_{\mathbb{R}^N}\Big(W^p-\sum_{i=1}^{k}U_{x_i}^p\Big)\phi\Big| = k\Big|\int_{\Omega_1}\Big(W^p-\sum_{i=1}^{k}U_{x_i}^p\Big)\phi\Big|\\ &\le Ck\int_{\Omega_1}U_{x_1}^{p-1}\sum_{j=2}^{k}U_{x_j}|\phi| \le Cke^{-\eta|x_1|\frac{\pi}{k}}\int_{\Omega_1}U_{x_1}^{p-1}\sum_{j=2}^{k}e^{-(1-\eta)|y-x_j|}|\phi|\\ &\le Ck^{\frac{1}{2}}e^{-\eta|x_1|\frac{\pi}{k}}\sum_{j=2}^{k}\Big(\int_{\Omega_1}U_{x_1}^{2(p-1)}e^{-2(1-\eta)|y-x_j|}\Big)^{\frac{1}{2}}\|\phi\|.\end{aligned}$$

If $\eta \in (0, 1)$ is chosen so that $1-\eta < p-1$, we then have

$$\int_{\Omega_1}U_{x_1}^{2(p-1)}e^{-2(1-\eta)|y-x_j|} \le Ce^{-2(1-\eta)|x_j-x_1|}.$$

Thus, it follows that

$$\begin{aligned}\Big|\int_{\mathbb{R}^N}\Big(W^p-\sum_{i=1}^{k}U_{x_i}^p\Big)\phi\Big| &\le Ck^{\frac{1}{2}}e^{-\eta|x_1|\frac{\pi}{k}-(1-\eta)|x_1|\frac{2\pi}{k}}\|\phi\|\\ &= Ck^{\frac{1}{2}}e^{-(2-\eta)\frac{|x_1|\pi}{k}}\|\phi\|.\end{aligned} \tag{4.2.13}$$

Now we fix $\eta = 1 - \sigma$ with $\sigma \in (0, p-1)$. From the definition of S_k in (4.2.2), we see that for any $r \in S_k$,

$$\begin{aligned} e^{-(2-\eta)|x_1|\frac{2\pi}{k}} &= e^{-(1+\sigma)|x_1|\frac{2\pi}{k}} \\ &\le \frac{C}{k^{\frac{(1+\sigma)(m-\beta)}{2}}} \le \frac{C}{k^{\frac{(1+\sigma')m}{2}}}, \end{aligned}$$

where $\sigma' > 0$, since we can take $\beta > 0$ small. Putting this into (4.2.13), we obtain

$$\left| \int_{\mathbb{R}^N} \left(W^p - \sum_{i=1}^{k} U_{x_i}^p \right) \phi \right| \le \frac{C}{k^{\frac{m-1}{2}+\sigma}} \|\phi\|. \tag{4.2.14}$$

The result now follows from (4.2.12) and (4.2.14). □

From Lemma 4.2.4, we see that $\|l_k\|^2 = O\big(\frac{1}{k^{m-1+2\sigma}}\big)$, which is smaller than $\frac{k}{r^m}$, $r \in S_k$, in the expansion (4.2.7) for $I(W_r)$. Hence we conclude that the approximate solution W_r is sufficient.

4.3 The Reduction

We know the kernel of the linear operator $-\Delta\phi + \phi - pU^{p-1}\phi$ in $H^1(\mathbb{R}^N)$ is given by

$$span\left\{ \frac{\partial U}{\partial y_1}, \cdots, \frac{\partial U}{\partial y_N} \right\}.$$

For $j = 2, \cdots N$, $\frac{\partial U}{\partial y_j} = U'(|y|)\frac{y_j}{|y|}$, which is odd in y_j. So $\frac{\partial U}{\partial y_j} \notin H_s$ for $j = 2, \cdots, N$, since any function u in H_s is even in y_j for $j \ge 2$. Therefore, in H_s, the kernel of this linear operator is $span\{\frac{\partial U}{\partial y_1}\}$.

The norm of $H^1(\mathbb{R}^N)$ is defined as follows:

$$\|v\| = \sqrt{\langle v, v\rangle}, \quad v \in H^1(\mathbb{R}^N),$$

where

$$\langle v_1, v_2\rangle = \int_{\mathbb{R}^N} \big(\nabla v_1 \nabla v_2 + V(|y|)v_1 v_2\big).$$

Let

$$Z = \frac{\partial W_r}{\partial r}.$$

Define

$$E = \big\{v \colon v \in H_s,\ \langle Z, v\rangle = 0\big\}.$$

It is easy to check that

$$\int_{\mathbb{R}^N} \big(\nabla v_1 \nabla v_2 + V(|y|) v_1 v_2 - p W_r^{p-1} v_1 v_2\big), \quad v_1, v_2 \in E,$$

is a bounded symmetric bi-linear functional in E. Thus there is a bounded linear operator L from E to E such that

$$\big\langle L v_1, v_2 \big\rangle = \int_{\mathbb{R}^N} \big(\nabla v_1 \nabla v_2 + V(|y|) v_1 v_2 - p W_r^{p-1} v_1 v_2\big), \quad v_1, v_2 \in E.$$

The next lemma shows that L is invertible in E.

Lemma 4.3.1 *There is a constant $\rho > 0$, independent of k, such that for any $r \in S_k$,*

$$\|Lv\| \geq \rho \|v\|, \quad v \in E.$$

Proof. We argue by contradiction. Suppose that there are $k \to +\infty$, $r_k \in S_k$, and $v_k \in E$ such that

$$\|L v_k\| = o(1)\|v_k\|.$$

Then it follows that

$$\big\langle L v_k, \varphi \big\rangle = o(1)\|v_k\|\|\varphi\|, \quad \forall\, \varphi \in E. \tag{4.3.1}$$

We may assume that $\|v_k\|^2 = k$.

Recall that

$$\Omega_j = \Big\{ y = (y', y'') \in \mathbb{R}^2 \times \mathbb{R}^{N-2} : \Big\langle \frac{(y', 0)}{|y'|}, \frac{x_j}{|x_j|} \Big\rangle \geq \cos \frac{\pi}{k} \Big\}.$$

By symmetry, we see from (4.3.1),

$$\begin{aligned} \int_{\Omega_1} \big(\nabla v_k \nabla \varphi + V(|y|) v_k \varphi - p W_r^{p-1} v_k \varphi\big) &= \frac{1}{k}\langle L v_k, \varphi\rangle \\ &= o\Big(\frac{1}{\sqrt{k}}\Big)\|\varphi\|, \quad \forall\, \varphi \in E. \end{aligned} \tag{4.3.2}$$

In particular, we have

$$\int_{\Omega_1} \big(|\nabla v_k|^2 + V(|y|) v_k^2 - p W_r^{p-1} v_k^2\big) = o(1),$$

and

$$\int_{\Omega_1} \big(|\nabla v_k|^2 + V(|y|) v_k^2\big) = 1. \tag{4.3.3}$$

Let $\bar{v}_k(y) = v_k(y - x_{1,k})$, where $x_{1,k} = \big(r_k, 0, 0\big)$. Then, for any $R > 0$, since

$$|x_{2,k} - x_{1,k}| = r \sin\frac{\pi}{k} \geq \frac{m}{4}\ln k,$$

we see that $B_R(x_{1,k}) \subset \Omega_1$ if k is sufficiently large. As a result, from (4.3.3), we find that for any $R > 0$, one has the inequality

$$\int_{B_R(0)} \big(|\nabla \bar{v}_k|^2 + V(|y - x_{1,k}|)\bar{v}_k^2\big) \leq 1.$$

Therefore, we may assume that there is a $v \in H^1(\mathbb{R}^N)$ such that as $k \to +\infty$,

$$\bar{v}_k \rightharpoonup v \quad \text{weakly in } H^1_{loc}(\mathbb{R}^N),$$

and

$$\bar{v}_k \to v \quad \text{strongly in } L^2_{loc}(\mathbb{R}^N).$$

Since $\bar{v}_k$ is even in y_h, $h = 2, \cdots, N$, it is easy to see that v is even in y_h, $h = 2, \cdots, N$. On the other hand, from

$$\big\langle Z, v_k\big\rangle = 0,$$

and the symmetry of the functions Z and v_k, we obtain

$$\int_{\mathbb{R}^N} \big(V(|y|) - 1\big)\frac{\partial U_{x_{1,k}}}{\partial y_1} v_k - (p-1)\int_{\mathbb{R}^N} U_{x_{1,k}}^{p-1}\frac{\partial U_{x_{1,k}}}{\partial y_1} v_k = 0,$$

which gives

$$\int_{\mathbb{R}^N} \big(V(|y - x_{1,k}|) - 1\big)\frac{\partial U}{\partial y_1}\bar{v}_k - (p-1)\int_{\mathbb{R}^N} U^{p-1}\frac{\partial U}{\partial y_1}\bar{v}_k = 0. \tag{4.3.4}$$

Letting $k \to +\infty$ in (4.3.4), we see that v satisfies

$$\int_{\mathbb{R}^N} U^{p-1}\frac{\partial U}{\partial y_1} v = 0. \tag{4.3.5}$$

Now, we claim that v satisfies

$$-\Delta v + v - pU^{p-1}v = 0, \quad \text{in } \mathbb{R}^N. \tag{4.3.6}$$

For any $R > 0$, let $\varphi \in C_0^\infty(B_R(0))$ be such that $\varphi(y)$ is even in y_h, $h = 2, \cdots, N$. Then $\varphi_k(y) := \varphi(y - x_{1,k}) \in C_0^\infty(B_R(x_{1,k}))$ and $\sum_{j=1}^k \varphi(y - x_{j,k}) \in H_s$.

We choose b_k such that

$$\bar{\varphi}_k(y) = \sum_{j=1}^k \varphi(y - x_{j,k}) - b_k Z \in E.$$

Then one has

$$b_k = \frac{\left\langle \sum_{j=1}^k \varphi(y - x_{j,k}), Z \right\rangle}{\|Z\|^2} = \frac{k\langle \varphi(y - x_{1,k}), Z\rangle}{\|Z\|^2}.$$

It is easy to check that

$$\|Z\|^2 = k\Big(\Big\| \frac{\partial U}{\partial y_1} \Big\|^2 + o(1) \Big).$$

Thus $|b_k| \le C$. This gives $\|\bar{\varphi}_k(y)\|^2 \le Ck$.

Inserting $\bar{\varphi}_k$ into (4.3.2), we obtain the following estimate:

$$\begin{aligned} \int_{\Omega_1} \big(\nabla v_k \nabla \bar{\varphi}_k + V(|y|) v_k \bar{\varphi}_k - p W_r^{p-1} v_k \bar{\varphi}_k \big) &= o\Big(\frac{1}{\sqrt{k}} \Big) \|\bar{\varphi}_k\| \\ &= o(1). \end{aligned} \tag{4.3.7}$$

Thus we have

$$\begin{aligned} &\int_{\Omega_1} \big(\nabla v_k \nabla \varphi_k + V(|y|) v_k \varphi_k - p W_r^{p-1} v_k \varphi_k \big) \\ &= \frac{1}{k} \int_{\mathbb{R}^N} \Big(\nabla v_k \nabla \sum_{j=1}^k \varphi(y - x_{j,k}) + \big(V(|y|) v_k - p W_r^{p-1} v_k \big) \sum_{j=1}^k \varphi(y - x_{j,k}) \Big) \\ &= \int_{\Omega_1} \big(\nabla v_k \nabla \bar{\varphi}_k + V(|y|) v_k \bar{\varphi}_k - p W_r^{p-1} v_k \bar{\varphi}_k \big) \\ &\quad + b_k \int_{\Omega_1} \big(\nabla v_k \nabla Z + V(|y|) v_k Z - p W_r^{p-1} v_k Z \big) \\ &= \int_{\Omega_1} \big(\nabla v_k \nabla \bar{\varphi}_k + V(|y|) v_k \bar{\varphi}_k - p W_r^{p-1} v_k \bar{\varphi}_k \big) - \gamma_k \langle \varphi_k, Z \rangle \\ &= o(1) + \gamma_k \langle \varphi_k, Z \rangle, \end{aligned} \tag{4.3.8}$$

for any $\varphi_k \in C_0^\infty(B_R(x_{1,k}))$, which is even in y_h, $h = 2, \cdots, N$, where

$$\gamma_k = \|Z\|^{-2} k \int_{\Omega_1} \big(\nabla v_k \nabla Z + V(|y|) v_k Z - p W_r^{p-1} v_k Z \big).$$

To estimate γ_k, let $\xi \in C_0^\infty(B_R(x_{1,k}))$ be such that $\xi(y) = \xi(|y - x_{1,k}|)$, $\xi = 1$ in $B_{\frac{1}{2}R}(x_{1,k})$, $|\nabla \xi| \le CR^{-1}$ and $|\nabla^2 \xi| \le CR^{-2}$. We take $\varphi_k = \xi Z$ in (4.3.8) to obtain

$$\begin{aligned}\gamma_k\langle \xi Z, Z\rangle &= \int_{\mathbb{R}^N} \big(\nabla v_k \nabla(\xi Z) + V(|y|)v_k\xi Z - pW_r^{p-1}v_k\xi Z\big) + o(1)\\ &= \int_{\mathbb{R}^N} \big(-\Delta(\xi Z) + V(|y|)\xi Z - pW_r^{p-1}\xi Z\big)v_k + o(1)\\ &= o(1),\end{aligned}$$

which in turn implies that $\gamma_k = o(1)$. In view of (4.3.8), we have established that

$$\int_{\mathbb{R}^N} \big(\nabla v_k \nabla \varphi_k + V(|y|)v_k\varphi_k - pW_r^{p-1}v_k\varphi_k\big) = o(1),$$

for any $\varphi_k \in C_0^\infty(B_R(x_{1,k}))$, which is even in y_h, $h = 2, \cdots, N$. Thus

$$\int_{\mathbb{R}^N} \big(\nabla \bar{v}_k \nabla \varphi + V(|y - x_{1,k}|)\bar{v}_k\varphi - p(W_r(|y - x_{1,k}|))^{p-1}\bar{v}_k\varphi\big) = o(1), \quad (4.3.9)$$

for any $\varphi \in C_0^\infty(B_R(0))$, which is even in y_h, $h = 2, \cdots, N$. Letting $k \to +\infty$ in (4.3.9), we see that

$$\int_{\mathbb{R}^N} \big(\nabla v \nabla \varphi + v\varphi - pU^{p-1}v\varphi\big) = 0, \qquad (4.3.10)$$

for any $\varphi \in C_0^\infty(B_R(0))$, which is even in y_h, $h = 2, \cdots, N$. For any $\varphi \in C_0^\infty(B_R(0))$, we let

$$\psi(y) = \frac{1}{2}\big(\varphi(y) + \varphi(y_1, -y_2, \cdots, -y_N)\big)$$

in (4.3.10). Since v is even in y_h for $h = 2, \cdots, N$, it follows that (4.3.10) holds for all $\varphi \in C_0^\infty(B_R(0))$. Thus we have proved (4.3.6).

On the other hand, since v is even in y_h, $h = 2, \cdots, N$, we have that $v = \alpha \frac{\partial U}{\partial y_1}$ for some constant α. From (4.3.5), $\alpha = 0$. Thus $v = 0$. As a result,

$$\int_{B_R(x_{1,k})} v_k^2 = o(1), \quad \forall\, R > 0.$$

Finally, it follows from Lemma 4.2.1 that for any small $\eta > 0$, there is a constant $C > 0$ such that

$$W_{r_k}(y) \le Ce^{-(1-\eta)|y - x_{1,k}|}, \quad y \in \Omega_1.$$

Thus

$$\begin{aligned} o(1) &= \int_{\Omega_1} \big(|\nabla v_k|^2 + V(|y|)v_k^2 - pW_{r_k}^{p-1}v_k^2\big)\\ &\ge \int_{\Omega_1} \Big(|\nabla v_k|^2 + V(|y|)v_k^2 - Ce^{-(p-1)(1-\eta)|y - x_{1,k}|}v_k^2\Big)\\ &= \int_{\Omega_1} \big(|\nabla v_k|^2 + V(|y|)v_k^2\big) + o(1).\end{aligned}$$

This is a contradiction to (4.3.3). Therefore, the proof of this lemma is completed.

□

Recall that

$$I(u) = \frac{1}{2}\int_{\mathbb{R}^N} \left(|\nabla u|^2 + V(|y|)u^2 - \frac{1}{p+1}\int_{\mathbb{R}^N} |u|^{p+1}.\right.$$

We define

$$J(\phi) = I(W_r + \phi), \quad \phi \in E.$$

Proposition 4.3.2 *Set $r = |x_1|$. Then there is an integer $k_0 > 0$ such that for each $k \geq k_0$, there exists a C^1 map ϕ_r from S_k to H_s that satisfies $\phi \in E$ and*

$$\langle J'(\phi_r), \varphi\rangle = 0, \quad \forall\, \varphi \in E.$$

Moreover, there is a small $\sigma > 0$ such that

$$\|\phi_r\| \leq \frac{C}{k^{\frac{m-1}{2}+\sigma}}. \tag{4.3.11}$$

Proof. Expand $J(\phi)$ as follows:

$$J(\phi) = J(0) + l(\phi) + \frac{1}{2}\langle L\phi, \phi\rangle + R(\phi), \quad \phi \in E,$$

where

$$l(\phi) = \sum_{i=1}^{k}\int_{\mathbb{R}^N} (V(|y|) - 1)U_{x_i}\phi + \int_{\mathbb{R}^N}\Big(W^p - \sum_{i=1}^{k} U_{x_i}^p\Big)\phi,$$

L is the bounded symmetric linear map from E to E in Lemma 4.3.1 and

$$R(\phi) = \frac{1}{p+1}\int_{\mathbb{R}^N}\Big(|W_r + \phi|^{p+1} - W_r^{p+1} - (p+1)W_r^p\phi - \frac{1}{2}(p+1)pW_r^{p-1}\phi^2\Big).$$

Since $l(\phi)$ is a bounded linear functional in E, we know that there is an $l_k \in E$ such that

$$l(\phi) = \langle l_k, \phi\rangle.$$

Thus finding a critical point for $J(\phi)$ in E is equivalent to solving

$$l_k + L\phi + R'(\phi) = 0. \tag{4.3.12}$$

By Lemma 4.3.1, L is invertible. Thus (4.3.12) can be rewritten as

$$\phi = A(\phi) := -L^{-1}l_k - L^{-1}R'(\phi).$$

Let

$$S = \left\{\phi : \phi \in E, \|\phi\| \le \frac{1}{k^{\frac{m-1}{2}}}\right\}.$$

If $p \le 2$, then it is easy to check that

$$\|R'(\phi)\| \le C\|\phi\|^p.$$

It then follows from Lemma 4.2.4 that

$$\begin{aligned} \|A(\phi)\| &\le C\|l_k\| + C\|\phi\|^p \\ &\le \frac{C}{k^{\frac{m-1}{2}+\sigma}} + \frac{C}{k^{\frac{p(m-1)}{2}}} \le \frac{1}{k^{\frac{m-1}{2}}}. \end{aligned} \tag{4.3.13}$$

Thus A maps S into S if $p \le 2$.

On the other hand, if $p \le 2$, then

$$\|R''(\phi)\| \le C\|\phi\|^{p-1}.$$

Thus one has

$$\begin{aligned} \|A(\phi_1) - A(\phi_2)\| &= \|L^{-1}R'(\phi_1) - L^{-1}R'(\phi_2)\| \\ &\le C\big(\|\phi_1\|^{p-1} + \|\phi_2\|^{p-1}\big)\|\phi_1 - \phi_2\| \le \frac{1}{2}\|\phi_1 - \phi_2\|. \end{aligned}$$

In conclusion, we have proved that if $p \le 2$, A is a contraction map from S to S. Hence the result follows from the contraction mapping theorem.

It remains to deal with the case $p > 2$. Since

$$\begin{aligned} \big|\langle R'(\phi), \xi\rangle\big| &\le C\int_{\mathbb{R}^N} W_r^{p-2}|\phi|^2|\xi| \\ &\le C\Big(\int_{\mathbb{R}^N} \big(W_r^{p-2}|\phi|^2\big)^{\frac{p+1}{p}}\Big)^{\frac{p}{p+1}}\|\xi\|, \end{aligned}$$

we see that

$$\|R'(\phi)\| \le C\Big(\int_{\mathbb{R}^N} \big(W_r^{p-2}|\phi|^2\big)^{\frac{p+1}{p}}\Big)^{\frac{p}{p+1}}.$$

On the other hand, it follows from Lemma 4.2.1 that W_r is bounded. Since $2 < \frac{2(p+1)}{p} < p+1$, we obtain

$$\|R'(\phi)\| \le C\Big(\int_{\mathbb{R}^N} |\phi|^{\frac{2(p+1)}{p}}\Big)^{\frac{p}{p+1}} \le \|\phi\|^2.$$

For the estimate of $\|R''(\phi)\|$, we have

$$\begin{aligned}|R''(\phi)(\xi,\eta)| &\le C\int_{\mathbb{R}^N} W_r^{p-2}|\phi||\xi||\eta| \le C\int_{\mathbb{R}^N}|\phi||\xi||\eta|\\ &\le C\Big(\int_{\mathbb{R}^N}|\phi|^3\Big)^{\frac{1}{3}}\Big(\int_{\mathbb{R}^N}|\xi|^3\Big)^{\frac{1}{3}}\Big(\int_{\mathbb{R}^N}|\eta|^3\Big)^{\frac{1}{3}} \le C\|\phi\|\|\xi\|\|\eta\|,\end{aligned}$$

since $2 < 3 < p+1$. So

$$\|R''(\phi)\| \le C\|\phi\|.$$

Thus we have

$$\|R'(\phi)\| \le \frac{C}{k^{m-1}} \le \frac{C}{k^{\frac{m-1}{2}+\sigma}}.$$

As a result,

$$\|A(\phi)\| \le C\|l_k\| + C\|R'(\phi)\| \le \frac{C}{k^{\frac{m-1}{2}+\sigma}} \le \frac{1}{k^{\frac{m-1}{2}}}. \tag{4.3.14}$$

Thus A maps S to S.

On the other hand,

$$\|R''(\phi)\| \le C\|\phi\| \le \frac{C}{k^{\frac{m-1}{2}}},$$

which implies that A is a contraction map from S to S, and the result follows from the contraction mapping theorem.

Finally, (4.3.11) follows from (4.3.13) and (4.3.14). □

4.4 The Finite Dimensional Problem

We are now ready to prove Theorem 4.1.2. Let ϕ_r be the map obtained in Proposition 4.3.2. Define

$$F_r(r) = I(W_r + \phi_r), \quad \forall\, r \in S_k.$$

Then, similar to Chapter 2, we know that if r is a critical point of $F(r)$, then $W_r + \phi_r$ is a solution of (4.0.3).

Proof of Theorem 4.1.2. It follows from Propositions 4.3.2 and 4.2.3 that

$$\begin{aligned}F_r(r) &= I(W) + l(\phi_r) + \frac{1}{2}\langle L\phi_r, \phi_r\rangle + R(\phi_r)\\ &= I(W) + O\big(\|l_k\|\|\phi_r\| + \|\phi_r\|^2\big) = I(W) + O\Big(\frac{1}{k^{m-1+\sigma}}\Big)\\ &= k\Big(A + \frac{B_1}{r^m} - B_2(1+o(1))e^{-\frac{2\pi r}{k}}\Big(\frac{2\pi r}{k}\Big)^{-\frac{N-1}{2}} + O\Big(\frac{1}{k^{m+\sigma}}\Big)\Big).\end{aligned}$$

Consider

$$\max\{F_k(r) : r \in S_k\}, \tag{4.4.1}$$

where S_k is defined in (4.2.2). Then (4.4.1) is achieved by some r_k. Now if we prove that r_k is in the interior of S_k, then r_k is a critical point of $F(r)$. As a result, $W_{r_k} + \phi_{r_k}$ is a solution of (4.0.3).

Take $\bar{r}_k = \frac{m}{2\pi} k \ln k$. Then one has

$$F_k(\bar{r}_k) = k\Big(A + \frac{B_1 (2\pi)^m}{m^m (k \ln k)^m} + o\Big(\frac{1}{(k \ln k)^m}\Big)\Big).$$

If $r_k = (\frac{m}{2\pi} - \beta) k \ln k$, then

$$\begin{aligned} F_k(r_k) &= k\Big(A - \frac{B_2(1 + o(1))\big((m - 2\pi\beta)\ln k\big)^{-\frac{N-1}{2}}}{k^{m-2\pi\beta}} + O\Big(\frac{1}{(k \ln k)^m}\Big)\Big) \\ &< F(\bar{r}_k), \end{aligned}$$

which is impossible.

If $r_k = (\frac{m}{2\pi} + \beta) k \ln k$, then we have

$$F_k(r_k) = k\Big(A + \frac{B_1 (2\pi)^m}{(m + 2\pi\beta)^m (k \ln k)^m} + O\Big(\frac{1}{k^{m+2\pi\beta}} + \frac{1}{k^{m+\sigma}}\Big)\Big) < F(\bar{r}_k),$$

which is also impossible.

Hence r_k should be in the interior of S_k, and we have completed the proof. □

4.5 Further Results and Comments

The mountain pass lemma has been extensively used in the search for nontrivial solutions of superlinear elliptic equations. For example, by using the symmetric mountain pass lemma in the radially symmetric space, we know that (4.0.3) has infinitely many solutions.

We now consider a more general problem:

$$-\Delta u + V(|y|)u = f(u), \ \text{in } \mathbb{R}^N, \ u \in H^1(\mathbb{R}^N), \tag{4.5.1}$$

where $f(t) = t^p$ for $t \geq 0$, $f(t) = |t|^{q-1}t$ for $t < 0$, and $1 < p, q < \frac{N+2}{N-2}$. If $p \neq q$, then $f(u)$ is not an odd function and the functional corresponding to (4.5.1) is not even. Therefore, the symmetric mountain pass lemma cannot be used to discuss the multiplicity of solutions for (4.5.1). In this case, Cao and

Zhu [42] proved that (4.5.1) still has infinitely many solutions via the construction of a radially symmetric solution that changes sign exactly k times for any positive integer k.

On the other hand, if V satisfies

$$V(r) = 1 - \frac{a}{r^m} + O\Big(\frac{1}{r^{m+\theta}}\Big), \quad \text{as } r \to +\infty, \tag{4.5.2}$$

then we can use the same method to construct infinitely many sign-changing solutions for (4.5.1). More precisely, if $V(r)$ satisfies (4.5.2), we just need to replace W_r by the function

$$\tilde{W}_r = \sum_{j=1}^{k} \big(U_{x_{2j}} - W_{x_{2j+1}}\big),$$

where

$$x_j = \Big(r\cos\frac{(j-1)\pi}{k}, r\sin\frac{(j-1)\pi}{k}, 0\Big),$$

$U(y) = U(|y|) \in H^1(\mathbb{R}^N)$ is the positive solution of $-\Delta u + u = u^p$ in $\mathbb{R}^N$, and $W(y) = W(|y|) \in H^1(\mathbb{R}^N)$ is the positive solution of $-\Delta u + u = u^q$ in $\mathbb{R}^N$. Noting that the contribution to the energy from the interaction of positive bumps and negative bumps is positive, it then follows from a similar argument as Proposition 4.2.3 that we have the following energy expansion for $\tilde{W}_r$:

Proposition 4.5.1 *There is a small constant $\sigma > 0$ such that*

$$I(\tilde{W}_r) = k\Big(A - \frac{B_1}{r^m} + B_2(1+o(1))e^{-\frac{\pi r}{k}}\Big(\frac{\pi r}{k}\Big)^{-\frac{N-1}{2}} + O\Big(\frac{1}{k^{m+\sigma}}\Big)\Big), \tag{4.5.3}$$

where A, B_1 and $B_2 > 0$ are some positive constants.

The main term in (4.5.3) is given by

$$-\frac{B_1}{r^m} + B_2 e^{-\frac{\pi r}{k}}\Big(\frac{\pi r}{k}\Big)^{-\frac{N-1}{2}},$$

which has a minimum in

$$\tilde{S}_k := \Big[\Big(\frac{m}{\pi} - \beta\Big)k\ln k, \Big(\frac{m}{\pi} + \beta\Big)k\ln k\Big].$$

So we can prove that if k is large, then (4.5.1) has a solution $u_k \approx \tilde{W}_{r_k}$ for some $r_k \in \tilde{S}_k$. Thus we have the following result:

Theorem 4.5.2 *If $V(r)$ satisfies* (4.5.2), *then problem (4.5.1) has infinitely many non-radial sign-changing solutions.*

Remark 4.5.3 The condition that $V(y)$ is radially symmetric can be weaken. In fact, if $V(y)$ is radially symmetric in (y_1, y_2), and is even in $y_i, i = 3, \cdots, N$, then we can prove Theorem 4.5.2 by using the same argument. Under such weaker assumption on $V(y)$, the method used in [42] does not work.

The contents of this chapter are taken from [148]. The idea to use the number of the bubbles or bumps as a parameter to construct solutions was first introduced in [149], where the prescribed scalar curvature problem was studied. Such ideas can find wide applications for many non-compact elliptic problems. The coefficient $V(|x|)$ in (4.0.3) plays a crucial role in the construction of solutions with many bumps.

A typical counterpart for (4.0.3) in the critical growth problems is the following Schrödinger equation with critical Sobolev exponent

$$\begin{cases} -\Delta u + V(|x|)u = u^{2^*-1}, \quad u > 0, \quad \text{in } \mathbb{R}^N, \\ u \in D^{1,2}(\mathbb{R}^N). \end{cases} \tag{4.5.4}$$

Similar to (4.0.3), we want to construct a solution u for (4.5.4) such that

$$u \approx W_{r,\mu}(y) := \sum_{j=1}^{k} U_{x_j,\mu}(y)$$

and

$$\begin{aligned} H_s = \Big\{ u : \ & u \in H^1(\mathbb{R}^N), u \text{ is even in } y_h, h = 2, \cdots, N, \\ & u(r\cos\theta, r\sin\theta, y'') \\ & = u\Big(r\cos\Big(\theta + \frac{2\pi j}{k}\Big), r\sin\Big(\theta + \frac{2\pi j}{k}\Big), y''\Big)\Big\}. \end{aligned}$$

where $U_{x,\mu}(y)$ is the positive solution of $-\Delta u = u^{2^*-1}$ in $\mathbb{R}^N$ with $U_{x,\mu}(x) = \max_{y\in\mathbb{R}^N} U_{x,\mu}(y)$,

$$x_j = \Big(r\cos\frac{2(j-1)\pi}{k}, r\sin\frac{2(j-1)\pi}{k}, 0\Big), \quad j = 1, \cdots, k,$$

and 0 is the zero vector in $\mathbb{R}^{N-2}$. We calculate the following energy expansion for W:

$$\begin{aligned} & \frac{1}{2}\int_\Omega \big(|\nabla W|^2 + V(|y|)W^2\big) - \frac{1}{2^*}\int_\Omega |W|^{2^*} \\ & = k\Big(A + \frac{B_1 V(r)}{\mu^2} - \frac{B_2 k^{N-2}}{r^{N-2}\mu^{N-2}} + O\Big(\frac{1}{\mu^{2+\sigma}} + \frac{k^{N-2+\sigma}}{\mu^{N-2+\sigma}}\Big)\Big), \end{aligned}$$

where A, B_1 and B_2 are some positive constants, and $\sigma > 0$ is a small constant. Now, the main terms in the above expansion are given by

$$F(r,\mu) = \frac{B_1 V(r)}{\mu^2} - \frac{B_2 k^{N-2}}{r^{N-2}\mu^{N-2}}.$$

We will now find the critical point for $F(r,\mu)$. Letting

$$\frac{\partial F}{\partial \mu} = -\frac{2B_1 V(r)}{\mu^3} + \frac{(N-2)B_2 k^{N-2}}{r^{N-2}\mu^{N-1}} = 0,$$

we get

$$\mu = \mu(r) = \Big(\frac{(N-2)B_2}{2B_1 r^{N-2} V(r)}\Big)^{\frac{1}{N-4}} k^{\frac{N-2}{N-4}}.$$

Thus

$$F(r,\mu(r)) = \frac{2B_1(N-4)}{N-2}\Big(\frac{2B_1}{(N-2)B_2}\Big)^{\frac{2}{N-4}} k^{\frac{-2(N-2)}{N-4}} \big(r^2 V(r)\big)^{\frac{N-2}{N-4}}.$$

Therefore, if $r^2V(r)$ has a critical point $r_0 > 0$, then $F(x,\mu)$ has a critical point $(r_0, \mu(r_0))$. If we require r_0 to be a stable critical point of $r^2V(r)$, then r_0 must be a local maximum point or a local minimum point of $r^2V(r)$.

In [47, 121], the following result is proved:

Theorem 4.5.4 *Assume that $N \geq 5$. Suppose that $V \geq 0$ is bounded and belongs to C^1. If $r^2V(r)$ has a strict local maximum point, or a strict local minimum point $r_0 > 0$ with $V(r_0) > 0$, then there exists a large $k_0 > 0$ such that for any $k \geq k_0$, problem* (4.5.4) *has a solution $u \approx W_{r_k,\mu_k}$ with $r_k \to r_0$ and $\mu_k \sim k^{\frac{N-2}{N-4}}$ as $k \to +\infty$. In particular,* (4.5.4) *has infinitely many non-radial solutions.*

It follows from the Pohozaev identity that any solution u of (4.5.4) satisfies

$$\int_{\mathbb{R}^N} \Big(V(|y|) + \frac{1}{2}|y|V'(|y|)\Big)u^2 = 0.$$

This shows that (4.5.4) has no solution if $r^2V(r)$ is always non-decreasing or non-increasing. So the condition that $r^2V(r)$ has a strict local maximum point, or a strict local minimum point $r_0 > 0$, is necessary to obtain at least one solution for (4.5.4).

For problems with critical growth, the bubble $U_{x,\mu}(y)$ decays at the rate $\frac{1}{|y|^{N-2}}$ as $|y| \to +\infty$. This algebraic decay makes the reduction much more difficult than the subcritical problem. In [149], the reduction is carried out in a space with a weighted maximal norm, which catches the concentration properties of the bubbling solutions.

Existence results that can be obtained by similar techniques include the prescribed scalar curvature problems [149], the nonlinear Neumann problems with critical growth [139, 140], the super-critical growth problems [110, 150], the elliptic problems of higher order [86], and some nonlinear Schrödinger systems [102, 118, 119].

5

A Compactness Theorem and Application

In this chapter, we will study the multiplicity of solutions for the following problem:

$$\begin{cases} -\Delta u = |u|^{2^*-2}u + \lambda u, & \text{in } \Omega, \\ u = 0, & \text{on } \partial\Omega, \end{cases} \tag{5.0.1}$$

where Ω is a bounded domain in $\mathbb{R}^N$ and $\lambda > 0$ is a positive constant. The functional corresponding to (5.0.1) is

$$I(u) = \frac{1}{2}\int_\Omega \left(|\nabla u|^2 - \lambda u^2\right) - \frac{1}{2^*}\int_\Omega |u|^{2^*}, \quad u \in H_0^1(\Omega).$$

In Chapter 1, we show that $I(u)$ satisfies the (P.-S.)$_c$ condition if $c < \frac{1}{N}S^{\frac{N}{2}}$. Since $I(u)$ usually does not satisfy the (P.-S.)$_c$ condition for any $c \geq \frac{1}{N}S^{\frac{N}{2}}$, we can not apply the symmetric mountain pass lemma to prove that (5.0.1) has infinitely many solutions. Therefore, instead of working on (5.0.1) directly, we can approximate the critical nonlinearity term $|u|^{2^*-2}u$ by an odd function $f_\varepsilon(u)$, which is subcritical at infinity for each $\varepsilon > 0$. We then consider the following perturbed problem:

$$\begin{cases} -\Delta u = f_\varepsilon(u) + \lambda u, & \text{in } \Omega, \\ u = 0, & \text{on } \partial\Omega. \end{cases} \tag{5.0.2}$$

In order to apply the symmetric mountain pass lemma to obtain infinitely many solutions for (5.0.2), we need to make sure that $f_\varepsilon(u)$ satisfies the following conditions:

(i) $\lim\limits_{t\to 0}\dfrac{f_\varepsilon(t)}{t} = \lim\limits_{|t|\to+\infty}\dfrac{f_\varepsilon(t)}{|t|^{2^*-1}} = 0.$

(ii) There exists $p > 2$, such that

$$0 < pF_\varepsilon(t) \le f_\varepsilon(t)t, \quad \forall\, |t| \ge T,$$

where $F_\varepsilon(t) = \int_0^t f_\varepsilon(s)\,ds$ and $T > 0$ is a large constant.

An obvious choice of f_ε is $f_\varepsilon(t) = |t|^{2*-2-\varepsilon}t$. This approximation was used by Devillannova and Solimini [71]. Another convenient choice is given by

$$f_\varepsilon(t) = \begin{cases} |t|^{2^*-2}t, & |t| \le \varepsilon^{-1}, \\ \varepsilon^{-\varepsilon}|t|^{2^*-2-\varepsilon}t, & |t| \ge \varepsilon^{-1}. \end{cases} \tag{5.0.3}$$

The advantage of using the approximation in (5.0.3) is that if a solution u_ε of (5.0.2) satisfies $|u_\varepsilon| \le \varepsilon^{-1}$, then u_ε is also a solution of (5.0.1).

The functional corresponding to (5.0.2) becomes

$$I_\varepsilon(u) = \frac{1}{2}\int_\Omega \left(|\nabla u|^2 - \lambda u^2\right)dx - \int_\Omega F_\varepsilon(u)dx, \quad u \in H_0^1(\Omega).$$

Now for each $\varepsilon > 0$, $I_\varepsilon(u)$ is an even functional, which satisfies the (P.-S.) condition. Thus, by the symmetric mountain pass lemma, $I_\varepsilon(u)$ has infinitely many critical values $c_{\varepsilon,k}$, $k = 1, 2\cdots$, with $c_{\varepsilon,k} \to +\infty$ as $k \to +\infty$. Consequently, (5.0.2) possesses solutions $u_{\varepsilon,k}$ that satisfy $I_\varepsilon(u_{\varepsilon,k}) = c_{\varepsilon,k}$.

By the characteristic of the value $c_{\varepsilon,k}$ (see (1.3.5) in Chapter 1), it is easy to prove that $c_{\varepsilon,k}$ is bounded for each fixed k. Thus we conclude that $u_{\varepsilon,k}$ is a bounded sequence in $H_0^1(\Omega)$ for each fixed k. Now we want to study the compactness of the sequence of $u_{\varepsilon,k}$ in $H_0^1(\Omega)$ as $\varepsilon \to 0$.

For any $u \in H_0^1(\Omega)$, we extend $u(x)$ to the whole of $\mathbb{R}^N$ by setting $u(x) = 0$ if $x \notin \Omega$. For any $\mu > 0$ and $x \in \mathbb{R}^N$, we define

$$\rho_{x,\mu}(u) = \mu^{\frac{N-2}{2}} u\big(\mu(\cdot - x)\big).$$

Let u_n be a solution of (5.0.2) with $\varepsilon = \varepsilon_n \to 0$, which satisfies $\|u_n\| \le C$ for some positive constant C independent of n. Then it follows from Lemma 1.4.7 that there exist at most countable $x_n^{(j)} \in \bar{\Omega}$, $\mu_n^{(j)} \to +\infty$ and $v_j \ne 0$ for $j = 1, 2, \cdots$, such that

$$\frac{\mu_n^{(j)}}{\mu_n^{(i)}} + \frac{\mu_n^{(i)}}{\mu_n^{(j)}} + \mu_n^{(j)}\mu_n^{(i)}|x_n^{(i)} - x_n^{(j)}|^2 \to +\infty, \quad \forall\, j \ne i,$$

$$(\mu_n^{(j)})^{-\frac{N-2}{2}} u_n\big((\mu_n^{(j)})^{-1}y + x_n^{(j)}\big) \rightharpoonup v_j(y) \text{ in } D^{1,2}(\mathbb{R}^N), \text{ as } n \to \infty,$$

$$\|u_n\|^2 = \|u\|^2 + \sum_{j=1}^{\infty} \|v_j\|^2_{D^{1,2}(\mathbb{R}^N)} + \|r_n\|^2_{D^{1,2}(\mathbb{R}^N)} + o_n(1), \tag{5.0.4}$$

where $u_n \rightharpoonup u$, and $r_n = u_n - u - \sum_{j=1}^{\infty} \rho_{x_n^{(j)},\mu_n^{(j)}}(v_j)$ satisfies

$$\|r_n\|_{L^{2^*}(\mathbb{R}^N)} \to 0, \quad \text{as } n \to +\infty.$$

Our objective is to show that there is no bubble in (5.0.4) and $\|r_n\| \to 0$. We will achieve this by first showing that the number of bubbles in (5.0.4) is at most finite and $\|r_n\| \to 0$. Similar to the discussion of the (P.-S.) sequence, this step is not difficult because u_n is a solution of (5.0.2), which results in $\|v_j\|_{D^{1,2}(\mathbb{R}^N)} \geq c_0 > 0$. The most difficult part is to prove that there is no bubble in (5.0.4). Generally speaking, a (P.-S.) sequence of $I(u)$ at a very high level may have bubbles. But since u_n is a solution of (5.0.2), it satisfies the following Pohozaev identity (see Theorem 6.2.1 in the Appendix):

$$\begin{aligned}
&\int_{B_n} \Big(N F_{\varepsilon_n}(u_n) - \frac{N-2}{2} u_n f_{\varepsilon_n}(u_n)\Big) + \lambda \int_{B_n} |u_n|^2 \\
&= \int_{\partial B_n} F_{\varepsilon_n}(u_n)(x - x_0) \cdot \nu + \frac{\lambda}{2} \int_{\partial B_n} |u_n|^2 (x - x_0) \cdot \nu \\
&+ \int_{\partial B_n} (\nabla u_n \cdot (x - x_0))(\nabla u_n \cdot \nu) \\
&- \frac{1}{2} \int_{\partial B_n} |\nabla u_n|^2 (x - x_0) \cdot \nu d\sigma + \frac{N}{2^*} \int_{\partial B_n} \nabla u_n \cdot \nu u_n,
\end{aligned} \tag{5.0.5}$$

where ν is the unit outward normal vector of ∂B_n and $x_0 \in \mathbb{R}^N$. We will choose a suitable small set B_n such that B_n contains a blow-up point $x_n^{(j)}$ of u_n but ∂B_n does not contain any blow-up point of u_n. Intuitively, u_n is uniformly bounded on ∂B_n, as there is no any blow-up point of u_n on ∂B_n. Therefore, all the surface integrals on the right-hand side of (5.0.5) are 'small'. On the other hand, by scaling, it is not very difficult to prove that whenever there is a bubble in (5.0.4), the volume integrals on the left-hand side of (5.0.5) have a positive lower bound that is bigger than the upper bound on the right-hand side.

We will prove the following compactness result for the sequence of solutions u_n.

Theorem 5.0.5 (Devillannova and Solimini [71], 2002) *Suppose that $\lambda > 0$ and $N \geq 7$. Let u_n be a solution of* (5.0.2) *with $\varepsilon = \varepsilon_n \to 0$, which satisfies $\|u_n\| \leq C$ for some constant C independent of n. Then u_n converges strongly in $H_0^1(\Omega)$ as $n \to +\infty$. Consequently, $\|u_n\|_{L^\infty(\Omega)} \leq C'$.*

As a direct application of Theorem 5.0.5, we prove the following multiplicity result:

Theorem 5.0.6 (Devillannova and Solimini [71], 2002) *Suppose that* $\lambda > 0$ *and* $N \ge 7$. *Then* (5.0.1) *has infinitely many solutions.*

5.1 The Decomposition

For subsequent discussion, we take

$$f_\varepsilon(t) = \begin{cases} |t|^{2*-2}t, & |t| \le \varepsilon^{-1}, \\ \varepsilon^{-\varepsilon}|t|^{2*-2-\varepsilon}t, & |t| \ge \varepsilon^{-1}. \end{cases} \tag{5.1.1}$$

In this section, we will prove that the number of the bubbles in (5.0.4) is at most finite. As a start, we have the following.

Proposition 5.1.1 *Let* u_n *be a solution of* (5.0.2) *with* $\varepsilon = \varepsilon_n \to 0$, *and which satisfies* $\|u_n\| \le C$ *for some positive constant* C *independent of* n. *Then there exists an integer* $k \ge 0$, $x_n^{(j)} \in \bar{\Omega}$, $\mu_n^{(j)} \to +\infty$ *and* $v_j \ne 0$ *for* $j = 1, 2, \cdots, k$, *such that*

$$\frac{\mu_n^{(j)}}{\mu_n^{(i)}} + \frac{\mu_n^{(i)}}{\mu_n^{(j)}} + \mu_n^{(j)}\mu_n^{(i)}|x_n^{(i)} - x_n^{(j)}|^2 \to +\infty, \quad \forall\, j \ne i,$$

$$(\mu_n^{(j)})^{-\frac{N-2}{2}} u_n\big((\mu_n^{(j)})^{-1}y + x_n^{(j)}\big) \rightharpoonup v_j(y) \ \ in\ D^{1,2}(\mathbb{R}^N),\ as\ n \to \infty,$$

$$\|u_n\|^2 = \|u\|^2 + \sum_{j=1}^{k} \|v_j\|^2_{D^{1,2}(\mathbb{R}^N)} + o_n(1),$$

$$\Big\|u_n - u - \sum_{j=1}^{k} \rho_{x_n^{(j)},\mu_n^{(j)}}(v_j)\Big\| \to 0, \quad as\ n \to +\infty.$$

Moreover, one has the following estimate:

$$|v_j(y)| \le \frac{C}{(1+|y|)^{N-2}}. \tag{5.1.2}$$

Proof. To prove that the number of the bubbles in (5.0.4) is at most finite, we just need to prove that there is an $\alpha > 0$ such that $\|v_j\|_{D^{1,2}(\mathbb{R}^N)} \ge \alpha$ for all j. Let

$$\tilde{u}_n(y) = (\mu_n^{(j)})^{-\frac{N-2}{2}} u_n\big((\mu_n^{(j)})^{-1}y + x_n^{(j)}\big).$$

Then, one has

$$-\Delta \tilde{u}_n = (\mu_n^{(j)})^{-\frac{N+2}{2}} f_{\varepsilon_n}\big((\mu_n^{(j)})^{\frac{N-2}{2}} \tilde{u}_n\big) + \frac{\lambda}{(\mu_n^{(j)})^2} \tilde{u}_n.$$

Note that $|f_\varepsilon(t)| \le |t|^{2^*-1}$. From

$$\tilde{u}_n \rightharpoonup v_j \ \text{ in } D^{1,2}(\mathbb{R}^N), \text{ as } n \to \infty,$$

we see that $w = \pm v_j$ satisfies

$$\int_D \nabla w \nabla \phi \le \int_D |w|^{2^*-1} \phi, \quad \forall\, \phi \in C_0^\infty(D),\ \phi \ge 0,$$

where $D = \mathbb{R}^N$ if $\mu_n^{(j)} d(x_n^{(j)}, \partial\Omega) \to +\infty$, or $D = \mathbb{R}_+^N$ (after suitable rotation and translation) if $\mu_n^{(j)} d(x_n^{(j)}, \partial\Omega) \le C < \infty$.

Let $w^+ = w$ if $w > 0$, $w^+ = 0$ if $w \le 0$, and let $D^+ = \{y : y \in D,\, w(y) > 0\}$. Then, for any $\phi \in C_0^\infty(D)$ with $\phi \ge 0$, we have

$$\begin{aligned}\int_D \nabla w^+ \nabla \phi &= \int_{D^+} \nabla w^+ \nabla \phi = \int_{\partial D^+} \frac{\partial w^+}{\partial \nu} \phi - \int_{D^+} \Delta w^+ \phi \\ &\le - \int_{D^+} \Delta w^+ \phi \le \int_{D^+} |w|^{2^*-1} \phi.\end{aligned}$$

This shows that $|v_j| = v_j^+ + (-v_j)^+$ satisfies

$$\int_D \nabla |v_j| \nabla \phi \le \int_D |v_j|^{2^*-1} \phi, \quad \forall\, \phi \in C_0^\infty(D),\ \phi \ge 0. \tag{5.1.3}$$

In particular, we obtain

$$\int_D |\nabla |v_j||^2 \le \int_D |v_j|^{2^*} \le C\Big(\int_D |\nabla |v_j||^2\Big)^{\frac{2^*}{2}},$$

which gives

$$\int_D |\nabla v_j|^2 \ge \alpha > 0.$$

Finally, from

$$-\Delta |v_j| \le |v_j|^{2^*-1},$$

we can prove (5.1.2) by using the Kelvin transformation. □

Remark 5.1.2 For the function $f_\varepsilon(t)$ defined in (5.1.1), it is not easy to determine the weak limit for $(\mu_n^{(j)})^{-\frac{N+2}{2}} f_{\varepsilon_n}((\mu_n^{(j)})^{\frac{N-2}{2}}\tilde{u}_n)$. Therefore, the exact limit equation for v_j cannot be found easily. Here we are able to obtain the differential inequality (5.1.3) for v_j, from which we can derive the inequality $\|v_j\| \geq \alpha > 0$ and the decay estimate for v_j.

5.2 Some Integral Estimates

To prove the strong convergence of u_n in $H^1(\Omega)$, we just need to show that the bubbles $\rho_{x_n^{(j)},\lambda_n^{(j)}}(v_j)$ do not appear in the following decomposition:

$$u_n = u + \sum_{j=1}^{k} \rho_{x_n^{(j)},\mu_n^{(j)}}(v_j) + r_n, \tag{5.2.1}$$

where $\|r_n\| \to 0$ as $n \to \infty$.

Our objective is to use the Pohozaev identity (5.0.5) to obtain a contradiction if bubbles appear in (5.2.1). Intuitively, if we ignore the error term r_n in the decomposition of u_n, or in other words, if we replace u_n by $u + \sum_{j=1}^{k} \rho_{x_n^{(j)},\mu_n^{(j)}}(v_j)$ in the Pohozaev identity, all those integrals can be estimated by the size of B_n and the height $\mu_n^{(j)}$ of the bubbles $\rho_{x_n^{(j)},\mu_n^{(j)}}(v_j)$. To illustrate this idea, let us do some quick calculations for both sides of (5.0.5), with

$$u_n = u + \rho_{x_n,\mu_n}(v)$$

and

$$\rho_{x_n,\mu_n}(v) = \frac{C\mu_n^{\frac{N-2}{2}}}{(1 + \mu_n^2|y - x_n|^2)^{\frac{N-2}{2}}}.$$

Here, we also assume that $x_n \to x^* \in \Omega$.

Initially, we take $B_n = B_{\sigma_n}(x_n)$ for some $\sigma_n \to 0$. Since $|u| \leq C$, it is convenient to choose $\sigma_n = \mu_n^{-\frac{1}{2}}$. As a consequence, we have $|\rho_{x_n,\mu_n}(v)| \leq C$ on $\partial B_{\mu_n^{-\frac{1}{2}}}(x_n)$. Now, take $B_n = B_{\mu_n^{-\frac{1}{2}}}(x_n)$ and $x_0 = x_n$ in (5.0.5). On $\partial B_{\mu_n^{-\frac{1}{2}}}(x_n)$, we have $\rho_{x_n,\mu_n}(v) \leq C$ and $|\nabla \rho_{x_n,\mu_n}(v)| \leq C\mu_n^{\frac{1}{2}}$. It is easy to check that

$$\text{RHS of (5.0.5)} \leq \frac{C}{\mu_n^{\frac{N-2}{2}}},$$

while

$$\text{LHS of (5.0.5)} \geq \lambda \int_{B_n} |u_n|^2 \geq \frac{\lambda}{2} \int_{B_n} |\rho_{x_n,\mu_n}(v)|^2 + O\Big(\frac{1}{\mu_n^{\frac{N}{2}}}\Big)$$
$$\geq C\mu_n^{-2} + O\Big(\frac{1}{\mu_n^{\frac{N}{2}}}\Big).$$

Thus we obtain the inequality $\mu_n^{-2} \leq C\mu_n^{-\frac{N-2}{2}}$, which is a contradiction if $N \geq 7$.

The main difficulty to make the above argument work comes from the error term r_n. Therefore, some subtle estimates for u_n are needed. In this section, we will derive some integral estimates for u_n by using the height $\mu_n^{(j)}$ of the bubbles $\rho_{x_n^{(j)},\mu_n^{(j)}}(v_j)$. Among all the bubbles $\rho_{x_n^{(j)},\mu_n^{(j)}}(v_j)$, we can choose a bubble $\rho_{x_n^{(i)},\mu_n^{(i)}}(v_i)$ such that this bubble has the slowest concentration rate. That is, the height $\mu_n^{(i)}$ of this bubble is the lowest order infinity among all $\mu_n^{(j)}$. For simplicity, we denote by μ_n the slowest concentration rate and x_n the corresponding concentration point.

In the following, we use $\|u\|_p$ to denote $\|u\|_{L^p(\Omega)}$. Roughly, the functions we study have the decomposition $u + \rho_{x_n,\mu_n}(v)$. The good part u satisfies $\|u\|_p \leq C$ for any $p > 1$, while the concentration part $\rho_{x_n,\mu_n}(v)$ has the property $\|\rho_{x_n,\mu_n}(v)\|_p \leq C\mu_n^{\frac{N}{2^*}-\frac{N}{p}}$, which is small if $p < 2^*$. To capture this property for functions of the form $u+\rho_{x_n,\mu_n}(v)$, we can define a norm for such functions as follows.

For any $p_2 < 2^* < p_1$, $\alpha > 0$ and $\mu > 0$, we consider the following relation:

$$\begin{cases} \|u_1\|_{p_1} \leq \alpha, \\ \|u_2\|_{p_2} \leq \alpha\mu^{\frac{N}{2^*}-\frac{N}{p_2}}. \end{cases} \tag{5.2.2}$$

We define

$$\|u\|_{p_1,p_2,\mu} = \inf\big\{\alpha > 0 : \text{there are } u_1 \text{and } u_2, \text{ such that (5.2.2) holds and } |u| \leq u_1 + u_2\big\}.$$

The main result of this section is the following proposition:

Proposition 5.2.1 *For any $p_1, p_2 \in \left(\frac{2^*}{2}, +\infty\right)$ with $p_2 < 2^* < p_1$, there is a constant C, depending on p_1 and p_2, such that*

$$\|u_n\|_{p_1,p_2,\mu_n} \leq C.$$

Note that $|f_\varepsilon(u)| \le |u|^{2^*-1} + \lambda|u|$. We let $u_n(x) = 0$ if $x \notin \Omega$. To avoid the estimate for u_n near the boundary of Ω, we take a domain D such that $\bar{\Omega} \subset D$ and such that $B_1(y) \subset D$ for $y \in \Omega$. Let $w_n \in H_0^1(D)$ be a solution of

$$-\Delta w_n = |u_n|^{2^*-1} + \lambda|u_n|, \quad \text{in } D. \tag{5.2.3}$$

Then one has $w_n \ge |u_n|$ in D. From (5.2.3), we find that $\int_D |\nabla w_n|^2 \le C$. Moreover, it follows from the estimate $|u_n| \le w_n$ that

$$-\Delta w_n \le (|u_n|^{2^*-2} + \lambda)w_n, \quad \text{in } D. \tag{5.2.4}$$

To prove Proposition 5.2.1, we only need to prove the following result:

Proposition 5.2.2 *Let w_n be a solution of (5.2.3). For any $p_1, p_2 \in \left(\frac{2^*}{2}, +\infty\right)$ with $p_2 < 2^* < p_1$, there is a constant C, depending on p_1 and p_2, such that*

$$\|w_n\|_{p_1,p_2,\mu_n} \le C.$$

Let us point out that in Proposition 5.2.2, the norm is defined in D.

First, let us discuss what we need to do to prove Proposition 5.2.2. By Proposition 5.1.1, we have

$$u_n = u + \sum_{j=1}^{k} \rho_{x_n^{(j)},\mu_n^{(j)}}(v_j) + r_n,$$

where r_n satisfies $\|r_n\| \to 0$ as $n \to +\infty$. Then

$$|u_n|^{2^*-2} + \lambda \le a_0 + a_1 + a_2,$$

where

$$a_0(x) = A|u|^{2^*-2} + \lambda,\ a_1 = A\left|\sum_{j=1}^{k} \rho_{x_n^{(j)},\mu_n^{(j)}}(v_j)\right|^{2^*-2},\ a_2(x) = A|r_n|^{2^*-2},$$

for some large $A > 0$. By (5.2.4), we have

$$-\Delta w_n \le (a_0 + a_1 + a_2)w_n, \quad \text{in } D.$$

Let $w = Gv$ be the solution of

$$\begin{cases} -\Delta w = v, & \text{in } D, \\ w = 0, & \text{on } \partial D. \end{cases}$$

By comparison, one has

$$w_n \le G(a_0 w_n) + G(a_1 w_n) + G(a_2 w_n). \tag{5.2.5}$$

To prove Proposition 5.2.2, we need to estimate $\|G(a_i w_n)\|_{p_1,p_2,\mu_n}$, $i = 1, 2, 3$. For this purpose, we will derive some L^p estimates for the solution w of the following linear problem:

$$\begin{cases} -\Delta w = a(x)v, & \text{in } D, \\ w = 0, & \text{on } \partial D, \end{cases}$$

where $v \in H^1(D)$ is a given function. Here we need to discuss three cases for $a(x)$. (i) $|a(x)| \le C$. (ii) $a(x)$ satisfies $\|a(x)\|_s \le C\mu_n^{2-\frac{N}{s}} \to 0$ for $s \in (\frac{N}{4}, \frac{N}{2})$. (iii) $\|a\|_{\frac{N}{2}}$ is small. We will obtain different estimates according to the condition that $a(x)$ satisfies. Since these results are of independent interest, we will put them in Section 5.5.

We will prove Proposition 5.2.2 in two steps: First, we prove that the result holds for some $\frac{2^*}{2} < p_2 < 2^* < p_1$. Then we prove the result for all $p_1, p_2 \in \left(\frac{2^*}{2}, +\infty\right)$ with $p_2 < 2^* < p_1$.

Lemma 5.2.3 *There are constants $C > 0$ and $p_1, p_2 \in \left(\frac{N}{N-2}, +\infty\right)$ with $p_2 < 2^* < p_1$ such that*

$$\|w_n\|_{p_1,\,p_2,\,\mu_n} \le C.$$

Proof. By (5.2.5), we have

$$\|w_n\|_{p_1,p_2,\mu_n} \le \|G(a_0 w_n)\|_{p_1,p_2,\mu_n} + \|G(a_1 w_n)\|_{p_1,p_2,\mu_n} + \|G(a_2 w_n)\|_{p_1,p_2,\mu_n}.$$

We will first treat the term $G(a_0 w_n)$. Let $p \in (\frac{2N}{N+2}, 2^*)$ be a fixed constant. Then $p_1 =: \frac{Np}{N-2p} > 2^*$. It follows from Lemma 5.5.3 that

$$\|G(a_0 w_n)\|_{p_1} \le C\|a_0 w_n\|_p \le C\|w_n\|_p \le C.$$

Consequently, we may conclude that

$$\|G(a_0 w_n)\|_{p_1,p_2,\mu_n} \le \|G(a_0 w_n)\|_{p_1} \le C. \tag{5.2.6}$$

Next, we treat the term $G(a_1 w_n)$. Let $p_2 \in \left(\frac{N}{N-2}, 2^*\right)$ be a constant. It follows from Lemma 5.5.4 that

$$\|G(a_1 w_n)\|_{p_2} \le C\|a_1\|_r \|w_n\|_{2^*} \le C\|a_1\|_r,$$

where r is determined by the equation $\frac{1}{p_2} = \frac{1}{r} + \frac{1}{2^*} - \frac{2}{N}$. But

$$\int_D |\rho_{x_n^{(j)},\mu_n^{(j)}}(v_j)|^{\frac{4r}{N-2}} = (\mu_n^{(j)})^{-N+2r} \int_{D_{x_n^{(j)},\mu_n^{(j)}}} |v_j|^{\frac{4r}{N-2}},$$

where $D_{x,\mu} = \{y : x + \mu^{-1} y \in D\}$.

For $j = 1, \cdots, k$, it follows from (5.1.2) that

$$|v_j(x)| \le \frac{C}{1 + |x|^{N-2}}.$$

Therefore, one has

$$\int_{D_{x_n^{(j)}, \mu_n^{(j)}}} |v_j|^{\frac{4r}{N-2}} \le C,$$

for any $r \in \left(\frac{N}{4}, \frac{N}{2}\right)$. It is easy to check that if $r < \frac{N}{2}$, then $p_2 < 2^*$. Thus we have proved that there is a $p_2 < 2^*$ such that

$$\|G(a_1 w_n)\|_{p_2} \le C \mu_n^{2 - \frac{N}{r}} = C \mu_n^{\frac{N}{2^*} - \frac{N}{p_2}},$$

which in turn implies that

$$\|G(a_1 w_n)\|_{p_1, p_2, \mu_n} \le C. \tag{5.2.7}$$

Finally, we treat the term $G(a_2 w_n)$. It follows from Lemma 5.5.2 that

$$\|G(a_2 w_n)\|_{p_1, p_2, \mu_n} \le C \|a_2\|_{\frac{N}{2}} \|w_n\|_{p_1, p_2, \mu_n} \le \frac{1}{2} \|w_n\|_{p_1, p_2, \mu_n}, \tag{5.2.8}$$

since $\|a_2\|_{\frac{N}{2}} = \|r_n\|_{2^*}^{\frac{4}{N-2}} \to 0$ as $n \to \infty$.

Combining (5.2.6), (5.2.7) and (5.2.8), we obtain the following estimates:

$$\begin{aligned} \|w_n\|_{p_1, p_2, \mu_n} &\le \|G(a_0 w_n)\|_{p_1, p_2, \mu_n} + \|G(a_1 w_n)\|_{p_1, p_2, \mu_n} + \|G(a_2 w_n)\|_{p_1, p_2, \mu_n} \\ &\le C + \frac{1}{2} \|w_n\|_{p_1, p_2, \mu_n}. \end{aligned}$$

Hence we have completed the proof of the lemma.

□

Lemma 5.2.3 holds for $p_1 \le \frac{Np}{N-p}$ and $p \in (\frac{2N}{N+2}, 2^*)$. In other words, it is valid for $p_1 < \frac{2^*}{N-2^*}$. To prove that the result also holds for all $p_1 > 2^*$, we need the following lemma:

Lemma 5.2.4 *Let $w \ge 0$ be a weak solution of*

$$\begin{cases} -\Delta w = 2v^{2^*-1} + A, & \text{in } D, \\ w = 0, & \text{on } \partial D, \end{cases}$$

where $v \ge 0$ is a function satisfying $\|v\|_{p_1, p_2, \mu} < \infty$ for some $p_1, p_2 \in \left(\frac{N+2}{N-2}, \frac{N}{2}\frac{N+2}{N-2}\right)$ with $p_2 < 2^* < p_1$, and $\mu > 0$. Let q_i be given by

$$\frac{1}{q_i} = \frac{N+2}{(N-2)p_i} - \frac{2}{N}, \quad i = 1, 2. \tag{5.2.9}$$

Then there is a constant $C = C(p_1, p_2)$ *such that*

$$\|w\|_{q_1,q_2,\mu} \le C\|v\|_{p_1,p_2,\mu}^{\frac{N+2}{N-2}} + C.$$

Proof. For any small $\theta > 0$, let $v_1 \ge 0$ and $v_2 \ge 0$ be the functions such that $v \le v_1 + v_2$ and (5.2.2) holds with $\alpha = \|v\|_{p_1,p_2,\mu} + \theta$. Consider

$$\begin{cases} -\Delta w_1 = C v_1^{\frac{N+2}{N-2}} + A, & \text{in } D, \\ w_1 = 0, & \text{on } \partial D, \end{cases}$$

and

$$\begin{cases} -\Delta w_2 = C v_2^{\frac{N+2}{N-2}}, & \text{in } D, \\ w_2 = 0, & \text{on } \partial D, \end{cases}$$

where $C > 0$ is a large constant. By the maximum principle, one then has $w \le w_1 + w_2$.

Let $\hat{p}_i = p_i \frac{N-2}{N+2}$. Then we have $q_i = \frac{N\hat{p}_i}{N-2\hat{p}_i}$ for $i = 1, 2$. If $p_i \in \left(\frac{N+2}{N-2}, \frac{N}{2}\frac{N+2}{N-2}\right)$, then $\hat{p}_i \in \left(1, \frac{N}{2}\right)$.

By Lemma 5.5.3, we have

$$\begin{aligned} \|w_1\|_{q_1} &\le C' \|v_1^{\frac{N+2}{N-2}} + A\|_{\hat{p}_1} \\ &\le C\left(\|v_1\|_{p_1}^{\frac{N+2}{N-2}} + 1\right) \\ &\le C\left(\left(\|v_1\|_{p_1,p_2,\mu} + \theta\right)^{\frac{N+2}{N-2}} + 1\right) \end{aligned}$$

and

$$\begin{aligned} \|w_2\|_{q_2} &\le C\|v_2\|_{p_2}^{\frac{N+2}{N-2}} \\ &\le C\left(\left(\|v_2\|_{p_1,p_2,\mu} + \theta\right)\mu^{\frac{N}{2^*} - \frac{N}{p_2}}\right)^{\frac{N+2}{N-2}} \\ &= C\left(\|v_2\|_{p_1,p_2,\mu} + \theta\right)^{\frac{N+2}{N-2}} \mu^{\frac{N}{2^*} - \frac{N}{q_2}}. \end{aligned}$$

Therefore, we have proved the result. □

Proof of Proposition 5.2.2. Applying Lemma 5.2.3, we see that there exist $p_1, p_2 \in \left(\frac{N}{N-2}, +\infty\right)$ with $p_2 < 2^* < p_1$ such that

$$\|w_n\|_{p_1,p_2,\mu_n} \le C.$$

From (5.2.4), we then have $w_n \le w_n^*$, where $w_n^* \in H_0^1(D)$ is the solution of

$$-\Delta w_n^* = 2 w_n^{2^*-1} + A, \quad \text{in } D.$$

By Lemma 5.2.4, one has

$$\|w_n\|_{q_1,q_2,\mu_n} \le \|w_n^*\|_{q_1,q_2,\mu_n} \le C\|w_n\|_{p_1,p_2,\mu_n}^{\frac{N+2}{N-2}} + C,$$

where q_i, $i = 1, 2$, is given as in (5.2.9). Let us point out that $q_1 > p_1$ and $q_2 < p_2$ since $p_2 < 2^* < p_1$. By applying Lemma 5.2.4 iteratively, we can prove that $\|w_n\|_{q_1,q_2,\mu}$ holds for all $p_1, p_2 \in \left(\frac{2^*}{2}, +\infty\right)$ with $p_2 < 2^* < p_1$.

□

5.3 Estimates on Safe Regions

As was pointed out at the beginning of this chapter, we need to find a region B_n such that ∂B_n contains no blow-up points of u_n. When using the Pohozaev identity to obtain a contradiction, it is essential to estimate u_n near ∂B_n. This is the estimates of u_n on a safe region.

Since the number of the bubbles of u_n is finite, we may always find a constant $\bar{C} > 0$, independent of n, such that the region

$$\mathcal{A}_n^1 = \left(B_{(\bar{C}+5)\mu_n^{-\frac{1}{2}}}(x_n) \setminus B_{\bar{C}\mu_n^{-\frac{1}{2}}}(x_n)\right) \cap \Omega$$

does not contain any concentration points of u_n for any n. We call this region a *safe region* for u_n.

Let

$$\mathcal{A}_n^2 = \left(B_{(\bar{C}+4)\mu_n^{-\frac{1}{2}}}(x_n) \setminus B_{(\bar{C}+1)\mu_n^{-\frac{1}{2}}}(x_n)\right) \cap \Omega.$$

In this section, we will prove the following result:

Proposition 5.3.1 *Let w_n be a weak solution of* (5.2.3). *Then, there is a constant $C > 0$, independent of n, such that*

$$\left(\int_{\mathcal{A}_n^2} |w_n|^p\right)^{\frac{1}{p}} \le C\mu_n^{-\frac{N}{2p}},$$

where $p \ge 2$ is any constant.

To prove Proposition 5.3.1, we need the following lemma:

Lemma 5.3.2 *Suppose w_n satisfies* (5.2.3). *Then there is a constant $C > 0$, independent of n, such that*

$$\frac{1}{r^{N-1}}\int_{\partial B_r(y)} w_n \le C, \quad \forall\, y \in \Omega,$$

for all $r \in [\bar{C}\mu_n^{-1/2}, (\bar{C}+5)\mu_n^{-1/2}]$.

Proof. From $\int_{B_1(y)} w_n \le C$, we can find a $r_n \in \left[\frac{1}{2}, 1\right]$ such that

$$\frac{1}{r_n^{N-1}}\int_{\partial B_{r_n}(y)} w_n \le C.$$

Let us denote by (r, ω) the spherical coordinate centered at y. Since

$$\begin{aligned}\int_{B_t(y)} \Delta w_n &= \int_{\partial B_t(y)} \frac{\partial w_n}{\partial \nu} = \int_{\partial B_t(y)} \frac{\partial w_n(t,\omega)}{\partial t}\\ &= t^{N-1}\int_{|\omega|=1} \frac{\partial w_n(t,\omega)}{\partial t} = t^{N-1}\frac{d}{dt}\int_{|\omega|=1} w_n(t,\omega)\\ &= t^{N-1}\frac{d}{dt}\Big(t^{1-N}\int_{\partial B_t(y)} w_n\Big),\end{aligned}$$

we see that

$$\frac{d}{dt}\Big(\frac{1}{t^{N-1}}\int_{\partial B_t(y)} w_n\Big) = \frac{1}{t^{N-1}}\int_{B_t(y)} \Delta w_n.$$

It then follows that

$$\begin{aligned}\frac{1}{r^{N-1}}\int_{\partial B_r(y)} w_n &= \frac{1}{r_n^{N-1}}\int_{\partial B_{r_n}(y)} w_n + \int_{r_n}^{r} \frac{d}{dt}\Big(\frac{1}{t^{N-1}}\int_{B_t(y)} w_n\Big)\,dt\\ &= \frac{1}{r_n^{N-1}}\int_{\partial B_{r_n}(y)} w_n + \int_{r}^{r_n} \frac{1}{t^{N-1}}\int_{B_t(y)} (-\Delta w_n)\,dt\\ &\le C + C\int_{r}^{r_n} \frac{1}{t^{N-1}}\int_{B_t(y)} \big(w_n^{2^*-1}+1\big)\,dt\\ &\le C' + C\int_{r}^{r_n} \frac{1}{t^{N-1}}\int_{B_t(y)} w_n^{2^*-1}\,dt.\end{aligned} \tag{5.3.1}$$

By Proposition 5.2.1, we know that $\|w_n\|_{p_1,p_2,\mu_n} \le C$, for any $p_1, p_2 \in \left(\frac{2^*}{2}, +\infty\right)$, $p_2 < 2^* < p_1$.

Let $p_2 = \frac{N+2}{N-2}$, and let $p_1 > \frac{N(N+2)}{(N-2)}$. Then we can choose $v_{1,n}$, and $v_{2,n}$, such that $w_n \le v_{1,n} + v_{2,n}$ and

$$\|v_{1,n}\|_{p_1} \le C, \quad \|v_{2,n}\|_{p_2} \le C\mu_n^{\frac{N}{2^*}-\frac{N}{p_2}}.$$

Thus one has

$$\begin{aligned}
&\int_r^{r_n} \frac{1}{t^{N-1}} \int_{B_t(y)} v_{1,n}^{\frac{N+2}{N-2}}\, dt \\
&\leq \int_r^{r_n} \frac{1}{t^{N-1}} \left(\int_{B_t(y)} |v_{1,n}|^{p_1} \right)^{\frac{N+2}{(N-2)p_1}} t^{N(1-\frac{N+2}{p_1(N-2)})}\, dt \qquad (5.3.2)\\
&\leq C \int_r^{r_n} t^{1-\frac{N(N+2)}{p_1(N-2)}}\, dt \leq C'
\end{aligned}$$

and

$$\begin{aligned}
\int_r^{r_n} \frac{1}{t^{N-1}} \int_{B_t(y)} v_{2,n}^{\frac{N+2}{N-2}}\, dt \leq& C \int_r^{r_n} \frac{1}{t^{N-1}} \mu_n^{\left(\frac{N}{2^*}-\frac{N}{p_2}\right)\frac{N+2}{N-2}}\, dt \\
\leq& C\mu_n^{\frac{2-N}{2}} \int_r^{r_n} t^{-N+1}\, dt \qquad (5.3.3)\\
\leq& C\mu_n^{\frac{2-N}{2}} \left(C + r^{2-N}\right) \leq C,
\end{aligned}$$

for all $r \in [\bar{C}\mu_n^{-1/2}, (\bar{C}+5)\mu_n^{-1/2}]$.

Combining (5.3.2) and (5.3.3), we obtain

$$\begin{aligned}
&\int_r^{r_n} \frac{1}{t^{N-1}} \int_{B_t(y)} w_n^{\frac{N+2}{N-2}}\, dt \\
\leq& C \int_r^{r_n} \frac{1}{t^{N-1}} \int_{B_t(y)} v_{1,n}^{\frac{N+2}{N-2}}\, dt + C \int_r^{r_n} \frac{1}{t^{N-1}} \int_{B_t(y)} v_{2,n}^{\frac{N+2}{N-2}}\, dt \qquad (5.3.4)\\
\leq& C.
\end{aligned}$$

The result now follows from (5.3.1) and (5.3.4). □

Proof of Proposition 5.3.1. It follows from Lemma 5.3.2 that

$$\int_{\partial B_r(x_n)} w_n \leq C r^{N-1}, \quad r \in [\bar{C}\mu_n^{-1/2}, (\bar{C}+5)\mu_n^{-1/2}].$$

Integrating from $\bar{C}\mu_n^{-1/2}$ to $(\bar{C}+5)\mu_n^{-1/2}$ yields

$$\int_{\mathcal{A}_n^1} w_n \leq C\mu_n^{-\frac{N}{2}}.$$

In view of the inclusion $B_{\mu_n^{-\frac{1}{2}}}(y) \subset \mathcal{A}_n^1$, we then have that

$$\int_{B_{\mu_n^{-\frac{1}{2}}}(y)} |w_n| \leq C\mu_n^{-\frac{N}{2}} \qquad (5.3.5)$$

for any $y \in \mathcal{A}_n^2$.

Let

$$v_n(x) = w_n(\mu_n^{-\frac{1}{2}} x + y), \quad x \in \Omega_n,$$

where $\Omega_n = \{x : \mu_n^{-\frac{1}{2}} x + y \in \Omega\}$. By (5.2.4), v_n satisfies

$$\begin{cases} -\Delta v_n \leq \mu_n^{-1}\big(|u_n(\mu_n^{-\frac{1}{2}} x + y)|^{2^*-2} + \lambda\big) v_n, & \text{in } \Omega_n, \\ v_n = 0, & \text{on } \partial\Omega_n. \end{cases}$$

Since $B_{\mu_n^{-\frac{1}{2}}}(y)$ does not contain any concentration point of u_n for $y \in \mathcal{A}_n^2$, we can deduce that

$$\int_{B_1(0)} \left|\mu_n^{-1}\big(|u_n(\mu_n^{-\frac{1}{2}} x + y)|^{2^*-2} + \lambda\big)\right|^{\frac{N}{2}} \leq C \int_{B_{\mu_n^{-\frac{1}{2}}}(y)} |u_n|^{2^*} + C\mu_n^{-\frac{N}{2}} \to 0$$

as $n \to +\infty$. Thus, by Lemma 6.4.6 in the Appendix and (5.3.5), we see that for any $p > 2^*$,

$$\begin{aligned} \|v_n\|_{L^p(B_{\frac{1}{2}}(0))} \leq & C\Big(\int_{B_1(0)} |v_n| + 1\Big) \\ = & C\Big(\mu_n^{\frac{N}{2}} \int_{B_{\mu_n^{-\frac{1}{2}}}(y)} |w_n| + 1\Big) \leq C. \end{aligned}$$

As a result, we have

$$\mu_n^{\frac{N}{2p}} \Big(\int_{B_{\mu_n^{-\frac{1}{2}}}(y)} |w_n|^p\Big)^{\frac{1}{p}} \leq C, \quad \forall\, y \in \mathcal{A}_n^2.$$

This in turn gives

$$\int_{\mathcal{A}_n^2} |w_n|^p \leq C\mu_n^{-\frac{N}{2}}.$$

For $2 \leq p \leq 2^*$, take $\bar{p} > 2^*$. Then, one has

$$\begin{aligned} \int_{\mathcal{A}_n^2} |w_n|^p \leq & C\Big(\int_{\mathcal{A}_n^2} |w_n|^{\bar{p}}\Big)^{\frac{p}{\bar{p}}} \mu_n^{-\frac{N}{2}(1-\frac{p}{\bar{p}})} \\ \leq & C\mu_n^{-\frac{N}{2}\frac{p}{\bar{p}} - \frac{N}{2} + \frac{pN}{2\bar{p}}} = C\mu_n^{-\frac{N}{2}}. \end{aligned}$$

□

Let

$$\mathcal{A}_n^3 = \Big(B_{(\bar{C}+3)\mu_n^{-\frac{1}{2}}}(x_n) \setminus B_{(\bar{C}+2)\mu_n^{-\frac{1}{2}}}(x_n)\Big) \cap \Omega.$$

Proposition 5.3.3 *We have*

$$\int_{\mathcal{A}_n^3} |\nabla u_n|^2 \le C \int_{\mathcal{A}_n^2} \left(|u_n|^{2^*} + 1\right) + C\mu_n \int_{\mathcal{A}_n^2} |u_n|^2. \tag{5.3.6}$$

In particular,

$$\int_{\mathcal{A}_n^3} |\nabla u_n|^2 \le C\mu_n^{\frac{2-N}{2}}. \tag{5.3.7}$$

Proof. Let $\phi_n \in C_0^2(\mathcal{A}_n^2)$ be a function with $\phi_n = 1$ in $\mathcal{A}_n^3$, $0 \le \phi_n \le 1$ and $|\nabla \phi_n| \le C\mu_n^{\frac{1}{2}}$.

From

$$\int_\Omega \nabla u_n \nabla(\phi_n^2 u_n) \le \int_\Omega \left(2|u_n|^{\frac{N+2}{N-2}} + A\right)\phi_n^2 |u_n|,$$

we can prove (5.3.6). By (5.3.6) and Proposition 5.3.1, we have

$$\int_{\mathcal{A}_n^3} |\nabla u_n|^2 \le C\mu_n^{-\frac{N}{2}} + C\mu_n^{1-\frac{N}{2}} \le C'\mu_n^{\frac{2-N}{2}}.$$

□

5.4 Proof of the Main Results

We are now ready to use the estimates in Propositions 5.3.1 and 5.3.3 and the Pohozaev identity to prove Theorem 5.0.5. Take a $t_n \in [\bar{C}+2, \bar{C}+3]$, which satisfies

$$\begin{aligned} &\int_{\partial B_{t_n\mu_n^{-\frac{1}{2}}}(x_n)} \left(|u_n|^{2^*} + |u_n|^2 + \mu_n^{-1}|\nabla u_n|^2\right) \\ &\le C\mu_n^{\frac{1}{2}} \int_{\mathcal{A}_n^3} \left(|u_n|^{2^*} + |u_n|^2 + \mu_n^{-1}|\nabla u_n|^2\right). \end{aligned} \tag{5.4.1}$$

By Proposition 5.3.1, (5.4.1) and (5.3.7), we obtain the following inequality:

$$\int_{\partial B_{t_n\mu_n^{-\frac{1}{2}}}(x_n)} \left(|u_n|^{2^*} + |u_n|^2 + \mu_n^{-1}|\nabla u_n|^2\right) \le C\mu_n^{-\frac{N-1}{2}}. \tag{5.4.2}$$

Proof of Theorem 5.0.5. We have two different cases.

(i) $B_{t_n\mu_n^{-\frac{1}{2}}}(x_n) \cap (\mathbb{R}^N \setminus \Omega) \ne \emptyset$.

(ii) $B_{t_n\mu_n^{-\frac{1}{2}}}(x_n) \subset \Omega$.

We have the following local Pohozaev identity for u_n in $B_n = B_{t_n \mu_n^{-\frac{1}{2}}}(x_n) \bigcap \Omega$:

$$
\begin{aligned}
&\int_{B_n}\Big(NF_{\varepsilon_n}(u_n) - \frac{N-2}{2}u_n f_{\varepsilon_n}(u_n)\Big) + \lambda\int_{B_n}|u_n|^2 \\
&= \int_{\partial B_n} F_{\varepsilon_n}(u_n)(x-x_0)\cdot\nu + \frac{\lambda}{2}\int_{\partial B_n}|u_n|^2(x-x_0)\cdot\nu \\
&+ \int_{\partial B_n}(\nabla u_n\cdot(x-x_0))(\nabla u_n\cdot\nu) \\
&- \frac{1}{2}\int_{\partial B_n}|\nabla u_n|^2(x-x_0)\cdot\nu d\sigma + \frac{N}{2^*}\int_{\partial B_n}\nabla u_n\cdot\nu u_n,
\end{aligned}
\tag{5.4.3}
$$

where ν is the outward normal to ∂B_n. Here, the point x_0 in (5.4.3) is chosen as follows. In case (i), we take $x_0 \in \mathbb{R}^N \setminus \Omega$ with $|x_0 - x_n| \le 2t_n\mu_n^{-\frac{1}{2}}$ and $\nu\cdot(x-x_0)\le 0$ in $\partial\Omega\cap B_n$. In case (ii), we take $x_0 = x_n$.

For the function $f_\varepsilon(t)$, defined in (5.1.1), one has

$$
NF_\varepsilon(t) - \frac{N-2}{2}tf_\varepsilon(t) \ge 0, \quad \forall\, t\in\mathbb{R}.
$$

We thus obtain from (5.4.3) that

$$
\begin{aligned}
\lambda\int_{B_n}|u_n|^2 \le &\int_{\partial B_n} F_{\varepsilon_n}(u_n)(x-x_0)\cdot\nu + \frac{\lambda}{2}\int_{\partial B_n}|u_n|^2(x-x_0)\cdot\nu \\
&+ \int_{\partial B_n}(\nabla u_n\cdot(x-x_0))(\nabla u_n\cdot\nu) \\
&- \frac{1}{2}\int_{\partial B_n}|\nabla u_n|^2(x-x_0)\cdot\nu + \frac{N}{2^*}\int_{\partial B_n}\nabla u_n\cdot\nu u_n.
\end{aligned}
\tag{5.4.4}
$$

Now decompose ∂B_n into

$$
\partial B_n = \partial_i B_n \cup \partial_e B_n,
$$

where $\partial_i B_n = \partial B_n \cap \Omega$ and $\partial_e B_n = B_n \cap \partial\Omega$.

Noting $u_n = 0$ on $\partial\Omega$, it follows from our choice of x_0 that

$$
\begin{aligned}
&\int_{\partial_e B_n} F_{\varepsilon_n}(u_n)(x-x_0)\cdot\nu + \frac{\lambda}{2}\int_{\partial_e B_n}|u_n|^2(x-x_0)\cdot\nu \\
&+ \int_{\partial_e B_n}(\nabla u_n\cdot(x-x_0))(\nabla u_n\cdot\nu) \\
&- \frac{1}{2}\int_{\partial_e B_n}|\nabla u_n|^2(x-x_0)\cdot\nu d\sigma + \frac{N}{2^*}\int_{\partial_e B_n}\nabla u_n\cdot\nu u_n \\
&= \frac{1}{2}\int_{\partial_e B_n}|\nabla u_n|^2(x-x_0)\cdot\nu \le 0.
\end{aligned}
$$

So we can rewrite (5.4.4) as

$$\begin{aligned}\lambda\int_{B_n}|u_n|^2 \le & \int_{\partial_i B_n} F_{\varepsilon_n}(u_n)(x-x_0)\cdot\nu + \frac{\lambda}{2}\int_{\partial_i B_n}|u_n|^2(x-x_0)\cdot\nu \\ & + \int_{\partial_i B_n}(\nabla u_n\cdot(x-x_0))(\nabla u_n\cdot\nu) \\ & -\frac{1}{2}\int_{\partial_i B_n}|\nabla u_n|^2(x-x_0)\cdot\nu + \frac{N}{2^*}\int_{\partial_i B_n}\nabla u_n\cdot\nu u_n.\end{aligned} \tag{5.4.5}$$

Using (5.4.2) and noting that $|x-x_0|\le C\mu_n^{-\frac{1}{2}}$ for $x\in\partial_i B_n$, we see that

$$\begin{aligned}&\text{RHS of (5.4.5)}\\ \le & C\mu_n^{-\frac{1}{2}}\int_{\partial_i B_n}\left(|u_n|^{2^*}+u_n^2+|\nabla u_n|^2\right)+C\int_{\partial_i B_n}|\nabla u_n||u_n|, \qquad (5.4.6)\\ \le & C\mu_n^{-\frac{N-2}{2}}.\end{aligned}$$

On the other hand, let $\rho_{x_n^{(i)},\mu_n^{(i)}}(v_i)$ be the bubble which has the slowest concentration rate. Let

$$\tilde{u}_n(y)=(\mu_n^{(i)})^{-\frac{N-2}{2}}u_n((\mu_n^{(i)})^{-1}y+x_n^{(i)}).$$

Then $\tilde{u}_n\rightharpoonup v_i$ weakly in $D^{1,2}(\mathbb{R}^N)$. Denote $B_n'=B_{R\mu_n^{-1}}(x_n)\cap\Omega$, where $R>0$ is a large constant. Then we have

$$\int_{B_n}|u_n|^2\ge\int_{B_n'}|u_n|^2=(\mu_n^{(i)})^{-2}\int_{B_R(0)}|\tilde{u}_n|^2\ge c'\mu_n^{-2},$$

where $c'>0$ is some constant, since by Fatou's lemma,

$$\lim_{n\to+\infty}\int_{B_R(0)}|\tilde{u}_n|^2\ge\int_{B_R(0)}|v_i|^2>0.$$

Therefore, we have proved

$$\text{LHS of (5.4.5)}\ge\lambda c'\mu_n^{-2}. \tag{5.4.7}$$

From (5.4.6) and (5.4.7), we see that

$$\mu_n^{-2}\le C\mu_n^{-\frac{N-2}{2}}.$$

This is a contradiction if $N\ge 7$.

□

Proof of Theorem 5.0.6. Given any positive integer k, it follows from Theorem 1.3.8 that for any $\varepsilon>0$ small, (5.0.2) has k different solutions $u_{1,\varepsilon},\cdots,u_{k,\varepsilon}$ such that $I_\varepsilon(u_{j,\varepsilon})=c_{j,\varepsilon}$ for $j=1,\cdots,k$. It is easy to check that for any fixed k, there is a constant $C > 0$, independent of $\varepsilon > 0$, such that $|c_{j,\varepsilon}| \le C$, for all $\varepsilon > 0$ and $j = 1, \cdots, k$. Thus $u_{j,\varepsilon}$ is bounded in $H^1(\Omega)$. In view of Theorem 5.0.5, we may assume that $u_{j,\varepsilon}$ converges strongly in $H^1(\Omega)$ as $\varepsilon \to 0$ for all $j = 1, \cdots, k$.

By Moser iteration, we can prove that $\|u_{j,\varepsilon}\|_{L^\infty(\Omega)} \le C'$ for some $C' > 0$, which is independent of ε. Thus, if $\varepsilon \in (0, \frac{1}{C'})$, then $f_\varepsilon(u_{j,\varepsilon}) = |u_{j,\varepsilon}|^{2^*-2} u_{j,\varepsilon}$. This shows that $u_{1,\varepsilon}, \cdots, u_{k,\varepsilon}$ are k different solutions for (5.0.1). Since $k > 0$ is an arbitrary integer, we conclude that (5.0.1) has infinitely many solutions.

□

Remark 5.4.1 The cut-off techniques were introduced in [156]. Such techniques make the application of Theorem 5.0.5 easier, as can be seen in the proof of Theorem 5.0.6.

5.5 Estimates for Linear Problems

In this section, we deduce some elementary estimates for solutions of some linear elliptic problems.

Lemma 5.5.1 *Let w be a solution of*

$$\begin{cases} -\Delta w = a(x)v, & \text{in } D, \\ w = 0, & \text{on } \partial D, \end{cases}$$

where $a(x) \ge 0$ and $v \ge 0$ are functions that satisfy $a,\ v \in C^2(\bar{D})$. Then, for any $p > \frac{N}{N-2}$, there is a constant $C = C(p)$ such that

$$\|w\|_p \le C\|a\|_{\frac{N}{2}} \|v\|_p.$$

Proof. By the hypothesis of the lemma, we have $w \in C^2(\bar{D})$.

Let $q = \frac{p}{2^*}$. Then $q > \frac{1}{2}$ since $p > \frac{N}{N-2}$.

First, we assume $p \ge 2^*$. In this case, $q \ge 1$. Let $\eta = w^{2q-1} \in H_0^1(D)$. We then have

$$\int_D \nabla w \nabla (w^{2q-1}) = \int_D a(x) v w^{2q-1}. \tag{5.5.1}$$

Let $\bar{\eta} = w^q$. Then (5.5.1) gives

$$\frac{2q-1}{q^2} \int_D |\nabla \bar{\eta}|^2 = \int_D a(x) v \eta.$$

Thus we obtain

$$\left(\int_D |\bar{\eta}|^{2^*}\right)^{\frac{2}{2^*}} \le C \int_D a(x)v\eta. \tag{5.5.2}$$

On the other hand, one has

$$\begin{aligned}\int_D a(x)v\eta \le& \|v\|_p \left(\int_D (a(x)\eta)^{\frac{2^*q}{2^*q-1}}\right)^{\frac{2^*q-1}{2^*q}}\\ \le& \|v\|_p \|a\|_{\frac{N}{2}} \left(\int_D \eta^{\frac{2^*q}{2q-1}}\right)^{\frac{2q-1}{2^*q}}\\ \le& C\|v\|_p \|a\|_{\frac{N}{2}} \left(\int_D \bar{\eta}^{2^*}\right)^{\frac{2q-1}{2^*q}}.\end{aligned} \tag{5.5.3}$$

Combining (5.5.2) and (5.5.3), we deduce that

$$\left(\int_D \bar{\eta}^{2^*}\right)^{\frac{1}{2^*q}} \le C\|v\|_p \|a\|_{\frac{N}{2}}.$$

This proves the result for $p \ge 2^*$.

Now we consider the case $q \in (\frac{1}{2}, 1)$. In this case, w^{2q-1} may not be in $H_0^1(\Omega)$, so we need to proceed differently.

By the maximum principle, we know that $w \ge 0$ in D. For any $\theta > 0$ small, let $\eta = (w+\theta)^{2q-1}\xi^2$, where $\xi \ge 0$ is a function satisfying $\xi = 0$ on ∂D, $\xi > 0$ in D, $\xi = 1$ in $D_\theta = \{x : x \in D, d(x, \partial D) \ge \theta\}$ and $|\nabla \xi| \le 2\theta^{-1}$. Then $\eta \in H_0^1(D)$ and

$$\nabla \eta = (2q-1)\xi^2 (w+\theta)^{2(q-1)}\nabla w + (w+\theta)^{2q-1}\nabla \xi^2.$$

Thus it follows that

$$\int_D \nabla w \nabla \eta = \int_D a(x)v(w+\theta)^{2q-1}\xi^2. \tag{5.5.4}$$

On the other hand, we have

$$\begin{aligned}&\int_D \nabla w \nabla \eta\\ =&(2q-1)\int_D \xi^2 (w+\theta)^{2(q-1)}|\nabla(w+\theta)|^2 + \int_D (w+\theta)^{2q-1}\nabla(w+\theta)\nabla\xi^2\\ =&(2q-1)\frac{1}{q^2}\int_D \xi^2|\nabla(w+\theta)^q|^2 + \int_D (w+\theta)^{2q-1}\nabla(w+\theta)\nabla\xi^2\\ =&(2q-1)\frac{1}{q^2}\int_D |\nabla(\xi(w+\theta)^q)|^2\\ &-2(2q-1)\frac{1}{q^2}\int_D q(w+\theta)^{2q-1}\nabla\xi\nabla(w+\theta)\\ &-(2q-1)\frac{1}{q^2}\int_D (w+\theta)^{2q}|\nabla\xi|^2 + \int_D (w+\theta)^{2q-1}\nabla(w+\theta)\nabla\xi^2.\end{aligned} \tag{5.5.5}$$

From $w \in C^2(\bar{D})$, we see that

$$w(x) \le Cd(x, \partial D) \le C\theta, \quad |\nabla w| \le C, \quad \forall\, x \in D \setminus D_\theta.$$

As a result, (5.5.5) becomes

$$\begin{aligned} \int_D \nabla w \nabla \eta &= (2q-1)\frac{1}{q^2}\int_D |\nabla(\xi(w+\theta)^q)|^2 + O\Big(\int_{D\setminus D_\theta} \theta^{2q-2}\Big) \\ &= (2q-1)\frac{1}{q^2}\int_D |\nabla(\xi(w+\theta)^q)|^2 + O(\theta^{2q-1}). \end{aligned} \tag{5.5.6}$$

Combining (5.5.4) and (5.5.6), we obtain

$$\begin{aligned} &\frac{2q-1}{q^2}\int_D |\nabla(\xi(w+\theta)^q)|^2 + O(\theta^{2q-1}) \\ &= \int_D a(x)v(w+\theta)^{2q-1}\xi^2, \end{aligned}$$

which gives

$$\Big(\int_D (\xi(w+\theta))^{q2^*}\Big)^{\frac{2}{2^*}} + O(\theta^{2q-1}) \le C\int_D a(x)v\xi^2(w+\theta)^{2q-1}. \tag{5.5.7}$$

Letting $\theta \to 0$ in (5.5.7), we obtain

$$\Big(\int_D w^{q2^*}\Big)^{\frac{2}{2^*}} \le \int_D a(x)vw^{2q-1} \le \|v\|_p \|a\|_{\frac{N}{2}} \|w\|_p^{2q-1},$$

and the result follows. □

A direct consequence of Lemma 5.5.1 is the following:

Lemma 5.5.2 *Let w be a solution of*

$$\begin{cases} -\Delta w = a(x)v, & \text{in } D, \\ w = 0, & \text{on } \partial D, \end{cases}$$

where $a(x) \ge 0$ and $v \ge 0$ are functions which satisfy $a,\ v \in C^2(\bar{D})$. Then, for any $p_1 > 2^* > p_2 > \frac{N}{N-2}$ and $\mu > 0$, one has

$$\|w\|_{p_1,p_2,\mu} \le C\|a\|_{\frac{N}{2}} \|v\|_{p_1,p_2,\mu}.$$

Proof. For any small $\theta > 0$, let $v_1 \ge 0$ and $v_2 \ge 0$ be the functions such that $v \le v_1 + v_2$, and that (5.2.2) holds with $\alpha = \|v\|_{p_1,p_2,\mu} + \theta$. Consider

$$\begin{cases} -\Delta w_i = a(x)v_i, & \text{in } D, \\ w_i = 0, & \text{on } \partial D. \end{cases}$$

It follows from Lemma 5.5.1 that

$$\|w_i\|_{p_i} \le C\|a\|_{\frac{N}{2}}\|v_i\|_{p_i}, \quad i = 1, 2.$$

On the other hand, by the maximum principle, we can deduce that $w \le w_1 + w_2$. Thus we have proved this result. □

Lemma 5.5.3 *Let w be a solution of*

$$\begin{cases} -\Delta w = f(x), & \text{in } D, \\ w = 0, & \text{on } \partial D. \end{cases}$$

Suppose that $f \in L^p(D)$ and is non-negative. Then we have

$$\|w\|_{\frac{Np}{N-2p}} \le C\|f\|_p.$$

Proof. This lemma can be proved by using the following L^p estimate for the elliptic equation

$$\|w\|_{W^{2,p}(D)} \le C\|f\|_p$$

and the Sobolev embedding theorem. Here, we give a direct proof that can be generalized to other elliptic problems.

By a similar argument to that of the proof of Lemma 5.5.1, we deduce that if $q > \frac{1}{2}$, then

$$\|w\|_{2^*q}^{2q} \le C\int_D |f(x)|w^{2q-1}.$$

For any $\frac{N}{2} > p \ge 1$, we see that $2p - 2^*(p-1) > 0$. Let $q = \frac{p}{2p-2^*(p-1)} > \frac{1}{2}$. Then it follows that

$$2^*q = \frac{2^*p}{2p - 2^*(p-1)} = \frac{Np}{N-2p}$$

and

$$\begin{aligned} \left|\int_D |f(x)|w^{2q-1}\right| &\le \|f\|_p \left(\int_\Omega w^{\frac{(2q-1)p}{p-1}}\right)^{1-\frac{1}{p}} \\ &\le \|f\|_p \|w\|_{2^*q}^{2q-1}. \end{aligned}$$

Thus we have

$$\|w\|_{2^*q}^{2q} \le \|f\|_p \|w\|_{2^*q}^{2q-1},$$

which implies that

$$\|w\|_{2^*q} \le C\|f\|_p,$$

as required. □

Lemma 5.5.4 *Let w be a solution of*

$$\begin{cases} -\Delta w = a(x)v, & \text{in } D, \\ w = 0, & \text{on } \partial D, \end{cases}$$

where $a(x) \geq 0$ and $v \geq 0$ are functions that satisfy $a,\ v \in C^2(\bar{D})$. Then, for any $\frac{N}{N-2} < p_2 < 2^*$, there is a constant $C = C(p_2)$ such that

$$\|w\|_{p_2} \leq C\|a\|_r\|v\|_{2^*},$$

where r is determined by the equation $\frac{1}{p_2} = \frac{1}{r} + \frac{1}{2^*} - \frac{2}{N}$.

Proof. Let $q = \frac{2p_2}{2^*}$. Since $p_2 > \frac{N}{N-2}$, we see that $q > 1$. Let $t = \frac{2N}{N+2}$. Via a similar argument to that of the proof of Lemma 5.5.1, we obtain

$$\begin{aligned} \|w\|_{p_2}^q \leq & C\int_D avw^{q-1} \leq \|v\|_{2^*}\Big(\int_D (aw^{q-1})^t\Big)^{\frac{1}{t}} \\ \leq & \|v\|_{2^*}\|a\|_r\Big(\int_D w^{(q-1)tr/(r-t)}\Big)^{\frac{1}{t}-\frac{1}{r}}. \end{aligned}$$

By definition, we can check that

$$\frac{(q-1)tr}{r-t} = \frac{2q}{2^*} = p_2.$$

As a result, one has

$$\|w\|_{p_2}^q \leq C\|v\|_{2^*}\|a\|_r\|w\|_{p_2}^{p_2(\frac{1}{t}-\frac{1}{r})}.$$

On the other hand, it is easy to check

$$p_2\Big(\frac{1}{t} - \frac{1}{r}\Big) = \frac{2p_2}{2^*} - 1 = q - 1.$$

Thus the result follows. □

5.6 Further Results and Comments

Since the publication of the celebrated work of Brezis and Nirenberg [24], the Brezis–Nirenberg problem has attracted considerable attention (for example, see [43, 45] and the references therein). However, the existence of infinitely many solutions remained an open problem in the 1980s and 1990s. Earlier results on this were obtained when the domain Ω is a ball [46, 133]. This open problem was solved by Devillannova and Solimini [71] in 2002.

In [41], Devillannova and Solimini's result is generalized to the following problem involving the Hardy potential:

$$\begin{cases} -\Delta u - \dfrac{\mu u}{|x|^2} = |u|^{2^*-2}u + \lambda u, & \text{in } \Omega, \\ u = 0, & \text{on } \partial\Omega, \end{cases} \tag{5.6.1}$$

where $N \geq 7$ and $\mu \in \left[0, \frac{(N-2)^2}{2} - 4\right)$. Note that the solutions of (5.6.1) are not regular at $x = 0$. This implies that for the corresponding perturbed problem of (5.6.1), the result in Proposition 5.3.1 cannot be true for any $p \geq 2$.

To show the compactness in $H^1(\Omega)$ of the approximate solutions, we use the Pohozaev identity (5.0.5) to prove that there is no bubble in the decomposition (5.0.4). For this purpose, it is essential to estimate $\int_{B_{\mu_m^{-\frac{1}{2}}}(y)} |w_n|^p$ for all y in a safe region, where $p \geq 2^*$ is a fixed constant. Such estimates can be obtained using the Moser iteration in Lemma 6.4.6.

Some of the estimates based on the Moser iteration presented in this chapter are taken from [41].

Theorem 5.0.5 was also generalized in [38] to the following quasi-linear elliptic problems:

$$\begin{cases} -\Delta_p u = |u|^{p^*-2}u + \lambda |u|^{p-2}u, & \text{in } \Omega, \\ u = 0, & \text{on } \partial\Omega, \end{cases}$$

where $1 < p < N$, $p^* = \frac{pN}{N-p}$, Ω is an open bounded domain in $\mathbb{R}^N$, λ is a positive constant and

$$\Delta_p u = \sum_{i=1}^{N} \frac{\partial\left(|\nabla u|^{p-2} \frac{\partial u}{\partial x_i}\right)}{\partial x_i}.$$

The following result was proved in [38].

Theorem 5.6.1 *Assume that $\lambda > 0$ and $N > p^2 + p$. Suppose that u_n $(n = 1, 2, \cdots)$ is a solution of*

$$\begin{cases} -\Delta_p u = |u|^{p^*-2-\varepsilon}u + \lambda |u|^{p-2}u, & \text{in } \Omega, \\ u = 0, & \text{on } \partial\Omega, \end{cases}$$

with $\varepsilon = \varepsilon_n \to 0$, and u_n satisfies $\|u_n\| \leq C$ for some constant C independent of n. Then $\{u_n\}_{n\geq 1}$ has a subsequence that converges strongly in $W_0^{1,p}(\Omega)$ as $n \to +\infty$.

First of all, let us point out that the estimates in Section 5.5 are proved by Moser iteration. Therefore, all these estimates can be generalized to p-Laplacian equations without any difficulty. On the other hand, to carry out similar estimates in Section 5.3 and Section 5.4 for the quasilinear case $p \neq 2$, some serious technical difficulties need to be overcome. Here, we point out some of the difficulties.

To obtain the integral estimate in Section 5.2, we need to derive the pointwise control of the solution w for

$$-\Delta_p w = f_1(x) + f_2(x), \quad \text{in } \Omega, \quad w = 0 \text{ on } \partial\Omega,$$

by the solution w_i $(i = 1, 2)$ of

$$-\Delta_p w_i = |f_i(x)|, \quad \text{in } \Omega, \quad w_i = 0 \text{ on } \partial\Omega.$$

If $p = 2$, then $|w(x)| \leq w_1(x) + w_2(x)$. For $p \neq 2$, such a simple relation does not exist. This difficulty can be solved by using the Wolff potential and some results from [91] to obtain a pointwise control of w.

To prove Lemma 5.2.3 for the p-Laplacian equation, we need to estimate the decay rate at infinity for the solutions of the following problem:

$$\begin{cases} -\Delta_p u = |u|^{p^*-2}u, & \text{in } \mathbb{R}^N, \\ u \in D^{1,p}(\mathbb{R}^N). \end{cases} \tag{5.6.2}$$

If $p = 2$, then one can employ the Kelvin transformation to obtain such estimate. For $p \neq 2$, there is no Kelvin transformation for (5.6.2). To overcome this difficulty, in [38], an iteration argument is developed to obtain the decay estimate for the solutions of (5.6.2).

For Laplacian operator, one has the mean value type identity (5.3.1), which is linear in w_n. Such identity cannot be true for the p-Laplacian operator if $p \neq 2$. Now we need to find a counterpart of the mean value type identity (5.3.1) for the following problem:

$$-\Delta_p u = f(x), \quad \text{in } \Omega. \tag{5.6.3}$$

By adapting the iteration scheme developed in [91], the following result was proved in [38].

Proposition 5.6.2 *There is a* $\gamma \in \left(p - 1, \frac{N(p-1)}{N-p+1}\right)$ *such that for any solution* $u \in W^{1,p}(\Omega)$ *of* (5.6.3), *the following inequality*

$$\left(r^{-N}\int_{B_r(x_0)} |u|^\gamma\right)^{\frac{1}{\gamma}} \leq C + C\int_t^{r_0}\left(\frac{1}{t^{N-p}}\int_{B_t(x_0)} |f|\right)^{\frac{1}{p-1}}\frac{dt}{t}$$

holds for any $0 < r < r_0 = d(x_0, \partial\Omega)$.

Multiplicity results on other nonlinear elliptic problems involving critical growth can be found in [70, 87, 120, 154–156].

Finally, let us mention a multiplicity result for the following nonlinear Schrödinger equation

$$\begin{cases} -\Delta u + V(x)u = |u|^{p-2}, & \text{in } \mathbb{R}^N, \\ u \in H^1(\mathbb{R}^N), \end{cases} \tag{5.6.4}$$

where $N > 2$, $2 < p < 2^*$ and $V(x)$ satisfies the following assumptions:

(V_1) $V \in C^1(\mathbb{R}^N, \mathbb{R})$, $\liminf\limits_{|x| \to +\infty} V(x) = V_\infty > 0$.

(V_2) $\dfrac{\partial V}{\partial \mathbf{x}}(x)e^{\alpha|x|} \to +\infty$ as $|x| \to +\infty$, $\forall \alpha > 0$, where $\mathbf{x} = \dfrac{x}{|x|}$, $\forall x \neq 0$.

(V_3) There exists a constant $c > 1$ such that

$$|\nabla_{\tau_x} V(x)| \leq c\frac{\partial V}{\partial \mathbf{x}}(x), \quad \forall\, x \in \mathbb{R}^N : |x| > c,$$

where $\nabla_{\tau_x} V(x)$ denotes the component of the gradient of $V(x)$ at x, which is in the hyperplane orthogonal to x and containing x.

In [44], Cerami, Devillanova and Solimini proved the following result:

Theorem 5.6.3 *If the potential* $V(x)$ *satisfies the assumptions* $(V_1) - (V_3)$, *then problem* (5.6.4) *has infinitely many solutions.*

6
Appendix

6.1 Some Elementary Estimates

First, it is easy to prove the following inequalities:

$$(1+t)^p = 1 + pt + O(t^p), \quad t \geq 0, \text{ if } p \in (1, 2],$$

$$(1+t)^p = 1 + pt + \frac{p(p-1)}{2}t^2 + O(t^p), \quad t \geq 0, \text{ if } p > 2,$$

and

$$(1+t)^p = 1 + O(t^p), \quad t \geq 0, \text{ if} p \in (0, 1].$$

From these inequalities, we obtain the following:

Lemma 6.1.1 *For any $a > 0$ and $b > 0$, one has*

$$(a+b)^p = a^p + pa^{p-1}b + O(b^p), \quad \text{if} p \in (1, 2],$$

$$(a+b)^p = a^p + pa^{p-1}b + \frac{p(p-1)}{2}a^{p-2}b^2 + O(b^p), \text{ if} p > 2, \quad (6.1.1)$$

and

$$(a+b)^p = a^p + O(b^p), \quad \text{if} p \in (0, 1]. \quad (6.1.2)$$

Lemma 6.1.2 *Suppose that $p > 1$. Then, for any $a > 0$ and $b > 0$, we have*

$$(a+b)^{p+1} - a^{p+1} - b^{p+1} - (p+1)a^p b - (p+1)ab^p \quad (6.1.3)$$
$$= \begin{cases} O\left(a^{\frac{p+1}{2}} b^{\frac{p+1}{2}}\right), & p \leq 2, \\ O\left(a^{p-1}b^2 + a^{\frac{p+1}{2}} b^{\frac{p+1}{2}}\right), & p > 2, \end{cases}$$

and

$$(a+b)^p - a^p - b^p = \begin{cases} O\big(a^{\frac{p}{2}} b^{\frac{p}{2}}\big), & p \le 2, \\ O\big(a^{p-1}b + ab^{p-1}\big), & p > 2. \end{cases} \tag{6.1.4}$$

Proof. Without loss of generality, we may assume $a \ge b$. Let $t = \frac{b}{a} \le 1$. If $p \le 2$, then we have

$$\begin{aligned}&(1+t)^{p+1} - 1 - t^{p+1} - (p+1)t - (p+1)t^p \\ &\quad = (1+t)^{p+1} - 1 - (p+1)t + O(t^p) = O(t^2 + t^p) = O(t^p) = O(t^{\frac{p+1}{2}}),\end{aligned}$$

which in turn implies that

$$(a+b)^{p+1} - a^{p+1} - b^{p+1} - (p+1)a^p b - (p+1)ab^p = O\big(a^{\frac{p+1}{2}} b^{\frac{p+1}{2}}\big).$$

If $p > 2$, it then follows from the calculation

$$\begin{aligned}&(1+t)^{p+1} - 1 - t^{p+1} - (p+1)t - (p+1)t^p \\ &= O(t^2 + t^p) = O(t^2 + t^{\frac{p+1}{2}})\end{aligned}$$

that

$$(a+b)^{p+1} - a^{p+1} - b^{p+1} - (p+1)a^p b - (p+1)ab^p = O\big(a^{p-1}b^2 + a^{\frac{p+1}{2}} b^{\frac{p+1}{2}}\big).$$

Thus we have proved (6.1.3). By a similar argument, we can also prove (6.1.4). □

Lemma 6.1.3 *Suppose that $u, v : \mathbb{R}^N \to \mathbb{R}$ are two non-negative continuous functions satisfying*

$$u(x) \le Ce^{-b|x|},\ v(x) \le Ce^{-b'|x|}, \quad (|x| \to \infty),$$

where $b' > b > 0$. Let $\xi \in \mathbb{R}^N$ with $|\xi| \to +\infty$. Then, one has

$$\int_{\mathbb{R}^N} u(x+\xi)v(x) \le Ce^{-b|\xi|}.$$

Proof. We have

$$\begin{aligned}\int_{\mathbb{R}^N} u(x+\xi)v(x) &\le C \int_{\mathbb{R}^N} e^{-b|x-\xi|} e^{-b'|x|} \\ &\le Ce^{-b|\xi|} \int_{\mathbb{R}^N} e^{-(b'-b)|x|} \le C'e^{-b|\xi|}.\end{aligned}$$

□

Lemma 6.1.4 *Suppose that $u, v : \mathbb{R}^N \to \mathbb{R}$ are two positive continuous functions satisfying*

$$|x|^a e^{b|x|} u(x) \to c_1 > 0, \quad |x|^{a'} e^{b'|x|} v(x) \to c_2 > 0, \quad as\ |x| \to \infty,$$

where $a, a' \in \mathbb{R}$, $b' > b > 0$. Let $\xi \in \mathbb{R}^N$ with $|\xi| \to +\infty$. Then there exists a constant $a_0 > 0$ such that

$$\int_{\mathbb{R}^N} u(x-\xi)v(x) = \big(a_0 + o(1)\big)e^{-b|\xi|}|\xi|^{-a}.$$

Proof. By the hypotheses of the lemma, we have

$$u(x) \le C(1+|x|)^{-a}e^{-b|x|}, \quad v(x) \le C(1+|x|)^{-a'}e^{-b'|x|}.$$

Let $\delta > 0$ be a small constant. Then it follows that

$$\begin{aligned}&\int_{\mathbb{R}^N\setminus B_{\delta|\xi|}(0)} u(x-\xi)v(x)\\ \le& Ce^{-b|\xi|}\int_{\mathbb{R}^N\setminus B_{\delta|\xi|}(0)} (1+|x-\xi|)^{-a}|x|^{-a'}e^{-(b'-b)|x|} \le Ce^{-(b+\frac{(b'-b)\delta}{2})|\xi|}.\end{aligned}$$

By rotation, we can assume $\frac{\xi}{|\xi|} = (1, 0, \cdots, 0)$. On the other hand, for any $x \in B_{\delta|\xi|}(0)$, one has

$$|x-\xi| = |\xi|\Big(1 - \Big\langle x, \frac{\xi}{|\xi|^2}\Big\rangle + O\Big(\frac{|x|^2}{|\xi|^2}\Big)\Big).$$

As a result, we have the following calculations:

$$\begin{aligned}&\int_{B_{\delta|\xi|}(0)} u(x-\xi)v(x)\,dx = \int_{B_{\delta|\xi|}(0)} (c_1+o(1))|x-\xi|^{-a}e^{-b|x-\xi|}v(x)\\ &= |\xi|^{-a}e^{-b|\xi|}\int_{B_{\delta|\xi|}(0)} (c_1+o(1))v(x)\Big(1 - \Big\langle x, \frac{\xi}{|\xi|^2}\Big\rangle + O\Big(\frac{|x|^2}{|\xi|^2}\Big)\Big)^{-a} e^{-b\langle x, \frac{\xi}{|\xi|}\rangle}\\ &\times \Big(1 + O\Big(\frac{|x|^2}{|\xi|}\Big)\Big)\,dx\\ &= |\xi|^{-a}e^{-b|\xi|}\Big(\int_{\mathbb{R}^N} v(x)e^{-bx_1} + o(1)\Big).\end{aligned}$$

Hence this lemma is proved, since

$$\int_{\mathbb{R}^N} |v(x)|e^{-bx_1} \le C\int_{\mathbb{R}^N} (1+|x|)^{-a'}e^{-(b'-b)|x|} < +\infty.$$

□

6.2 Pohozaev Identities and Applications

Pohozaev identity [123] has been used widely in the study of the non-existence of solutions for some non-compact elliptic problems. In this section, we provide several types of Pohozaev identities, which are used in Chapter 2 and Chapter 3.

Suppose that $f(x,t) \in C(\overline{D} \times \mathbb{R}, \mathbb{R})$, where D is a domain in $\mathbb{R}^N$. Set

$$F(x,u) = \int_0^u f(x,t)dt.$$

In this section, we will prove the following Pohozaev identities:

Theorem 6.2.1 *Suppose that $u \in C^2(\bar{D})$ is a solution of*

$$-\Delta u = f(x,u), \quad \text{in } D. \tag{6.2.1}$$

Then, for any bounded domain $\Omega \subset D$, one has the following identity:

$$\begin{aligned}
&\int_{\partial\Omega} \Big((\nabla u \cdot \nu)((x - x_0) \cdot \nabla u) - \frac{|\nabla u|^2}{2}(x - x_0) \cdot \nu \Big) + \frac{N-2}{2} \int_{\partial\Omega} u \frac{\partial u}{\partial \nu} \\
&\quad + \int_{\partial\Omega} F(x,u)((x - x_0) \cdot \nu) \\
&= N \int_{\Omega} F(x,u) + \frac{2-N}{2} \int_{\Omega} f(x,u)u + \int_{\Omega} (x - x_0) \cdot \nabla_x F(x,u),
\end{aligned} \tag{6.2.2}$$

and, for $i = 1, \cdots, N$, one has

$$\int_{\partial\Omega} \frac{\partial u}{\partial x_i} \frac{\partial u}{\partial \nu} - \frac{1}{2} \int_{\partial\Omega} |\nabla u|^2 \nu_i + \int_{\partial\Omega} F(x,u)\nu_i = \int_{\Omega} F_{x_i}(x,u), \tag{6.2.3}$$

where ν denotes the unit outward normal to the boundary $\partial\Omega$.

Proof. Without loss of generality, we suppose that $x_0 = 0$. Multiplying (6.2.1) by $x \cdot \nabla u$ and integrating on Ω, we obtain

$$-\int_{\Omega} \Delta u x \cdot \nabla u = \int_{\Omega} f(x,u) x \cdot \nabla u. \tag{6.2.4}$$

By the divergence theorem, we see that

$$\begin{aligned}
\int_{\Omega} f(x,u) x \cdot \nabla u &= \int_{\Omega} \Big(x \cdot \nabla F(x,u) - x \cdot \nabla_x F(x,u) \Big) \\
&= \int_{\partial\Omega} F(x,u)(x \cdot \nu) - \int_{\Omega} (N F(x,u) + x \cdot \nabla_x F(x,u)),
\end{aligned} \tag{6.2.5}$$

and

$$-\int_\Omega \Delta u x\cdot\nabla u = -\int_{\partial\Omega}\frac{\partial u}{\partial\nu}(x\cdot\nabla u) + \int_\Omega\Big(|\nabla u|^2 + \frac{1}{2}x\cdot\nabla(|\nabla u|^2)\Big)$$
$$= -\int_{\partial\Omega}\frac{\partial u}{\partial\nu}x\cdot\nabla u + \frac{1}{2}\int_{\partial\Omega}(x\cdot\nu)|\nabla u|^2 - \frac{N-2}{2}\int_\Omega|\nabla u|^2. \tag{6.2.6}$$

Combining (6.2.4), (6.2.5) and (6.2.6), we obtain

$$\begin{aligned}&-\int_{\partial\Omega}\frac{\partial u}{\partial\nu}x\cdot\nabla u + \frac{1}{2}\int_{\partial\Omega}(x\cdot\nu)|\nabla u|^2 - \frac{N-2}{2}\int_\Omega|\nabla u|^2\\ &= \int_{\partial\Omega}F(x,u)(x\cdot\nu) - \int_\Omega(NF(x,u) + x\cdot\nabla_x F(x,u)).\end{aligned} \tag{6.2.7}$$

On the other hand, one has

$$\begin{aligned}\int_\Omega|\nabla u|^2 &= -\int_\Omega u\Delta u + \int_{\partial\Omega}u\frac{\partial u}{\partial\nu}\\ &= \int_\Omega f(x,u)u + \int_{\partial\Omega}u\frac{\partial u}{\partial\nu}.\end{aligned} \tag{6.2.8}$$

Inserting (6.2.8) into (6.2.7), we obtain (6.2.2).

To prove (6.2.3), we multiply (6.2.1) by $\frac{\partial u}{\partial x_i}$ and integrate on Ω to obtain

$$-\int_\Omega \Delta u\frac{\partial u}{\partial x_i} = \int_\Omega f(x,u)\frac{\partial u}{\partial x_i}.$$

Using the divergence theorem, we then find that

$$\int_\Omega f(x,u)\frac{\partial u}{\partial x_i} = \int_{\partial\Omega}F(x,u)\nu_i - \int_\Omega F_{x_i}(x,u),$$

and

$$\begin{aligned}-\int_\Omega \Delta u\frac{\partial u}{\partial x_i} &= -\int_{\partial\Omega}\frac{\partial u}{\partial x_i}\frac{\partial u}{\partial\nu} + \sum_{j=1}^N\int_\Omega\frac{\partial u}{\partial x_j}\frac{\partial^2 u}{\partial x_j\partial x_i}\\ &= -\int_{\partial\Omega}\frac{\partial u}{\partial x_i}\frac{\partial u}{\partial\nu} + \frac{1}{2}\int_{\partial\Omega}|\nabla u|^2\nu_i.\end{aligned} \tag{6.2.9}$$

So (6.2.8) follows. □

Corollary 6.2.2 *Suppose that $u\in C^2(\bar{\Omega})$ is a solution of*

$$-\Delta u = f(x,u), \quad \text{in } \Omega,$$

and $u = 0$ on $\partial\Omega$. In the event that Ω is unbounded, further assume that

$$\int_\Omega |\nabla u|^2, \quad \int_\Omega |F(x,u)|, \quad \int_\Omega |(x-x_0)\cdot\nabla_x F(x,u)| < +\infty. \tag{6.2.10}$$

Then we have

$$\begin{aligned}\frac{1}{2}\int_{\partial\Omega}\left(\frac{\partial u}{\partial \nu}\right)^2 (x-x_0)\cdot\nu \\ = N\int_\Omega F(x,u) + \frac{2-N}{2}\int_\Omega f(x,u)u + \int_\Omega (x-x_0)\cdot\nabla_x F(x,u).\end{aligned} \tag{6.2.11}$$

Proof. Suppose that Ω is bounded. Then, noting that $u = 0$ on $\partial\Omega$, (6.2.2) gives (6.2.11).

Now suppose that Ω is unbounded. To prove (6.2.11), we first note from (6.2.10) that there exists $R_n \to +\infty$, such that

$$R_n\int_{\Omega\cap\partial B_{R_n}(x_0)} |\nabla u|^2 \to 0, \quad R_n\int_{\Omega\cap\partial B_{R_n}(x_0)} |F(x,u)| \to 0. \tag{6.2.12}$$

Replacing Ω by $\Omega\cap B_{R_n}(x_0)$ in (6.2.7), letting $n\to+\infty$ and taking (6.2.12) into account, we obtain

$$\begin{aligned}-\frac{1}{2}\int_{\partial\Omega}\left(\frac{\partial u}{\partial \nu}\right)^2 (x-x_0)\cdot\nu - \frac{N-2}{2}\int_\Omega |\nabla u|^2 \\ = -\int_\Omega (NF(x,u) + (x-x_0)\cdot\nabla_x F(x,u)).\end{aligned}$$

Moreover, one has

$$\int_\Omega |\nabla u|^2 = \int_\Omega u f(x,u).$$

Hence the result follows.

□

We now use (6.2.9) to prove the following result:

Proposition 6.2.3 *Let $x\in\Omega$ and let $G(y,x)$ be the Green's function of $-\Delta$ in Ω with homogenous Dirichlet boundary condition. If $B_d(x)\subset\subset\Omega$, then we have*

$$\int_{\partial B_d(x)} \frac{\partial G(y,x)}{\partial\nu}\frac{\partial G(y,x)}{\partial y_i} - \frac{1}{2}\int_{\partial B_d(x)} \nu_i |\nabla G(y,x)|^2 = \left.\frac{\partial H(y,x)}{\partial y_i}\right|_{y=x}.$$

Proof. For any $\theta > 0$, $G(y,x)$ is harmonic in $B_d(x)\setminus B_\theta(x)$. So from (6.2.9), we obtain

$$
\begin{aligned}
0 &= -\int_{B_d(x)\setminus B_\theta(x)} \Delta G(y,x)\frac{\partial G(y,x)}{\partial y_i} \\
&= -\int_{\partial(B_d(x)\setminus B_\theta(x))} \frac{\partial G(y,x)}{\partial \nu}\frac{\partial G(y,x)}{\partial y_i} + \frac{1}{2}\int_{\partial(B_d(x)\setminus B_\theta(x))} \nu_i|\nabla G(y,x)|^2.
\end{aligned}
$$

Hence it follows that

$$
\begin{aligned}
&\int_{\partial B_d(x)} \frac{\partial G(y,x)}{\partial \nu}\frac{\partial G(y,x)}{\partial y_i} - \frac{1}{2}\int_{\partial B_d(x)} \nu_i|\nabla G(y,x)|^2 \\
&= \int_{\partial B_\theta(x)} \frac{\partial G(y,x)}{\partial \nu}\frac{\partial G(y,x)}{\partial y_i} - \frac{1}{2}\int_{\partial B_\theta(x)} \nu_i|\nabla G(y,x)|^2.
\end{aligned}
$$

Define the quadratic form as follows:

$$
L(u,v,\theta) = \int_{\partial B_\theta(x)} \frac{\partial u}{\partial \nu}\frac{\partial v}{\partial y_i} - \frac{1}{2}\int_{\partial B_\theta(x)} \nu_i\langle \nabla u, \nabla v\rangle.
$$

Then we have

$$
\begin{aligned}
&L(G(y,x),G(y,x),\theta) \\
&\quad = L(S(y,x)-H(y,x),S(y,x)-H(y,x),\theta) \\
&\quad = L(S(y,x),S(y,x),\theta) - L(S(y,x),H(y,x),\theta) \\
&\qquad - L(H(y,x),S(y,x),\theta) + L(H(y,x),H(y,x),\theta),
\end{aligned}
$$

where

$$
S(x,P) = \frac{1}{(N-2)\omega_{N-1}}\frac{1}{|x-P|^{N-2}}.
$$

Using the symmetry, we find that

$$
L(S(y,x),S(y,x),\theta) = 0, \tag{6.2.13}
$$

and that as $\theta \to 0$,

$$
L(H(y,x),H(y,x),\theta) \to 0. \tag{6.2.14}
$$

On the other hand, one has

$$
\begin{aligned}
&\int_{\partial B_\theta(x)} \frac{\partial S(y,x)}{\partial \nu}\frac{\partial H(y,x)}{\partial y_i} - \frac{1}{2}\int_{\partial B_\theta(x)} \nu_i\langle \nabla S(y,x), \nabla H(y,x)\rangle \\
&= \frac{1}{\omega_{N-1}}\Big(-\int_{\partial B_\theta(x)} \frac{1}{\theta^{N-1}}\frac{\partial H(y,x)}{\partial y_i} + \frac{1}{2}\int_{\partial B_\theta(x)} \nu_i\Big\langle \frac{y-x}{\theta^N}, \nabla H(y,x)\Big\rangle\Big)
\end{aligned}
$$

$$
\begin{aligned}
&= \frac{1}{\omega_{N-1}}\Big(-\int_{\partial B_\theta(x)} \frac{1}{\theta^{N-1}} \frac{\partial H(y,x)}{\partial y_i}\Big|_{y=x} \\
&\quad + \frac{1}{2}\int_{\partial B_\theta(x)} \nu_i \Big\langle \frac{y-x}{\theta^N}, \nabla_y H(y,x)\Big|_{y=x}\Big\rangle\Big) \\
&\quad + o\Big(\int_{\partial B_\theta(x)} \frac{|y-x|}{\theta^{N-1}} + \frac{|y-x|^2}{\theta^N}\Big) \\
&\to -\Big(1-\frac{1}{2N}\Big)\frac{\partial H(y,x)}{\partial y_i}\Big|_{y=x}, \quad \text{as } \theta \to 0,
\end{aligned}
\tag{6.2.15}
$$

since

$$
\int_{\partial B_\theta(x)} \nu_i \nu_j = 0, \quad i \neq j
$$

and

$$
\int_{\partial B_\theta(x)} \nu_i^2 = \frac{1}{N}\int_{\partial B_\theta(x)} |\nu|^2 = \frac{\omega_{N-1}\theta^{N-1}}{N}.
$$

Moreover, we have

$$
\begin{aligned}
&\int_{\partial B_\theta(x)} \frac{\partial H(y,x)}{\partial \nu}\frac{\partial S(y,x)}{\partial y_i} - \frac{1}{2}\int_{\partial B_\theta(x)} \nu_i \langle \nabla H(y,x), \nabla S(y,x)\rangle \\
&= \frac{1}{\omega_{N-1}}\Big(-\int_{\partial B_\theta(x)} \frac{\nu_i}{\theta^{N-1}}\langle \nabla H(y,x), \nu\rangle + \frac{1}{2}\int_{\partial B_\theta(x)} \nu_i \Big\langle \frac{y-x}{\theta^N}, \nabla H(y,x)\Big\rangle\Big) \\
&= -\frac{1}{2\omega_{N-1}}\int_{\partial B_\theta(x)} \frac{\nu_i^2}{\theta^{N-1}}\frac{\partial H(y,x)}{\partial y_i}\Big|_{y=x} + o(1) \\
&\to -\frac{1}{2N}\frac{\partial H(y,x)}{\partial y_i}\Big|_{y=x}, \quad \text{as } \theta \to 0.
\end{aligned}
\tag{6.2.16}
$$

Therefore, the result follows from (6.2.13)–(6.2.16). □

Using (6.2.6) and (6.2.8), we have that

$$
\begin{aligned}
&-\int_\Omega \Delta u x \cdot \nabla u dx - \frac{N-2}{2}\int_\Omega u\Delta u \\
&= -\int_{\partial\Omega} \frac{\partial u}{\partial \nu}(x-x_0)\cdot \nabla u + \frac{1}{2}\int_{\partial\Omega} (x-x_0)\cdot \nu |\nabla u|^2 - \frac{N-2}{2}\int_{\partial\Omega} u\frac{\partial u}{\partial \nu}.
\end{aligned}
\tag{6.2.17}
$$

Now we use (6.2.17) to prove the following result:

Proposition 6.2.4 *Let $x \in \Omega$ and let $G(y,x)$ be the Green's function of $-\Delta$ in Ω with homogenous Dirichlet boundary condition. If $B_d(x) \subset\subset \Omega$, then*

$$
\begin{aligned}
&-\int_{\partial B_d(x)} \frac{\partial G(y,x)}{\partial \nu}(y-x)\cdot\nabla G(y,x) + \frac{1}{2}\int_{\partial B_d(x)} (y-x)\cdot\nu|\nabla G(y,x)|^2 \\
&-\frac{N-2}{2}\int_{\partial B_d(x)} G(y,x)\frac{\partial G(y,x)}{\partial \nu} \\
=&-\frac{N-2}{2}H(x,x).
\end{aligned}
$$

Proof. For any $\theta > 0$, $G(y,x)$ is harmonic in $B_d(x)\setminus B_\theta(x)$. By (6.2.17), we obtain

$$
\begin{aligned}
&-\int_{\partial B_d(x)} \frac{\partial G(y,x)}{\partial \nu}(y-x)\cdot\nabla G(y,x) + \frac{1}{2}\int_{\partial B_d(x)} (y-x)\cdot\nu|\nabla G(y,x)|^2 \\
&-\frac{N-2}{2}\int_{\partial B_d(x)} G(y,x)\frac{\partial G(y,x)}{\partial \nu} \\
=&-\int_{\partial B_\theta(x)} \frac{\partial G(y,x)}{\partial \nu}(y-x)\cdot\nabla G(y,x) + \frac{1}{2}\int_{\partial B_\theta(x)} (y-x)\cdot\nu|\nabla G(y,x)|^2 \\
&-\frac{N-2}{2}\int_{\partial B_\theta(x)} G(y,x)\frac{\partial G(y,x)}{\partial \nu}.
\end{aligned}
$$

Direct calculations show that

$$
\begin{aligned}
&-\int_{\partial B_\theta(x)} \frac{\partial S(y,x)}{\partial \nu}(y-x)\cdot\nabla S(y,x) + \frac{1}{2}\int_{\partial B_\theta(x)} (y-x)\cdot\nu|\nabla S(y,x)|^2 \\
&-\frac{N-2}{2}\int_{\partial B_\theta(x)} S(y,x)\frac{\partial S(y,x)}{\partial \nu} \\
&= 0.
\end{aligned}
$$

Hence it follows that

$$
\begin{aligned}
&-\int_{\partial B_\theta(x)} \frac{\partial G(y,x)}{\partial \nu}(y-x)\cdot\nabla G(y,x) + \frac{1}{2}\int_{\partial B_\theta(x)} (y-x)\cdot\nu|\nabla G(y,x)|^2 \\
&\quad-\frac{N-2}{2}\int_{\partial B_\theta(x)} G(y,x)\frac{\partial G(y,x)}{\partial \nu} \\
&=\frac{N-2}{2}\int_{\partial B_\theta(x)} H(y,x)\frac{\partial S(y,x)}{\partial \nu} + o_\theta(1) \\
&\to -\frac{N-2}{2}H(x,x).
\end{aligned}
$$

□

We now discuss (6.2.9) further. Let

$$
Q(u,\Omega) = -\int_{\partial\Omega} \frac{\partial u}{\partial x_i}\frac{\partial u}{\partial \nu} + \frac{1}{2}\int_{\partial\Omega} |\nabla u|^2 \nu_i.
$$

Then we have

$$Q(u, \Omega) = -\int_\Omega \Delta u \frac{\partial u}{\partial x_i},$$

which gives

$$Q(u+v, \Omega) = -\int_\Omega \Delta(u+v)\frac{\partial(u+v)}{\partial x_i}.$$

So we obtain

$$Q(u, v, \Omega) = -\int_\Omega \Delta u \frac{\partial v}{\partial x_i} - \int_\Omega \Delta v \frac{\partial u}{\partial x_i}, \tag{6.2.18}$$

where

$$Q(u, v, \Omega) = -\int_{\partial\Omega} \frac{\partial u}{\partial x_i}\frac{\partial v}{\partial \nu} - \int_{\partial\Omega} \frac{\partial v}{\partial x_i}\frac{\partial u}{\partial \nu} + \int_{\partial\Omega} \langle \nabla u, \nabla v\rangle \nu_i.$$

For any small $d > 0$, we define

$$Q(u, v, d) = Q(u, v, B_d(x)).$$

We can now prove the following result:

Proposition 6.2.5 *We have*

$$Q(G(x,y), \partial_h G(x,y), d) = -\frac{1}{2}\frac{\partial^2 \varphi(x)}{\partial x_i \partial x_h},$$

where $\partial_h G(x,y) = \frac{\partial G(x,y)}{\partial x_h}$.

Proof. Since $G(x,y)$ and $\partial_h G(x,y)$ are harmonic for $y \in B_d(x) \setminus B_\theta(x)$, it follows from (6.2.18) that $Q(G(x,y), \partial_h G(x,y), d)$ is independent of d.

Note that as $d \to 0$, $Q(S(x,y), \partial_h S(x,y), d)$ approaches to either infinity or zero, since the singularity involved is of order $\frac{1}{|y-x|^{2N-1}}$. On the other hand, $Q(S(x,y), \partial_h S(x,y), d)$ is a constant if $d > 0$ is fixed. Thus $Q(S(x,y), \partial_h S(x,y), d) = 0$ (this can also be checked directly). Therefore, we have

$$\begin{aligned} &Q(G(x,y), \partial_h G(x,y), d)\\ &= -Q(S(x,y), \partial_h H(x,y), d) - Q(H(x,y), \partial_h S(x,y), d) + o_d(1). \end{aligned} \tag{6.2.19}$$

A direct calculation shows that

$$Q(S(x,y), \partial_h H(x,y), d) = D_{y_i}\partial_h H(x,y)\Big|_{y=x} + o_d(1). \tag{6.2.20}$$

On the other hand, in view of

$$\begin{aligned}&\int_{\partial B_d(x)} \langle \nabla_y \partial_h S(x,y), \nu\rangle D_{y_i} H(x,y)\big|_{y=x}\\&\quad+\int_{\partial B_d(x)} \langle \nabla_y H(x,y)\big|_{y=x}, \nu\rangle D_{y_i}\partial_h S(x,y)\\&\quad-\int_{\partial B_d(x)} \langle \nabla_y H(x,y)\big|_{y=x}, \nabla_y \partial_h S(x,y)\rangle \nu_i = 0,\end{aligned}$$

we see that

$$\begin{aligned}&Q(H(x,y), \partial_h S(x,y), d)\\&= -\int_{\partial B_d(x)} \langle \nabla_y \partial_h S(x,y), \nu\rangle \Big(D_{y_i} H(x,y) - D_{y_i} H(x,y)\big|_{y=x}\Big)\\&\quad-\int_{\partial B_d(x)} \Big(\langle \nabla_y H(x,y), \nu\rangle - \langle \nabla_y H(x,y)\big|_{y=x}, \nu\rangle\Big) D_{y_i}\partial_h S(x,y)\\&\quad+\int_{\partial B_d(x)} \langle \nabla_y H(x,y) - \nabla_y H(x,y)\big|_{y=x}, \nabla_y \partial_h S(x,y)\rangle \nu_i\\&= -\int_{\partial B_d(x)} \langle \nabla_y \partial_h S(x,y), \nu\rangle \langle \nabla_y D_{y_i} H(x,y)\big|_{y=x}, y-x\rangle\\&\quad-\int_{\partial B_d(x)} \langle \nabla_y^2 H(x,y)\big|_{y=x}(y-x), \nu\rangle D_{y_i}\partial_h S(x,y)\\&\quad+\int_{\partial B_d(x)} \langle \nabla_y^2 H(x,y)\big|_{y=x}(y-x), \nabla_y \partial_h S(x,y)\rangle \nu_i + o_d(1)\\&= J_1 + J_2 + J_3 + o_d(1).\end{aligned}$$

Recall that

$$S(y,x) = \frac{1}{(N-2)\omega_{N-1}} \frac{1}{|y-x|^{N-2}}.$$

Let $z = y - x$. We then have

$$\partial_h S(x,y) = \frac{1}{\omega_{N-1}} \frac{z_h}{|z|^N},$$

$$\langle \nabla_y \partial_h S(x,y), \nu\rangle = -\frac{N-1}{\omega_{N-1}} \frac{z_h}{|z|^{N+1}}$$

and

$$D_{y_t}\partial_h S(x,y) = \frac{1}{\omega_{N-1}}\Big(\frac{\delta_{ht}}{|z|^N} - \frac{N z_t z_h}{|z|^{N+2}}\Big).$$

Thus we obtain

$$J_1 = \frac{N-1}{\omega_{N-1}} \int_{|z|=d} \frac{z_h}{|z|^{N+1}} D^2_{y_h y_i} H(x,y)\big|_{y=x} z_h = \frac{N-1}{N} D^2_{y_h y_i} H(x,y)\big|_{y=x}. \tag{6.2.21}$$

On the other hand, one has

$$
\begin{aligned}
&J_2 + J_3 \\
&= -\frac{1}{\omega_{N-1}} \int_{|z|=d} \sum_{l=1}^{N} \sum_{t=1}^{N} D^2_{y_l y_t} H(x, y)\big|_{y=x} \frac{z_l z_t}{|z|} \left(\frac{\delta_{hi}}{|z|^N} - \frac{N z_i z_h}{|z|^{N+2}} \right) \\
&\quad + \frac{1}{\omega_{N-1}} \int_{|z|=d} \sum_{l=1}^{N} \sum_{t=1}^{N} D^2_{y_l y_t} H(x, y)\big|_{y=x} z_l \left(\frac{\delta_{ht}}{|z|^N} - \frac{N z_t z_h}{|z|^{N+2}} \right) \frac{z_i}{|z|} \\
&= -\frac{1}{\omega_{N-1}} \int_{|z|=d} \sum_{l=1}^{N} \sum_{t=1}^{N} D^2_{y_l y_t} H(x, y)\big|_{y=x} \frac{\delta_{hi} z_l z_t}{|z|^{N+1}} \\
&\quad + \frac{1}{\omega_{N-1}} \int_{|z|=d} \sum_{l=1}^{N} \sum_{t=1}^{N} D^2_{y_l y_t} H(x, y)\big|_{y=x} \frac{\delta_{ht} z_l z_i}{|z|^{N+1}}.
\end{aligned}
$$

If $i \neq h$, then we have

$$
\begin{aligned}
J_2 + J_3 &= \frac{1}{\omega_{N-1}} \int_{|z|=d} \sum_{l=1}^{N} \sum_{t=1}^{N} D^2_{y_l y_t} H(x, y)\big|_{y=x} \frac{\delta_{ht} z_l z_i}{|z|^{N+1}} \\
&= \frac{1}{\omega_{N-1}} \int_{|z|=d} D^2_{y_i y_h} H(x, y)\big|_{y=x} \frac{z_i^2}{|z|^{N+1}} = \frac{1}{N} D^2_{y_i y_h} H(x, y)\big|_{y=x}.
\end{aligned} \tag{6.2.22}
$$

Combining (6.2.21) and (6.2.22), we obtain

$$
J_1 + J_2 + J_3 = D^2_{y_i y_h} H(x, y)\big|_{y=x}, \quad i \neq h. \tag{6.2.23}
$$

If $i = h$, then

$$
\begin{aligned}
J_2 + J_3 &= -\frac{1}{\omega_{N-1}} \int_{|z|=d} \sum_{l=1}^{N} \sum_{t=1}^{N} D^2_{y_l y_t} H(x, y)\big|_{y=x} \frac{z_l z_t}{|z|^{N+1}} \\
&\quad + \frac{1}{\omega_{N-1}} \int_{|z|=d} \sum_{l=1}^{N} \sum_{t=1}^{N} D^2_{y_l y_t} H(x, y)\big|_{y=x} \frac{\delta_{ht} z_l z_i}{|z|^{N+1}} \\
&= -\frac{1}{N} \sum_{l=1}^{N} D_{y_l y_l} H(x, y)\big|_{y=x} + \frac{1}{N} D^2_{y_i y_h} H(x, y)\big|_{y=x} \\
&= \frac{1}{N} D^2_{y_i y_h} H(x, y)\big|_{y=x},
\end{aligned}
$$

since

$$
\sum_{l=1}^{N} D_{y_l y_l} H(x, y)\big|_{y=x} = \Delta_y H(x, y)\big|_{y=x} = 0, \quad \forall x \in \Omega.
$$

In conclusion, we have

$$J_1 + J_2 + J_3 = D^2_{y_i y_i} H(x, y)\big|_{y=x}, \quad i = h. \tag{6.2.24}$$

Combining (6.2.19), (6.2.20), (6.2.23) and (6.2.24), we have

$$\begin{aligned} & Q(G(x, y), \partial_h G(x, y), d) \\ = & - D^2_{y_i y_h} H(x, y)\big|_{y=x} - D_{y_i} \partial_h H(x, y)\big|_{y=x} \\ = & - \frac{1}{2} \frac{\partial^2 \varphi(x)}{\partial x_i \partial x_h}. \end{aligned}$$

□

6.3 Sobolev Spaces

In the section, we discuss briefly the Sobolev spaces. General reference to this topic can be found in works by Adams [1], Evans [74] or Gilbarg–Trudinger [77].

6.3.1 Sobolev Spaces

Let Ω be a domain in $\mathbb{R}^N$. For any multi-index $(\alpha_1, \ldots, \alpha_n)$, where each α_j is a non-negative integer, we write $|\alpha| = \sum_{j=1}^n \alpha_j$. Then for any function $u \in C^{|\alpha|}(\Omega)$, we have

$$\int_\Omega \varphi D^\alpha u = (-1)^{|\alpha|} \int_\Omega u D^\alpha \varphi, \quad \forall \varphi \in C_0^\infty(\Omega).$$

From this relation, we can define the weak derivatives for a function u as follows: Let u be a locally integrable function in Ω. If a locally integrable function v satisfies

$$\int_\Omega \varphi v = (-1)^{|\alpha|} \int_\Omega u D^\alpha \varphi, \quad \forall \varphi \in C_0^\infty(\Omega),$$

then v is said to be the αth order weak derivative of u, which we denote by $v = D^\alpha u$.

For any non-negative integer k and $1 \le p \le \infty$, we define the Sobolev space to be

$$W^{k,p}(\Omega) = \Big\{ u \in L^p(\Omega) : D^\alpha u \in L^p(\Omega) \text{ for all } \alpha : |\alpha| \le k \Big\},$$

with a norm given by

$$\|u\|_{W^{k,p}(\Omega)} = \Big(\sum_{|\alpha| \le k} \int_\Omega |D^\alpha u|^p \Big)^{\frac{1}{p}},$$

if $1 \le p < +\infty$, and

$$\|u\|_{W^{k,\infty}(\Omega)} = \sum_{|\alpha| \le k} \|D^\alpha u\|_{L^\infty},$$

if $p = \infty$.

From the properties of $L^p(\Omega)$, we can prove the following result:

Theorem 6.3.1 *For any non-negative integer k and* $1 \le p \le \infty$, $W^{k,p}(\Omega)$ *is a Banach space. Furthermore, the Banach space* $W^{k,p}(\Omega)$ *is reflexive if and only if* $1 < p < \infty$.

When $p = 2$, we let $H^k(\Omega)$ denote the space $W^{k,2}(\Omega)$. $H^k(\Omega)$ is a Hilbert space with inner product

$$\langle u, v \rangle_{H^k(\Omega)} = \sum_{|\alpha| \le k} \int_\Omega D^\alpha u D^\alpha v.$$

Finally, write $W_0^{k,p}(\Omega)$ for the closure of $C_0^\infty(\Omega)$ in $W^{k,p}(\Omega)$, and write $\mathcal{D}^{k,p}(\Omega)$ for the closure of $C_0^\infty(\Omega)$ with respect to the norm

$$\|u\|^p_{\mathcal{D}^{k,p}} = \sum_{|\alpha| = k} \|D^\alpha u\|^p_{L^p}.$$

To facilitate the discussion of the embedding theorem, we need to introduce the Hölder space. A function $u : \Omega \subset \mathbb{R}^N \to \mathbb{R}$ is Hölder continuous with exponent $\beta > 0$ if

$$[u]^{(\beta)} = \sup_{x \neq y \in \Omega} \frac{|u(x) - u(y)|}{|x - y|^\beta} < \infty.$$

For any non-negative integer m and a number β with $0 < \beta \le 1$, denote

$$C^{m,\beta}(\Omega) = \{u \in C^m(\Omega) : D^\alpha u \text{ is Hölder continuous with exponent } \beta \text{ for all } \alpha : |\alpha| = m\}.$$

If Ω is relatively compact, $C^{m,\beta}(\bar{\Omega})$ becomes a Banach space under the norm

$$\|u\|_{C^{m,\beta}} = \sum_{|\alpha| \le m} \|D^\alpha u\|_{L^\infty} + \sum_{|\alpha| = m} [D^\alpha u]^{(\beta)}.$$

6.3.2 Density Theorem

Now we turn to discuss the approximation of a function $u \in W^{k,p}$ by smooth functions. For this purpose, let us fix a function $\rho(x) = \rho(|x|) \ge 0$ such that

$\rho \in C_0^\infty(B_1(0))$, $\rho'(r) \le 0$ and $\int_{\mathbb{R}^N} \rho(x) = 1$. We define the operator from $L^1_{loc}(\mathbb{R}^N)$ to $C^\infty(\mathbb{R}^N)$ as follows:

$$u_h(x) = h^{-N} \int_{\mathbb{R}^N} \rho\Big(\frac{y-x}{h}\Big) u(y)\, dy, \quad u \in L^1_{loc}(\mathbb{R}^N). \tag{6.3.1}$$

For any $u \in L^1(\Omega)$, we can let $u(x) = 0$ for $x \in \mathbb{R}^N \setminus \Omega$. Note that for small $h > 0$, the function $h^{-N}\rho(\frac{y-x}{h})$ converges to the Dirac measure δ_x. So it is easy to prove the following result:

Lemma 6.3.2 *As $h \to 0$, the following statements hold.*

(i) If $u \in C^0(\Omega)$, then $u_h \to u$ uniformly in any $\Omega' \subset\subset \Omega$.
(ii) If $u \in L^p(\Omega)$, $p < +\infty$, then $u_h \to u$ in $L^p(\Omega)$.

For the proof, the readers can refer to Lemmas 7.1 and 7.2 in [77]. It is also easy to prove the following result (see Lemma 7.3 in [77]).

Lemma 6.3.3 *Suppose that $u \in L^1_{loc}(\Omega)$. If $D^\alpha u$ exists, then for any $x \in \Omega$ with $d(x, \partial\Omega) \ge h$, one has*

$$(D^\alpha u_h)(x) = (D^\alpha u)_h(x).$$

With Lemmas 6.3.2 and 6.3.3, we can prove the following density result (see Theorem 7.9 in [77]):

Theorem 6.3.4 *For any non-negative integer k and $1 \le p < \infty$, the subspace $W^{k,p} \cap C^\infty(\Omega)$ is dense in $W^{k,p}(\Omega)$.*

By Theorem 6.3.4, Sobolev functions can be approximated by functions enjoying any degree of smoothness in the interior of Ω. Some regularity condition on the boundary $\partial\Omega$ is necessary if smoothness up to the boundary is required.

Theorem 6.3.5 *Let k be a non-negative integer and $1 \le p < \infty$. Suppose that $\Omega \subset \mathbb{R}^N$ is a bounded domain of class C^1. Then $C^\infty(\bar{\Omega})$ is dense in $W^{k,p}(\Omega)$.*

For the proof, the reader can refer to Theorem 3.18 in [1].

6.3.3 Embedding Theorems

By definition, $L^p(\Omega) \subset W^{k,p}(\Omega)$. The following example shows that a function in $W^{k,p}(\Omega)$ may have better property. Let $u(x) = \frac{1}{|x|^\alpha}$, $x \in B_1(0)$. Then

$u \in W^{1,p}(B_1(0))$ if and only if $\alpha < \frac{N-p}{p}$. Consequently, for any $q \leq \frac{Np}{N-p}$, we have $u \in L^q(B_1(0))$, since $q\alpha < N$. The following embedding theorem states that this property is true for all the function in $W^{1,p}(\Omega)$.

Theorem 6.3.6 *(Sobolev embedding theorem) Let $\Omega \subset \mathbb{R}^N$ be a domain in $\mathbb{R}^N$ with a Lipschitz boundary. Then the following holds:*
(i) If $kp < N$, then $W^{k,p}(\Omega) \subset L^q(\Omega)$ for $1 \leq q \leq \frac{Np}{N-kp}$, and there exists a constant $C > 0$ such that

$$\|u\|_{L^p(\Omega)} \leq C\|u\|_{W^{k,p}(\Omega)}. \tag{6.3.2}$$

(ii) If $0 \leq m < k - \frac{N}{p} < m+1$, then $W^{k,p}(\Omega) \subset C^{m,\alpha}(\Omega)$ for $0 \leq \alpha \leq k - m - \frac{N}{p}$. If Ω is bounded, then $\|u\|_{C^{m,\alpha}(\Omega)} \leq C\|u\|_{W^{k,p}(\Omega)}$.

For the proof, see Theorem 5.4 in [1]. Note that by the density theorem, we only need to prove that (6.3.2) holds for any $u \in C^\infty(\Omega) \cap W^{k,p}(\Omega)$.

From the Sobolev embedding theorem, we have an embedding operator $i(u) = u$ from $W^{k,p}(\Omega)$ to $L^p(\Omega)$ for $1 \leq q \leq \frac{Np}{N-kp}$, if $kp < N$. If $kp > N$ and Ω is bounded, we have an embedding operator $i(u) = u$ from $W^{k,p}(\Omega)$ to $C^{m,\alpha}(\Omega)$. In both cases, the embedding operator is bounded.

Theorem 6.3.7 *(Compact Sobolev embedding theorem) Let $\Omega \subset \mathbb{R}^N$ be a bounded domain with a Lipschitz boundary. Then the following holds:*
(i) If $kp < N$, then the embedding operator i from $W^{k,p}(\Omega)$ to $L^q(\Omega)$ is compact for $q < \frac{Np}{N-kp}$.
(ii) If $0 \leq m < k - \frac{N}{p} < m+1$, then the embedding operator i from $W^{k,p}(\Omega)$ to $C^{m,\alpha}(\bar{\Omega})$ is compact for $\alpha < k - m - \frac{N}{p}$.

Compactness of the embedding $W^{k,p}(\Omega) \hookrightarrow L^q(\Omega)$ for $q < \frac{Np}{N-kp}$ is a consequence of Rellich-Kondrakov theorem. See [1, Theorem 6.2].

Theorem 6.3.6 is valid for $W_0^{k,p}(\Omega)$ -spaces on arbitrary bounded domains Ω. We can also prove the following inequality:

$$\|u\|_{L^{\frac{Np}{N-p}}(\mathbb{R}^N)} \leq C\|\nabla u\|_{L^p(\mathbb{R}^N)}, \quad \forall\, u \in C_0^\infty(\mathbb{R}^N),$$

whenever $p < N$ (see Theorem 7.16 in [77]). This shows that $\mathcal{D}^{1,p}(\Omega)$ is embedded in $L^{\frac{Np}{N-p}}(\mathbb{R}^N)$.

Before we end this subsection, it should be noted that in Theorem 6.3.7, the condition imposed is necessary.

Example 6.3.8 The embedding from $W^{k,p}(\mathbb{R}^N)$ to $L^q(\mathbb{R}^N)$ is not compact. To see this, we take two functions, $u,\ v \in C_0^\infty(B_1(0))$ and $v \neq 0$. We fix a vector $e \in \mathbb{R}^N$. We define

$$w_n(x) = u(x) + v(x + ne).$$

Then w_n is bounded in $W^{k,p}(\mathbb{R}^N)$. Suppose that w_n has a convergent subsequence in $L^q(\mathbb{R}^N)$. Then the limit must be u. But any subsequence of w_n does not converge strongly in $L^q(\mathbb{R}^N)$ to u. So w_n is not compact in $L^q(\mathbb{R}^N)$. In this case, the sequence w_n is not compact because part of w_n moves to infinity.

Example 6.3.9 The embedding from $W^{1,p}(\Omega)$ to $L^{\frac{Np}{N-p}}(\Omega)$ is not compact. To see this, we take a function $u \in C_0^\infty(B_1(0))$ and $u \neq 0$. Fix a point $x_0 \in \Omega$ and choose a positive sequence $\rho_n \to 0$. Define

$$w_n(x) = \rho_n^{-\frac{N-p}{p}} u\Big(\frac{x - x_0}{\rho_n}\Big).$$

Then for n large, we have $w_n \in W_0^{1,p}(\Omega)$ and

$$\|\nabla w_n\|_{L^p(\Omega)} = \|\nabla u\|_{L^p(B_1(0))}, \quad \|w_n\|_{L^{\frac{Np}{N-p}}(\Omega)} = \|u\|_{L^{\frac{Np}{N-p}}(B_1(0))}.$$

Note that $w_n \to 0$ for all $x \in \Omega \setminus \{x_0\}$. Suppose that w_n has a convergent subsequence in $L^{\frac{Np}{N-p}}(\Omega)$. Then the limit must be 0. This is impossible since $\|w_n\|_{L^{\frac{Np}{N-p}}(\Omega)} = \|u\|_{L^{\frac{Np}{N-p}}(B_1(0))}$. In conclusion, w_n is not compact in $L^{\frac{Np}{N-p}}(\Omega)$. In this example, the sequence w_n is not compact because w_n concentrates at a point $x_0 \in \Omega$.

6.3.4 The Trace of the Functions in $W_0^{1,p}(\Omega)$

Recall that $W_0^{1,p}(\Omega)$ is the closure of $C_0^\infty(\Omega)$ in $W^{1,p}(\Omega)$. Noting that any function $u \in C_0^\infty(\Omega)$ satisfies $D^\alpha u = 0$ on $\partial\Omega$, we can expect that $u = 0$ on $\partial\Omega$ in some sense if $u \in W_0^{1,p}(\Omega)$. The following theorem shows that one usually has $\nabla u \neq 0$ on $\partial\Omega$ if $u \in W_0^{1,p}(\Omega)$.

Theorem 6.3.10 *Let Ω be a bounded domain in $\mathbb{R}^N$ with C^1 boundary. For any function $u \in C^1(\Omega)$ satisfying $u = 0$ on $\partial\Omega$, there exists a sequence of $u_n \in C_0^\infty(\Omega)$ such that $u_n \to u$ in $W^{1,p}(\Omega)$. Thus, $u \in W_0^{1,p}(\Omega)$.*

Proof. Since $\partial\Omega$ is C^1, we can find a C^1 function ξ_h such that $\xi(x) = 1$ for $x \in \Omega$ with $d(x, \partial\Omega) \geq 2h > 0$, $\xi(x) = 0$ for $x \in \Omega$ with $d(x, \partial\Omega) \leq h$, $0 \leq$

$\xi_h \leq 1$ and $|\nabla \xi_h| \leq \frac{C}{h}$ for some constant $C > 0$. Define $u_h = \xi u \in C_0^1(\Omega)$. Then $u_h \to u$ strongly in $L^p(\Omega)$. Moreover, we have

$$\int_\Omega |\nabla u_h - \nabla u|^p \leq C \int_\Omega |1 - \xi_h|^p |\nabla u|^p + C \int_\Omega |\nabla \xi_h|^p |u|^p.$$

Note that $1 - \xi_h = 0$ and $\nabla \xi_h = 0$ for any $x \in \Omega$ with $d(x, \partial\Omega) \geq 2h$. Set $D_h = \{x : x \in \Omega, d(x, \partial\Omega) \leq 2h\}$. In D_h, it holds $|\nabla \xi_h| \leq \frac{C}{h}$ and $|u(x)| \leq C d(x, \partial\Omega) \leq 2Ch$. So, as $h \to 0$,

$$\int_\Omega |\nabla u_h - \nabla u|^p \leq C|D_h| \to 0.$$

Thus $u_h \to u$ strongly in $W^{1,p}(\Omega)$ as $h \to 0$. Now we can approximate each u_h by a $C_0^\infty(\Omega)$ function v_h to finish the proof. This approximation is possible by (6.3.1). □

In the following, we briefly discuss the trace of a function $u \in W^{1,p}(\Omega)$. Let Ω be a bounded domain in $\mathbb{R}^N$ with C^1 boundary and let $\nu = \nu(x)$ be the outward unit normal of $\partial\Omega$ at x. Take a small number $t > 0$ and define

$$\Omega_t = \left\{x : x \in \Omega, d(x, \partial\Omega) < t\right\}.$$

Take a function $\xi \geq 0$, which satisfies $\xi = 1$ in $\Omega_{\frac{t}{2}}$ and $\xi = 0$ in $\Omega \setminus \Omega_t$. For any $u \in C^1(\bar{\Omega})$ and $x \in \partial\Omega$, we have

$$\begin{aligned} |u(x)| = &|\xi(x) u(x)| = \left| - \int_0^t \frac{d(\xi u)(x - s\nu(x))}{ds} ds \right| \\ \leq &C \left(\int_0^t \left| \frac{d(\xi u)(x - s\nu(x))}{ds} \right|^p ds \right)^{\frac{1}{p}}. \end{aligned}$$

Thus

$$\int_{\partial\Omega} |u(x)|^p \leq C \int_{\partial\Omega} \int_0^t \left| \frac{d(\xi u)(x - s\nu(x))}{ds} \right|^p ds \leq C' \int_\Omega \left(|\nabla u|^p + |u|^p\right). \tag{6.3.3}$$

Now, for any function $u \in W^{1,p}(\Omega)$, it follows from Theorem 6.3.5 that we can find a sequence of $u_m \in C^1(\bar{\Omega})$ such that $u_m \to u$ strongly in $W^{1,p}(\Omega)$. By (6.3.3), this in turn implies that

$$\int_{\partial\Omega} |u_m(x) - u_n(x)|^p \leq C \int_\Omega \left(|\nabla(u_m - u_n)|^p + |u_m - u_n|^p\right) \to 0,$$

as $m, n \to +\infty$. Therefore, u_m converges strongly in $L^p(\partial\Omega)$. This limit is the trace of u on $\partial\Omega$. In particular, any $u \in W_0^{1,p}(\Omega)$ has trace zero on $\partial\Omega$, since $u_m = 0$ on $\partial\Omega$.

6.3.5 An Inequality and Applications

Let $L(u)$ be a bounded linear functional in $W^{1,p}(\Omega)$ with the property that $L(1) \neq 0$. In this subsection, we will prove the following result:

Theorem 6.3.11 *Suppose that $1 < p < +\infty$. For any bounded domain Ω of class C^1, there exists a constant $C = C(\Omega)$ such that for any $u \in W^{1,p}(\Omega)$, we have the following inequality:*

$$\|u\|_{L^p(\Omega)} \leq C\|\nabla u\|_{L^p(\Omega)} + C|L(u)|.$$

Proof. Suppose on the contrary that there exists a sequence $\{u_n\}$ in $W^{1,p}(\Omega)$ such that

$$\|u_n\|_{L^p(\Omega)} \geq n(\|\nabla u_n\|_{L^p(\Omega)} + |L(u_n)|). \tag{6.3.4}$$

By homogeneity, we may normalize

$$\|u_n\|_{L^p(\Omega)} = 1.$$

But then $\{u_n\}$ is bounded in $W^{1,p}(\Omega)$, and we may assume that $u_n \rightharpoonup u$ weakly. Moreover, by (i) of Theorem 6.3.7, $u_n \to u$ strongly in $L^p(\Omega)$. But (6.3.4) also implies that $\nabla u_n \to 0$ in $L^p(\Omega)$ and $L(u_n) \to 0$. Hence, by the weakly convergence of u_n in $W^{1,p}(\Omega)$, we have $\nabla u = 0$ and $L(u) = 0$. Let u_h be the operator defined as in (6.3.1). Then $\nabla u_h = 0$ in any compact subset D of Ω if $h > 0$ is small. It follows from this that $u_h = C_h$ in D for some constant C_h. Since $u_h \to u$ strongly in $L^p(D)$, we conclude that u is constant in any compact subset of Ω. This in turn implies that $u = C$ in Ω.

Since $\|u\|_{L^p(\Omega)} = 1$, we see that $u = C \neq 0$. As a result, we have $L(u) = L(C) = CL(1) \neq 0$. This is a contradiction.

□

Example 6.3.12 Let $L(u) = \int_\Omega u$. Then $L(1) \neq 0$. By Theorem 6.3.11, we then have

$$\|u\|_{L^p(\Omega)} \leq C\|\nabla u\|_{L^p(\Omega)} + C\left|\int_\Omega u\right|.$$

Define $\bar{u} = \frac{1}{|\Omega|}\int_\Omega u$. If we take $v = u - \bar{u}$ in the above inequality, we obtain the following Poincaré inequality:

$$\|u - \bar{u}\|_{L^p(\Omega)} \leq C\|\nabla u\|_{L^p(\Omega)}.$$

Example 6.3.13 Let $L(u) = \int_{\partial\Omega} u$. Then $L(1) \neq 0$. It then follows from Theorem 6.3.11 that

$$\|u\|_{L^p(\Omega)} \leq C\|\nabla u\|_{L^p(\Omega)} + C\left|\int_{\partial\Omega} u\right|.$$

In particular, if $u \in W_0^{1,p}(\Omega)$, we have

$$\|u\|_{L^p(\Omega)} \leq C\|\nabla u\|_{L^p(\Omega)}.$$

6.4 Some Fundamental Estimates for Elliptic Equations

In this section, we recall some fundamental estimates for elliptic equations. For simplicity, we just consider second-order elliptic differential operators of the form

$$Lu = -\Delta u + cu, \quad \text{in } \Omega,$$

where Ω is a domain in $\mathbb{R}^N$, and $c \in C(\bar{\Omega})$.

6.4.1 Maximum Principle and Comparison Theorem

If $u \in C^2(\Omega)$ solves $Lu = f$, then one has

$$\int_\Omega (\nabla u \cdot \nabla\varphi + cu\varphi) - \int_\Omega f\varphi = 0, \text{ for all } \varphi \in C_0^\infty(\Omega).$$

Based on the same spirit for the definition of the weak derivatives, we can make the following definition that a function $u \in H^1(\Omega)$ is said to weakly solve the equation $Lu = f$ if

$$\int_\Omega (\nabla u \cdot \nabla\varphi + cu\varphi) - \int_\Omega f\varphi = 0, \text{ for all } \varphi \in H_0^1(\Omega).$$

If $u \in H_0^1(\Omega)$, then this is equivalent to weakly solving the problem $Lu = f$ in Ω with boundary condition $u = 0$ on $\partial\Omega$. The integral

$$\mathcal{L}(u, \varphi) = \int_\Omega (\nabla u \cdot \nabla\varphi + cu\varphi)$$

is a symmetric bilinear form on $H_0^1(\Omega)$, which is called the Dirichlet form associated with the operator L.

Theorem 6.4.1 *Suppose that the Dirichlet form of $\mathcal{L}$ is positive definite on $H_0^1(\Omega)$ in the sense that*

$$\mathcal{L}(u, u) > 0, \textit{ for all } u \in H_0^1(\Omega) \textit{ with } u \neq 0.$$

If $u \in H_0^1(\Omega)$ weakly satisfies $Lu \geq 0$ in the sense that

$$\mathcal{L}(u, \varphi) \geq 0, \text{ for all non-negative } \varphi \in H_0^1(\Omega),$$

then $u \geq 0$ in Ω.

Proof. Choose $\varphi = u_- = \max\{-u, 0\} \in H_0^1(\Omega)$. Then

$$0 \leq \mathcal{L}(u, u_-) = -\mathcal{L}(u_-, u_-) \leq 0,$$

with equality if and only if $u_- \equiv 0$. This shows that $u \geq 0$. □

Consider

$$\inf\left\{\mathcal{L}(u, u) : u \in H_0(\Omega), \int_\Omega u^2 = 1\right\}. \tag{6.4.1}$$

Then (6.4.1) is achieved by a non-negative function φ_1, which weakly solves $L\varphi_1 = \lambda_1\varphi$ in Ω for some λ_1. The number λ_1 is the first eigenvalue of the operator L in Ω with homogenous Dirichlet boundary condition.

The condition in Theorem 6.4.1 is that the first eigenvalue of L is positive. Let us point out if one can find a function $h \in H_0^1(\Omega)$ such that $h > 0$ and

$$\mathcal{L}(h, \varphi) > 0 \text{ for all non-negative } \varphi \in H_0^1(\Omega),$$

then the first eigenvalue λ_1 must be positive. To see this, let $\varphi_1 \geq 0$ be the eigenfunction corresponding to λ_1. Then we have

$$0 < \mathcal{L}(h, \varphi_1) = \lambda_1 \int_\Omega h\varphi_1,$$

which gives $\lambda_1 > 0$.

Now we state the following classical maximum principle [77, Theorem 3.5].

Theorem 6.4.2 *Suppose that $u \in C^2(\Omega) \cap C^1(\bar{\Omega})$ satisfies*

$$Lu \geq 0, \text{ in } \Omega.$$

If $c \geq 0$ in Ω, then u cannot achieve its non-positive minimum in the interior of Ω unless u is a constant. In particular, if $u \geq 0$ on $\partial\Omega$, then $u \geq 0$ in Ω.

A direct consequence of the maximum principle is the following comparison theorem:

Theorem 6.4.3 *Suppose that $u, v \in C^2(\Omega) \cap C^1(\bar{\Omega})$, u satisfies $Lu \geq 0$ in Ω, $u \geq v$ on $\partial\Omega$, and v satisfies $Lv \leq 0$ in Ω. If $c \geq 0$ in Ω, then $u \geq v$ in Ω.*

6.4.2 Schauder Estimates

The following theorem can be found in [77, Theorems 6.2, 6.6].

Theorem 6.4.4 *Suppose that $u \in C^2(\Omega)$ solves $Lu = f$ and $f \in C^\alpha(\bar{\Omega})$. Then $u \in C^{2,\alpha}(\Omega)$, and for any $\Omega' \subset\subset \Omega$, we have*

$$\|u\|_{C^{2,\alpha}(\Omega')} \leq C(\|u\|_{L^\infty(\Omega)} + \|f\|_{C^\alpha(\bar{\Omega})}).$$

In addition, if Ω is of class $C^{2+\alpha}$ and $u \in C^0(\bar{\Omega})$ coincides with a function $u_0 \in C^{2+\alpha}(\bar{\Omega})$ on $\partial\Omega$, then $u \in C^{2,\alpha}(\bar{\Omega})$ and

$$\|u\|_{C^{2,\alpha}(\bar{\Omega})} \leq C(\|u\|_{L^\infty(\Omega)} + \|f\|_{C^\alpha(\bar{\Omega})} + \|u_0\|_{C^{2,\alpha}(\bar{\Omega})})$$

with constants C depending on L, Ω, n and α.

6.4.3 L^p Theory

Suppose that $f \in L^p(\Omega)$. The following estimates are the counterpart of the Schauder estimates for classical solutions. See Gilbarg–Trudinger [77, Theorems 9.11, 9.13].

Theorem 6.4.5 *Suppose that $u \in W^{2,p}_{loc}(\Omega)$ satisfies $Lu = f$ in Ω with $f \in L^p(\Omega)$, $1 < p < \infty$. Then for any $\Omega' \subset\subset \Omega$, we have*

$$\|u\|_{W^{2,p}(\Omega')} \leq C(\|u\|_{L^p(\Omega)} + \|f\|_{L^p(\Omega)}).$$

In addition, if Ω is of class $C^{1,1}$, and if there exists a function $u_0 \in W^{2,p}(\Omega)$ such that $u - u_0 \in W^{1,p}_0(\Omega)$, then

$$\|u\|_{W^{2,p}(\Omega)} \leq C(\|u\|_{L^p(\Omega)} + \|f\|_{L^p(\Omega)} + \|u_0\|_{W^{2,p}(\Omega)}).$$

The constants C may depend on L, Ω, n and p.

6.4.4 Moser Iteration

In this subsection, we consider the equation

$$-\Delta u = g(x, u), \quad x \in \Omega, \tag{6.4.2}$$

on a domain $\Omega \subset \mathbb{R}^N$, with a Carathéodory function $g : \Omega \times \mathbb{R} \to \mathbb{R}$, that is, $g(x, u)$ is measurable in $x \in \Omega$ and continuous in $u \in \mathbb{R}$. Moreover, we assume that g satisfies the growth condition

$$|g(x, u)| \leq C(1 + |u|^p), \tag{6.4.3}$$

where $p \le \frac{N+2}{N-2}$, if $N \ge 3$. We also assume that $g(x,0) = 0$. Write $a(x) = \frac{g(x,u)}{u}$. If $u \in H^1(\Omega)$, then $a(x) \in L^{\frac{N}{2}}_{loc}(\Omega)$.

If we want to estimate the L^p norm for the solution u, it is convenient to take $w = |u|$, and in many cases, we may end up with a differential inequality

$$-\Delta w \le a(x)w,$$

for some $a(x) \ge 0$ with $a(x) \in L^{\frac{N}{2}}_{loc}$.

Based on Moser's [109] iteration technique, we will prove the following result.

Lemma 6.4.6 *Suppose that $w \in H^1_{loc}(\mathbb{R}^N)$ satisfies $w \ge 0$, and is a weak solution of*

$$-\Delta w \le a(x)w, \quad in\ \mathbb{R}^N,$$

where $a(x) \ge 0$. For any $p > 2^*$, there is a small constant $\delta > 0$, depending on p, such that if

$$\int_{B_1(y)} |a|^{\frac{N}{2}} \le \delta,$$

then,

$$\|w\|_{L^p(B_{\frac{1}{2}}(y))} \le C\|w\|_{L^1(B_1(y))},$$

for some constant $C = C(p)$.

Proof. Let $1 \ge R > r > 0$. Let $\xi \in C^2_0(B_R(y))$, which has the property that $\xi = 1$ in $B_r(y)$, $0 \le \xi \le 1$ and $|\nabla \xi| \le 2(R-r)^{-1}$. Supposing that $w \in L^{2q}$ for some $q \ge \frac{2^*}{2}$, we define $\eta = \xi^2 w w_L^{2q-2}$, where $w_L = w$ if $w \le L$ and $w_L = L$ if $w > L$. We have

$$\int_{\mathbb{R}^N} \nabla w \nabla \eta \le \int_{\mathbb{R}^N} a(x) w \eta.$$

Firstly, we estimate

$$\int_{\mathbb{R}^N} a(x) w\eta \le \left(\int_{B_1(y)} |a|^{\frac{N}{2}}\right)^{\frac{2}{N}} \left(\int_{B_R(y)} \left(\xi w w_L^{q-1}\right)^{2^*}\right)^{\frac{2}{2^*}}. \tag{6.4.4}$$

On the other hand, we have

$$\nabla \eta = (2q-1)\xi^2 w^{2q-2}\nabla w + w^{2q-1}\nabla \xi^2, \quad \text{if } w \le L,$$

and

$$\nabla \eta = L^{2q-2}\left(\xi^2 \nabla w + w \nabla \xi^2\right), \quad \text{if } w > L.$$

Therefore, it follows that

$$
\begin{aligned}
\int_{\mathbb{R}^N} \nabla w \nabla \eta =& \int_{\{w \le L\}} \nabla w \nabla \eta + \int_{\{w > L\}} \nabla w \nabla \eta \\
=& \int_{\{w \le L\}} \Big((2q-1)\xi^2 w^{2q-2} |\nabla w|^2 + \nabla w w^{2q-1} \nabla \xi^2 \Big) \\
&+ \int_{\{w > L\}} \Big(L^{2q-2} (\xi^2 |\nabla w|^2 + \nabla w w \nabla \xi^2) \Big) \\
\ge& c' \int_{\{w \le L\}} |\nabla(\xi w^q)|^2 - C \int_{\{w \le L\}} w^{2q} |\nabla \xi|^2 \\
&+ c' \int_{\{w > L\}} |\nabla(\xi w L^{q-1})|^2 - C \int_{\{w > L\}} w^2 L^{2q-2} |\nabla \xi|^2 \\
=& c' \int_{\mathbb{R}^N} |\nabla(\xi w w_L^{q-1})|^2 - C \int_{\mathbb{R}^N} w^2 w_L^{2q-2} |\nabla \xi|^2 \\
\ge& c'' \Big(\int_{B_1(y)} (\xi w w_L^{q-1})^{2^*} \Big)^{\frac{2}{2^*}} - \frac{C}{(R-r)^2} \int_{B_R(y)} w^2 w_L^{2q-2}.
\end{aligned}
\tag{6.4.5}
$$

Here the constant $c'' > 0$ depends on q and c'' has a positive lower bound if $q \le p + 1$ for the given p.

Combining (6.4.4) and (6.4.5), we see that

$$
\begin{aligned}
& c'' \Big(\int_{B_1(y)} (\xi w w_L^{q-1})^{2^*} \Big)^{\frac{2}{2^*}} - \frac{C}{(R-r)^2} \int_{B_R(y)} w^2 w_L^{2q-2} \\
\le& \Big(\int_{B_1(y)} |a|^{\frac{N}{2}} \Big)^{\frac{N}{2}} \Big(\int_{B_R(y)} (\xi w w_L^{q-1})^{2^*} \Big)^{\frac{2}{2^*}}.
\end{aligned}
\tag{6.4.6}
$$

If $\int_{B_1(y)} |a|^{\frac{N}{2}} \le \delta$ is small enough, it follows from (6.4.6) that

$$
\Big(\int_{B_1(y)} (\xi w w_L^{q-1})^{2^*} \Big)^{\frac{2}{2^*}} \le \frac{C}{(R-r)^2} \int_{B_R(y)} w^2 w_L^{2q-2},
$$

which in turn implies that if $w \in L^{2q}_{loc}$, then $w \in L^{2^* q}_{loc}$ and

$$
\Big(\int_{B_r(y)} |w|^{q2^*} \Big)^{\frac{1}{q2^*}} \le \Big(\frac{C}{R-r} \Big)^{\frac{1}{q}} \Big(\int_{B_R(y)} |w|^{2q} \Big)^{\frac{1}{2q}}, \quad \forall\, 1 > R > r > 0.
\tag{6.4.7}
$$

Set $\gamma = \frac{N}{N-2} > 1$ and $p_j = 2\gamma^{j+1}$. We take the smallest positive integer k such that $p_k \ge p$. For any $R > r$, we let $R_j = R - \frac{j(R-r)}{k}$ for $j = 0, 1, \cdots, k$.

Then $R_k = r$ and $R_0 = R$. We also have $R_{j-1} - R_j = \frac{R-r}{k}$. Taking $r = R_j$, $R = R_{j-1}$ and $q = \gamma^j$ in (6.4.7), we obtain

$$\left(\int_{B_{R_j}(y)} |w|^{p_j}\right)^{\frac{1}{p_j}} \le \left(\frac{Ck}{R-r}\right)^{\frac{2}{p_{j-1}}} \left(\int_{B_{R_{j-1}}(y)} |w|^{p_{j-1}}\right)^{\frac{1}{p_{j-1}}}. \tag{6.4.8}$$

By iteration, we can obtain from (6.4.8) that

$$\begin{aligned}\left(\int_{B_{R_k}(y)} |w|^{p_k}\right)^{\frac{1}{p_k}} &\le \left(\frac{Ck}{R-r}\right)^{\sum_{j=0}^{k-1} \frac{2}{p_j}} \left(\int_{B_R(y)} |w|^{2^*}\right)^{\frac{1}{2^*}} \\ &\le \frac{C'}{(R-r)^{\sigma}} \left(\int_{B_R(y)} |w|^{2^*}\right)^{\frac{1}{2^*}},\end{aligned}$$

for some constants $C' > 0$ and $\sigma > 0$. Thus we have

$$\begin{aligned}\left(\int_{B_r(y)} |w|^{p}\right)^{\frac{1}{p}} &\le C\left(\int_{B_{R_k}(y)} |w|^{p_k}\right)^{\frac{1}{p_k}} \\ &\le \frac{C'}{(R-r)^{\sigma}} \left(\int_{B_R(y)} |w|^{2^*}\right)^{\frac{1}{2^*}}.\end{aligned}$$

But

$$\begin{aligned}\frac{C'}{(R-r)^{\sigma}} \left(\int_{B_R(y)} |w|^{2^*}\right)^{\frac{1}{2^*}} &\le \frac{C'}{(R-r)^{\sigma}} \|w\|^{\kappa}_{L^1(B_R(y))} \|w\|^{1-\kappa}_{L^p(B_R(y))} \\ &\le \frac{1}{2}\|w\|_{L^p(B_R(y))} + \frac{C}{(R-r)^{\sigma/\kappa}} \|w\|_{L^1(B_R(y))},\end{aligned}$$

where $\kappa = \frac{p-2^*}{p-1}$. Therefore, we obtain

$$\|w\|_{L^p(B_r(y))} \le \frac{1}{2}\|w\|_{L^p(B_R(y))} + \frac{C}{(R-r)^{\sigma/\kappa}} \|w\|_{L^1(B_R(y))}, \quad \forall\, 1 > R > r, > 0.$$

This gives

$$\|w\|_{L^p(B_r(y))} \le \frac{C'}{(1-r)^{\sigma/\kappa}} \|w\|_{L^1(B_1(y))}, \quad \forall\, 1 > r > 0,$$

by the following lemma.

□

Lemma 6.4.7 *Let $\phi(t)$ be non-negative bounded function defined in $[T_0, T_1]$ with $T_1 > T_0 \ge 0$. Assume that $\phi(t)$ satisfies*

$$\phi(t) \le \theta\phi(s) + \frac{A}{(s-t)^{\alpha}}, \quad \forall\, T_0 \le t < s \le T_1,$$

where $\theta \in (0,1)$, A and α are some positive constants. Then there exists a constant $C > 0$ such that

$$\phi(r) \le \frac{CA}{(R-r)^\alpha}, \qquad \forall\, T_0 \le r < R \le T_1.$$

Proof. Let $t_0 = r$ and $t_{j+1} = t_j + (1-\tau)\tau^j(R-r)$ for $j \ge 0$, where $\tau \in (0,1)$ is a constant satisfying $\theta\tau^{-\alpha} < 1$. By the assumption, one has

$$\phi(t_j) \le \theta\phi(t_{j+1}) + \frac{A}{[(1-\tau)\tau^j(R-r)]^\alpha}, \qquad j = 0, 1, \cdots.$$

By iteration, we obtain

$$\begin{aligned}\phi(r) = \phi(t_0) &\le \theta^k\phi(t_k) + A\sum_{j=1}^{k-1}\frac{\theta^j}{[(1-\tau)\tau^j(R-r)]^\alpha}\\ &= \theta^k\phi(t_k) + \frac{A}{[(1-\tau)(R-r)]^\alpha}\sum_{j=1}^{k-1}\frac{\theta^j}{\tau^{\alpha j}}.\end{aligned}$$

The conclusion of the lemma now follows by letting $k \to +\infty$ in the above inequality. □

To apply Lemma 4.2.1, note that if $u \in H^1_{loc}(\Omega)$ weakly solves (6.4.2), then $u \in L^q_{loc}(\Omega)$ for any $q < \infty$. From (6.4.3), $g(x,u) \in L^q_{loc}$ for any $q > 1$. Thus, by Theorem 6.4.5, $u \in W^{2,q}_{loc}(\Omega)$ for any $q < \infty$. Whence we have $u \in C^{1,\alpha}_{loc}(\Omega)$ by the Sobolev embedding theorem (Theorem 6.3.6) for any $\alpha < 1$. Now we may proceed using Schauder estimate (Theorem 6.4.4). In particular, if g is Hölder continuous in x, then $u \in C^2(\Omega)$.

6.5 The Kelvin Transformation

Suppose that $u \in C^2(\mathbb{R}^N)$ solves

$$-\Delta u = f(u), \quad \text{in } \mathbb{R}^N,$$

and $u(x) \to 0$ as $|x| \to +\infty$. In applications, we want to know the decay rate of u at infinity. If we want to prove

$$|x|^\alpha u(x) \to c_0 \ne 0,$$

as $|x| \to +\infty$, we can change this problem in infinity to a problem at the origin as follows.

We make the change of variable $y = \frac{x}{|x|^2}$. That is, if we use $\omega = \frac{x}{|x|}$ to denote the angle of x, then along the direction ω, we change $|x|$ to $|x|^{-1}$. Under this change of variable, we only need to prove

$$|y|^{-\alpha} u\Big(\frac{y}{|y|^2}\Big) \to c_0, \tag{6.5.1}$$

as $y \to 0$. In order to prove (6.5.1), we need to find the equation for $w(y) = |y|^{-\alpha} u\big(\frac{y}{|y|^2}\big)$.

Case 1. We assume that $u(x) = u(r)$, where $r = |x|$. Then,

$$w' = -r^{-\alpha-2} u'\Big(\frac{1}{r}\Big) - \alpha r^{-\alpha-1} u\Big(\frac{1}{r}\Big)$$

and

$$w'' = r^{-\alpha-4} u''\Big(\frac{1}{r}\Big) + (2\alpha+2) r^{-\alpha-3} u'\Big(\frac{1}{r}\Big) + \alpha(\alpha+1) r^{-\alpha-2} u\Big(\frac{1}{r}\Big).$$

Thus, it follows that

$$\begin{aligned}
\Delta w &= w'' + \frac{N-1}{r} w' \\
&= r^{-\alpha-4} u''\Big(\frac{1}{r}\Big) + \big(2\alpha + 2 - (N-1)\big) r^{-\alpha-3} u'\Big(\frac{1}{r}\Big) \\
&\quad + \alpha\big((\alpha+1) - (N-1)\big) r^{-\alpha-2} u\Big(\frac{1}{r}\Big).
\end{aligned}$$

Setting $\alpha = N - 2$, we have

$$\begin{aligned}
\Delta w &= r^{-N-2} u''\Big(\frac{1}{r}\Big) + (N-1) r^{-N-1} u'\Big(\frac{1}{r}\Big) \\
&= r^{-N-2}\Big(u''\Big(\frac{1}{r}\Big) + \frac{N-1}{r^{-1}} u'\Big(\frac{1}{r}\Big)\Big) \\
&= r^{-N-2} \Delta u\Big(\frac{1}{r}\Big).
\end{aligned}$$

Case 2. We write $u(x) = u(r, \omega)$ with $r = |x|$. Then,

$$\begin{aligned}
\Delta w &= \frac{\partial^2 w}{\partial r^2} + \frac{N-1}{r}\frac{\partial w}{\partial r} + \frac{1}{r^2}\Delta_\omega w \\
&= r^{-N-2}\Big(u''\Big(\frac{1}{r}\Big) + \frac{N-1}{r^{-1}} u'\Big(\frac{1}{r}\Big) + \frac{1}{r^{-2}}\Delta_\omega u\Big) \\
&= r^{-N-2}\Delta u\Big(\frac{y}{|y|^2}\Big).
\end{aligned}$$

In conclusion, we have proved

Proposition 6.5.1 *For $w(y) = |y|^{-(N-2)}u(\frac{y}{|y|^2})$, one has*

$$\Delta w = |y|^{-N-2}\Delta u\Big(\frac{y}{|y|^2}\Big).$$

The change $w(y) = |y|^{-(N-2)}u(\frac{y}{|y|^2})$ is called the Kelvin transformation. Now we consider the following equation

$$-\Delta u = |u|^{p-1}u, \quad \text{in } \mathbb{R}^N.$$

Then, under the Kelvin transformation, w satisfies

$$-\Delta w = |y|^{-N-2+p(N-2)}|w|^{p-1}w, \quad \text{in } \mathbb{R}^N \setminus \{0\}.$$

In particular, if $p = \frac{N+2}{N-2}$, then w satisfies

$$-\Delta w = |w|^{2^*-1}w, \quad \text{in } \mathbb{R}^N \setminus \{0\}. \tag{6.5.2}$$

Now we prove that w actually satisfies

$$-\Delta w = |w|^{2^*-1}w, \quad \text{in } \mathbb{R}^N.$$

First, we calculate that

$$\begin{aligned}
\int_{\mathbb{R}^N} |u|^{2^*} &= \int_{|\omega|=1}\int_0^{+\infty} r^{N-1}|u(r,\omega)|^{2^*}\,dr d\omega \\
&= \int_{|\omega|=1}\int_0^{+\infty} t^{-(N-1)}\Big|u\Big(\frac{1}{t},\omega\Big)\Big|^{2^*} t^{-2}\,dt d\omega \\
&= \int_{|\omega|=1}\int_0^{+\infty} t^{N-1}\Big|t^{-(N-2)}u\Big(\frac{1}{t},\omega\Big)\Big|^{2^*}\,dt d\omega \\
&= \int_{\mathbb{R}^N} |w|^{2^*}.
\end{aligned}$$

Take an $\xi(x) = \xi(|x|) \in C^1(\mathbb{R}^N)$ satisfying $\xi = 0$ in $B_1(0)$, $\xi(x) = 1$ if $|x| \geq 2$ and ξ is non-decreasing in $|x|$. Denote $\xi_h(x) = \xi(\frac{x}{h})$. For any $\varphi \in C_0^\infty(\mathbb{R}^N)$, it follows from (6.5.2) that

$$\int_{\mathbb{R}^N} \nabla w \nabla(\xi_h^2\varphi) = \int_{\mathbb{R}^N} |w|^{2^*-2}w\xi_h^2\varphi, \quad \varphi \in C_0^\infty(\mathbb{R}^N). \tag{6.5.3}$$

In particular, replacing φ by $w\varphi^2$ in (6.5.3), we obtain

$$\int_{\mathbb{R}^N} \nabla w \nabla(\xi_h^2 w\varphi^2) = \int_{\mathbb{R}^N} |w|^{2^*}\xi_h^2\varphi^2, \quad \varphi \in C_0^1(\mathbb{R}^N). \tag{6.5.4}$$

We then have

$$\begin{aligned}\int_{\mathbb{R}^N} \nabla w \nabla(\xi_h^2 w \varphi^2) &= \int_{\mathbb{R}^N} |\nabla w|^2 \xi_h^2 \varphi^2 + 2\int_{\mathbb{R}^N} \nabla w w \varphi \xi_h \nabla(\xi_h \varphi) \\ &\geq \frac{1}{2}\int_{\mathbb{R}^N} |\nabla w|^2 \xi_h^2 \varphi^2 - C\int_{\mathbb{R}^N} w^2 |\nabla(\xi_h \varphi)|^2.\end{aligned} \tag{6.5.5}$$

Since

$$\begin{aligned}\int_{\mathbb{R}^N} w^2 |\nabla(\xi_h \varphi)|^2 &\leq Ch^{-2}\int_{|x|\leq 2h} w^2 + C\int_{\mathbb{R}^N} w^2 |\nabla \varphi| \\ &\leq C\Big(\int_{|x|\leq 2h} w^{2^*}\Big)^{\frac{2}{2^*}} + C\int_{\mathbb{R}^N} w^2|\nabla\varphi| \leq C,\end{aligned}$$

we obtain from (6.5.4) and (6.5.5) that

$$\int_{B_R(0)} |\nabla w|^2 \leq C, \quad \forall R > 0. \tag{6.5.6}$$

From (6.5.6), we see that as $h \to 0$,

$$\begin{aligned}\Big|\int_{\mathbb{R}^N} \nabla w \xi_h \nabla \xi_h \varphi\Big| &\leq Ch^{-1}\int_{B_{2h}(0)} |\nabla w| \\ &\leq Ch^{\frac{N}{2}-1}\Big(\int_{B_{2h}(0)} |\nabla w|^2\Big)^{\frac{1}{2}} \to 0.\end{aligned}$$

Thus (6.5.3) gives

$$\int_{\mathbb{R}^N} \xi_h^2 \nabla w \nabla \varphi + o_h(1) = \int_{\mathbb{R}^N} |w|^{2^*-2} w \xi_h^2 \varphi, \quad \varphi \in C_0^\infty(\mathbb{R}^N). \tag{6.5.7}$$

Letting $h \to 0$ in (6.5.7), we obtain

$$\int_{\mathbb{R}^N} \nabla w \nabla \varphi = \int_{\mathbb{R}^N} |w|^{2^*-2} w \varphi, \quad \varphi \in C_0^\infty(\mathbb{R}^N).$$

Now we can use the result in Subsection 6.4.4 to conclude that w is continuous at $y = 0$. Hence we have proved

Theorem 6.5.2 *Let $u \in D^{1,2}(\mathbb{R}^N)$ be a solution of $-\Delta u = |u|^{2^*-2}u$. Then,*

$$|u(x)| \leq \frac{C}{(1+|x|)^{N-2}}.$$

6.6 The Kernel of a Linear Operator

Consider the following problem:

$$-\Delta u = f(u), \ u > 0, \ \text{in } \mathbb{R}^N. \tag{6.6.1}$$

We assume that the solution u of (6.6.1) decays fast enough at infinity, so u is radially symmetric and decreasing in $|x|$. Special cases are the following two problems:

The first one is

$$\begin{cases} -\Delta u = u^{2^*-1}, \ u > 0, \ \text{in } \mathbb{R}^N, \\ u \in D^{1,2}(\mathbb{R}^N). \end{cases} \tag{6.6.2}$$

The second one is

$$\begin{cases} -\Delta u + u = u^p, \ u > 0, \ \text{in } \mathbb{R}^N, \\ u \in H^1(\mathbb{R}^N), \end{cases} \tag{6.6.3}$$

where $1 < p < 2^* - 1$.

Let U be a radially symmetric solution of (6.6.1). In this section, we will study the kernel of the following linear operator:

$$Lu := -\Delta u - f'(U)u = 0, \quad \text{in } \mathbb{R}^N. \tag{6.6.4}$$

Here u is either in $D^{1,2}(\mathbb{R}^N)$ or in $H^1(\mathbb{R}^N)$. We will use the spherical coordinate systems and the separation of variables to deal with this problem. Denote $r = |x|$ and let $\omega \in \mathbb{S}^{N-1}$.

Let $\mu_0 = 0$, $\mu_1 = \cdots = \mu_N = N-1$, $\mu_j > N-1$, $j \geq N+1$ be all the eigenvalues of the Laplace operator $-\Delta_\omega$ on $\mathbb{S}^{N-1}$. The eigenfunction corresponding to μ_j is $\frac{x_j}{|x|}$ for $j = 1, \cdots, N$. We let φ_j denote the eigenfunction corresponding to μ_j.

Let u be a solution of (6.6.4). We decompose u as follows:

$$u(r,\omega) = \sum_{j=0}^{\infty} v_j(r)\varphi_j(\omega).$$

Then $v_j \in D^{1,2}(\mathbb{R}^N)$ and satisfies

$$-v_j'' - \frac{N-1}{r}v_j' + \frac{\mu_j}{r^2}v_j - f'(U)v_j = 0, \quad r > 0. \tag{6.6.5}$$

We thus have:

Theorem 6.6.1 *For $j > N$, $v_j = 0$ and $v_j = c_j U'(r)$ for some constant c_j for $j = 1, \cdots, N$.*

Proof. First, we differentiate (6.6.1) to find U' satisfies (6.6.5) with $j = 1, \cdots, N$.

To prove $v_j = 0$ for $j > N$, we compare v_j with U'. We write (6.6.5) as

$$-(r^{N-1}v_j')' + \mu_j r^{N-3}v_j - r^{N-1}f'(U)v_j = 0, \quad r > 0.$$

So we have

$$-(r^{N-1}v_j')'U' + \mu_j r^{N-3}v_jU' - r^{N-1}f'(U)v_jU' = 0, \quad r > 0, \quad (6.6.6)$$

and

$$-(r^{N-1}U_j'')'v_j + (N-1)r^{N-3}U'v_j - r^{N-1}f'(U)U'v_j = 0, \quad r > 0. \quad (6.6.7)$$

Subtracting (6.6.7) from (6.6.6), we obtain

$$-(r^{N-1}v_j')'U' + (r^{N-1}U'')'v_j + \big(\mu_j - (N-1)\big)r^{N-3}U'v_j = 0, \quad r > 0. \quad (6.6.8)$$

Suppose that $v_j \not\equiv 0$. We may assume that $v_j(0) > 0$.

Case 1. Suppose that the first zero of v_j is $r_0 > 0$. Then, by integrating (6.6.8) from 0 to r_0, we obtain

$$-r^{N-1}v_j'U'\big|_0^{r_0} + r^{N-1}U''v_j\big|_0^{r_0} + \int_0^{r_0} \big(\mu_j - (N-1)\big)r^{N-3}U'v_j = 0. \quad (6.6.9)$$

Since $v_j(r_0) = 0$, $U' < 0$, we see that each of the terms on the left-hand side of (6.6.9) is non-positive and that the last term is negative. This is impossible.

Case 2. $v_j > 0$ for all $r > 0$. Since

$$\int_0^{+\infty} r^{N-1}\big((v_j')^2 + |v_j|^{2^*}\big) < +\infty,$$

we can find a sequence $\{r_n\} \to +\infty$ as $n \to +\infty$, such that

$$r_n^{N-1}\big((v_j'(r_n))^2 + |v_j(r_n)|^{2^*}\big) \to 0.$$

Similar to (6.6.9), we can integrate (6.6.8) from 0 to r_n and then let $n \to +\infty$ to obtain a contradiction. Here we use the fact that $|U''(r)| \le Cr^{-N}$ as $r \to +\infty$. This is true in many problems we study. So we have proved the first part of this theorem.

For $j = 1, \cdots, N$, we have that U' is a solution of (6.6.5). By a standard theory for the second-order linear differential equations, the second solution that is linear independent with U' can be obtained by integration. In the present situation, this solution is not in $D^{1,2}(\mathbb{R}^N)$. It then follows that $v_j = c_jU'$. See the discussion at the end of this section. □

Now we consider (6.6.2). We consider the linear operator

$$Lu := -\Delta u - (2^* - 1)U_{0,1}^{2^*-2} u = 0, \quad \text{in } \mathbb{R}^N, \tag{6.6.10}$$

and $u \in D^{1,2}(\mathbb{R}^N)$.

By Theorem 6.6.1, we only need to study the following equation (which corresponds to $j = 0$):

$$-v_0'' - \frac{N-1}{r} v_0' - (2^* - 1)U_{0,1}^{2^*-2} v_0 = 0, \quad \text{in } \mathbb{R}^N. \tag{6.6.11}$$

It is obvious that (6.6.11) has a solution $\frac{\partial U_{0,\mu}}{\partial \mu}\big|_{\mu=1}$. The second solution, which is linear independent with $\frac{\partial U_{0,\mu}}{\partial \mu}\big|_{\mu=1,}$ is not in $D^{1,2}(\mathbb{R}^N)$. Therefore, $v_0 = c_0 \frac{\partial U_{0,\mu}}{\partial \mu}\big|_{\mu=1}$, and we have proved the following:

Theorem 6.6.2 *The kernel of the linear operator* (6.6.10) *in* $D^{1,2}(\mathbb{R}^N)$ *is*

$$span\Big\{\frac{\partial U_{0,1}}{\partial x_1}, \dots, \frac{\partial U_{0,1}}{\partial x_N}, \frac{\partial U_{0,\mu}}{\partial \mu}\Big|_{\mu=1}\Big\}.$$

Now we turn to (6.6.3). Consider the linear operator

$$Lu := -\Delta u + u - pU^{p-1} u = 0, \quad \text{in } \mathbb{R}^N, \tag{6.6.12}$$

where $u \in H^1(\mathbb{R}^N)$. We need to study the following equation:

$$-v_0'' - \frac{N-1}{r} v_0' + v_0 - pU^{p-1} v_0 = 0, \quad \text{in } \mathbb{R}^N, \tag{6.6.13}$$

where $v_0 \in H^1(\mathbb{R}^N)$. Problem (6.6.13) is much more difficult than (6.6.11). One can prove that (6.6.13) has only zero solution. Thus we state the following theorem without a complete proof:

Theorem 6.6.3 *The kernel of the linear operator* (6.6.12) *in* $H^1(\mathbb{R}^N)$ *is*

$$span\Big\{\frac{\partial U}{\partial x_1}, \dots, \frac{\partial U}{\partial x_N}\Big\}.$$

For the proof of the above theorem, the readers can refer to sections 4 and 5 in [92].

Finally, we briefly discuss the following second-order linear differential equation:

$$v'' + p(r)v' + q(r)v = 0. \tag{6.6.14}$$

Let v_1 and v_2 be two solutions of (6.6.14). The Wronkian is defined by $W = v_1 v_2' - v_2 v_1'$. Then W satisfies the equation

$$W' + p(r)W = 0.$$

Thus $W(r) = W(0)e^{-\int_0^r p(s)\,ds}$. Now if v_1 is found, then to find a second solution v_2, we can solve

$$v_1 v_2' - v_2 v_1' = W\Big(= W(0)e^{-\int_0^r p(s)\,ds}\Big),$$

which is equivalent to

$$\Big(\frac{v_2}{v_1}\Big)' = \frac{W}{v_1^2}.$$

This in turn yields

$$v_2(r) = v_1(r) \int_1^r \frac{W(s)}{v_1^2(s)} + C.$$

For problem (6.6.5), we have $p(r) = \frac{N-1}{r}$. It then follows that

$$W(r) = \frac{1}{r^{N-1}}.$$

Thus, if (6.6.5) has a good solution v^*, then the second one is given by

$$v(r) = v^*(r) \int_1^r \frac{1}{s^{N-1}(v^*)^2(s)} + C.$$

This function is not in $H^1(B_1(0))$.

6.7 The Estimates for the Green's Function

Let $G(y, x)$ be the Green's function for $-\Delta$ in a bounded C^1 domain in $\mathbb{R}^N$ with homogeneous Dirichlet boundary condition. Write

$$G(y, x) = \frac{1}{(N-2)\omega_{N-1}} \frac{1}{|y-x|^{N-2}} - H(y, x),$$

where ω_{N-1} is the area of the the unit sphere $\mathbb{S}^{N-1}$ in $\mathbb{R}^N$ and $H(y, x)$ is the regular part of the Green's function $G(y, x)$ satisfying

$$\begin{cases} \Delta H(y, x) = 0, & \text{in } \Omega, \\ H(y, x) = \dfrac{1}{(N-2)\omega_{N-1}} \dfrac{1}{|y-x|^{N-2}}, & \text{on } \partial\Omega. \end{cases}$$

We define the Robin function $\varphi(x)$ as $\varphi(x) = H(x, x)$.

In this section, we will prove the following result:

Proposition 6.7.1 *Let $d = d(x, \partial\Omega)$ for $x \in \Omega$. Then as $d \to 0$,*

$$\varphi(x) = \frac{1}{(N-2)\omega_{N-1}} \frac{1}{(2d)^{N-2}} \big(1 + O(d)\big), \tag{6.7.1}$$

and

$$\nabla\varphi(x) = \frac{2}{\omega_{N-1}} \frac{1}{(2d)^{N-1}} \frac{x' - x}{d} + O\Big(\frac{1}{d^{N-2}}\Big), \tag{6.7.2}$$

where $x' \in \partial\Omega$ is the unique point, satisfying $d(x, \partial\Omega) = |x - x'|$.

Proof. Since Ω is smooth, there exists $d_0 > 0$ such that for any $x \in \Omega$ with $d(x, \partial\Omega) < d_0$, there exists a unique $x' \in \partial\Omega$, satisfying $d(x, \partial\Omega) = |x - x'|$. By translation and rotation, we assume that $x = (0, d)$, $x' = 0$ and there is a C^2 function $\phi(y')$ such that $\phi(0) = 0$, $\nabla\phi(0) = 0$,

$$\partial\Omega \cap B_\delta(0) = \big\{y : \ y_N = \phi(y')\big\} \cap B_\delta(0),$$

$$\Omega \cap B_\delta(0) = \big\{y : \ y_N > \phi(y')\big\} \cap B_\delta(0),$$

where $\delta > 0$ is a small constant.

Let $x'' = (0, -d)$ be the reflection of x with respect to the boundary of Ω. For d_0 small enough, $x'' \notin \Omega$. The function $\frac{1}{|y - x''|^{N-2}}$ is harmonic in Ω. Consequently,

$$F(y) := H(y, x) - \frac{1}{(N-2)\omega_{N-1}} \frac{1}{|y - x''|^{N-2}}$$

is harmonic in Ω.

For any $y \in \partial\Omega \cap B_\delta(0)$, we have

$$|y - x| = \sqrt{|y|^2 + d^2 - 2dy_N} = \sqrt{|y|^2 + d^2}\Big(1 + O\Big(\frac{dy_N}{|y|^2 + d^2}\Big)\Big)$$

and

$$|y - x''| = \sqrt{|y|^2 + d^2 + 2dy_N} = \sqrt{|y|^2 + d^2}\Big(1 + O\Big(\frac{dy_N}{|y|^2 + d^2}\Big)\Big).$$

So for any $y \in \partial\Omega \cap B_\delta(0)$, in view of $|y_N| = |\phi(y')| = O(|y'|^2)$, one has

$$\begin{aligned}\frac{1}{|y - x|^{N-2}} - \frac{1}{|y - x''|^{N-2}} &= (|y|^2 + d^2)^{-\frac{N-2}{2}} O\Big(\frac{dy_N}{|y|^2 + d^2}\Big) \\ &= (|y|^2 + d^2)^{-\frac{N-2}{2}} O(d) = O\big(d^{-N+3}\big).\end{aligned}$$

On the other hand, for any $y \in \partial\Omega \cap \big(\mathbb{R}^N \setminus B_\delta(0)\big)$, we have

$$\frac{1}{|y - x|^{N-2}} - \frac{1}{|y - x''|^{N-2}} = O(1) = O\big(d^{-N+3}\big).$$

Thus, using the maximum principle for $F(y)$, we see that

$$|F(y)| \le \max_{y\in\partial\Omega} |F(y)| = O\Big(\frac{1}{d^{N-3}}\Big), \quad y \in \Omega.$$

As a result,

$$H(y,x) = \frac{1}{(N-2)\omega_{N-1}} \frac{1}{|y-x''|^{N-2}} + O\Big(\frac{1}{d^{N-3}}\Big),$$

and (6.7.1) follows.

The proof of (6.7.2) is similar. Let us recall that $\frac{\partial H}{\partial x_i}(y,x)$ is a harmonic function in Ω, and on the boundary $\partial\Omega$, it is equal to

$$\frac{1}{(N-2)\omega_{N-1}} \frac{\partial}{\partial x_i}\Big(\frac{1}{|x-y|^{N-2}}\Big) = -\frac{1}{\omega_{N-1}} \frac{x_i - y_i}{|x-y|^N}.$$

We consider the function defined on Ω in the following way:

$$f_i(y) := \begin{cases} \dfrac{1}{\omega_{N-1}} \dfrac{y_i}{|x''-y|^N}, & i = 1, \cdots, N-1, \\ -\dfrac{1}{\omega_{N-1}} \dfrac{d+y_N}{|x''-y|^N}, & i = N. \end{cases}$$

It is easy to prove that for any $y \in \partial\Omega$,

$$\frac{\partial H}{\partial x_i}(y,x) - f_i(y) = O\Big(\frac{1}{d^{N-2}}\Big).$$

Then, the maximum principle shows that

$$\frac{\partial H}{\partial x_i}(y,x) = f_i(y) + O\Big(\frac{1}{d^{N-2}}\Big),$$

uniformly in $y \in \Omega$ as $d \to 0$, and we get that as $d \to 0$,

$$\frac{\partial \varphi}{\partial x_i}(x) = 2\frac{\partial H}{\partial x_i}(x,x) = O\Big(\frac{1}{d^{N-2}}\Big), \quad i = 1, \cdots, N-1,$$

and

$$\frac{\partial \varphi}{\partial x_N}(x) = 2\frac{\partial H}{\partial x_N}(x,x) = -\frac{2}{\omega_{N-1}} \frac{1}{(2d)^{N-1}} + O\Big(\frac{1}{d^{N-2}}\Big).$$

Thus (6.7.2) follows. □

Notations

$\mathbb{R}$: the set of all real numbers

$\mathbb{R}^+$: the set of all nonnegative real numbers

$\mathbb{R}^N$: N-dimensional Euclidean space

$\mathbb{R}^N_+ := \{(x_1, \cdots, x_N) \in \mathbb{R}^N : \ x_N \geq 0\}$

$(x_1, \cdots, x_N)^T$: transpose of the row vector $(x_1, \cdots, x_N)$

$o(t)$: any quantity satisfying $\lim_{t\to 0} \frac{o(t)}{t} = 0$

$O(t)$: any quantity satisfying $|O(t)| \leq C|t|$ for some constant $C > 0$ as $t \to 0$

$a \perp b$: a is perpendicular to b

$\emptyset$: empty set

$B_\rho(x)$: ball centered at x with radius ρ

$E^\perp$: set consisting of vectors perpendicular to the set E

Ω: domain of $\mathbb{R}^N$

$\partial\Omega$: boundary of domain Ω

$\overline{\Omega}$: closure of Ω

$\Delta u := \sum_{i=1}^N \frac{\partial^2 u}{\partial x_i^2}$

$\nabla u := (\frac{\partial u}{\partial x_1}, \cdots, \frac{\partial u}{\partial x_N})$

$u_+ := \max\{u, \ 0\}$

$u_- := \min\{u, \ 0\}$

$d(x, \ \Omega) := \inf_{y\in\Omega} |x - y|$

$|\Omega|$: measure of the set Ω

$\| \cdot \|$, $\| \cdot \|_X$: norm in the Banach space X

$C(X, \ \mathbb{R})$: space of continuous functionals on Banach space X

$C^k(X, \mathbb{R})$: space of k-times continuously differentiable functionals

$L^p(\Omega)$: space of measurable functions whose p-th power is integrable on Ω

$L^p_{loc}(\Omega)$: space of locally p-integrable functions on Ω

$L^\infty(\Omega)$: space of measurable functions bounded almost everywhere in Ω

$C_0^\infty(\Omega)$: space of C^∞-functions with compact support in Ω

$D^{1,2}(\mathbb{R}^N)$: completion of $C_0^\infty(\mathbb{R}^N)$ under the norm $(\int_{\mathbb{R}^N} |\nabla u|^2)^{\frac{1}{2}}$

$H^1(\Omega)$: Sobolev space of order 1

$H_0^1(\Omega)$: closure of $C_0^\infty(\Omega)$ in $H^1(\Omega)$

$W^{k,p}(\Omega)$, $W_0^{k,p}(\Omega)$, $H^k(\Omega)$, $\mathcal{D}^{k,p}(\Omega)$, $C^k(\Omega)$, $C^{k,\alpha}(\Omega)$: see Sobolev space 6.3.1

$span\{e_1, \cdots, e_n\}$: space spanned by the vectors $e_1, \cdots, e_n$

$\langle\, , \rangle$: scalar product in the Hilbert space H

$X \bigoplus Y$: direct sum of Hilbert spaces X and Y

$I'(u)$: the Fréchet derivative of functional I at u

References

[1] R. A. Adams, *Sobolev spaces.* Academic Press, New York, San Francisco, and London (1975).

[2] Adimurthi, G. Mancini, The Neumann problem for elliptic equations with critical nonlinearity, 'A tribute in honour of G. Prodi'. *Scu. Norm. Sup. Pisa* (1991), 9–25.

[3] Adimurthi, F. Pacella, S. L. Yadava, Interaction between the geometry of the boundary and positive solutions of a semilinear Neumann problem with critical nonlinearity. *J. Funct. Anal.* 113 (1993), 318–350.

[4] Adimurthi, F. Pacella, S. L. Yadava, Characterization of concentration points and L^{∞}-estimates for solutions of a semilinear Neumann problem involving the critical Sobolev exponent. *Diff. Integ. Equ.* 8 (1995), 42–68.

[5] A. Ambrosetti, A. Malchiodi, *Perturbation methods and semilinear elliptic problems on* $\mathbb{R}^n$. Birkhauser Verlag, Basel, Boston, and Berlin. (2006).

[6] A. Ambrosetti, A. Malchiodi, Concentration phenomena for NLS: Recent results and new perspectives, perspectives in nonlinear partial differential equations. *Contemp. Math.* 446 (2007), 19–30.

[7] A. Ambrosetti, A. Malchiodi, W.-M. Ni, Singularly perturbed elliptic equations with symmetry: existence of solutions concentrating on spheres. I. *Comm. Math. Phys.* 235 (2003), 427–466.

[8] A. Ambrosetti, A. Malchiodi, W.-M. Ni, Singularly perturbed elliptic equations with symmetry: existence of solutions concentrating on spheres. II. *Indiana Univ. Math. J.* 53 (2004), 297–329.

[9] A. Ambrosetti, P. H. Rabinowitz, Dual variational methods in critical point theory and applications. *J. Funct. Anal.* 14 (1973), 349–381.

[10] T. Aubin, Problèmes isopérimétriques et espaces de Sobolev. (French) *J. Differential Geometry* 11 (1976), no. 4, 573–598.

[11] N. Ba, Y. Deng, S. Peng, Multi-peak bound states for Schrödinger equations with compactly supported or unbounded potentials. *Ann. Inst. H. Poincaré Anal. Non Linéaire* 27 (2010), no. 5, 1205–1226.

[12] A. Bahri, *Critical points at infinity in some variational problems.* Longman Scientific and Technical, New York (1989).

[13] A. Bahri, J. M. Coron, On the nonlinear elliptic equation involving the critical Sobolev exponent, the effect of the topology of the domain. *Comm. Pure Appl. Math.* 41 (1988), 253–294.

[14] A. Bahri, Y. Y. Li, On a min-max procedure for the existence of a positive solution for certain scalar field equations in $\mathbb{R}^N$. *Rev. Mat. Iberoamericana* 6 (1990), no. 1–2, 1–15.

[15] A. Bahri, Y. Y. Li, O. Rey, On a variational problem with lack of compactness: the topological effect of the critical points at infinity. *Calc. Var. Partial Differential Equations* 3 (1995), no. 1, 67–93.

[16] A. Bahri, P. L. Lions, On the existence of a positive solution of semilinear elliptic equations in unbounded domains. *Ann. Inst. H. Poincaré Anal. Non Linéaire* 14 (1997), 365–413.

[17] T. Bartsch, T. D'Aprile, A. Pistoia, Multi-bubble nodal solutions for slightly subcritical elliptic problems in domains with symmetries. *Ann. Inst. H. Poincaré Anal. Non Linéaire* 30 (2013), no. 6, 1027–1047.

[18] T. Bartsch, S. Peng, Solutions concentrating on higher dimensional subsets for singularly perturbed elliptic equations. *I. Indiana Univ. Math. J.* 57 (2008), no. 4, 1599–1631.

[19] T. Bartsch, S. Peng, Solutions concentrating on higher dimensional subsets for singularly perturbed elliptic equations II. *J. Differential Equations* 248 (2010), no. 11, 2746–2767.

[20] V. Benci, G. Cerami, Positive solutions of some nonlinear elliptic problems in exterior domains. *Arch. Ration. Mech. Anal.* 99 (1987), 282–300.

[21] V. Benci, G. Cerami, Existence of positive solutions of the equation $-\Delta u + a(x)u = u^{\frac{N+2}{N-2}}$ in $\mathbb{R}^N$. *J. Funct. Anal.* 88 (1990), 90–117.

[22] H. Berestycki, P. L. Lions, Nonlinear Scalar field equations, I, II. *Arch. Ration. Mech. Anal.* 82 (1983), 313–375.

[23] H. Brézis, E. Lieb, A relation between pointwise convergence of functions and convergence of functionals. *Proc. Amer. Math. Soc.* 88 (1983), no. 3, 486–490.

[24] H. Brézis, L. Nirenberg, Positive solutions of nonlinear elliptic equations involving critical Sobolev exponents. *Comm. Pure Appl. Math.* 36 (1983), no. 4, 437–477.

[25] J. Byeon, Z.-Q. Wang, Standing waves with a critical frequency for nonlinear Schröinger equations. *Arch. Ration. Mech. Anal.* 165 (2002), no. 4, 295–316.

[26] J. Byeon, Z.-Q. Wang, Standing waves with a critical frequency for nonlinear Schrödinger equations. II. *Calc. Var. Partial Differential Equations* 18 (2003), no. 2, 207–219.

[27] J. Byeon, Z.-Q. Wang, Spherical semiclassical states of a critical frequency for Schrödinger equations with decaying potentials. *J.Eur. Math. Soc. (JEMS)* 8 (2006), no. 2, 217–228.

[28] D. Cao, Positive solution and bifurcation from the essential spectrum of a semilinear elliptic equation on $\mathbb{R}^N$. *Nonlinear Anal.* 15 (1990), no. 11, 1045–1052.

[29] D. Cao, N. E. Dancer, E. S. Noussair, S. Yan, On the existence and profile of multi-peaked solutions to singularly perturbed semilinear Dirichlet problems. *Discrete Contin. Dynam. Systems* 2 (1996), no. 2, 221–236.

[30] D. Cao, Y. Guo, S. Peng, S. Yan, Local uniqueness for vortex patch problem in incompressible planar steady flow. *J. Math. Pures Appl.* 131 (2019), 251–289.

[31] D. Cao, S. Li, P. Luo, Uniqueness of positive bound states with multi-bump for nonlinear Schrödinger equations. *Calc. Var. Partial Differential Equations* 54 (2015), no. 4, 4037–4063.

[32] D. Cao, E. S. Noussair, Multi-bump standing waves with a critical frequency for nonlinear Schrödinger equations. *J. Differential Equations* 203 (2004), no. 2, 292–312.

[33] D. Cao, E. S. Noussair, S. Yan, Multiscale-bump standing waves with a critical frequency for nonlinear Schrödinger equations. *Trans. Amer. Math. Soc.* 360 (2008), no. 7, 3813–3837.

[34] D. Cao, E. S. Noussair, S. Yan, Existence and uniqueness results on single-peaked solutions of a semilinear problem. *Ann. Inst. H. Poincaré Anal. Non Linéaire* 15 (1998), 73–111.

[35] D. Cao, E. S. Noussair, S. Yan, Solutions with multiple peaks for nonlinear elliptic equations. *Proc. Roy. Soc. Edinburgh Sect. A* 129 (1999), no. 2, 235–264.

[36] D. Cao, S. Peng, Multi-bump bound states of Schrödinger equations with a critical frequency. *Math. Ann.* 336 (2006), no. 4, 925–948.

[37] D. Cao, S. Peng, Semi-classical bound states for Schrödinger equations with potentials vanishing or unbounded at infinity. *Comm. Partial Differential Equations* 34 (2009), no. 10–12, 1566–1591.

[38] D. Cao, S. Peng, S. Yan, Infinitely many solutions for p-Laplacian equation involving critical Sobolev growth. *J. Funct. Anal.* 262 (2012), no. 6, 2861–2902.

[39] D. Cao, S. Peng, S. Yan, Multiplicity of solutions for the plasma problem in two dimensions. *Adv. Math.* 225 (2010), no. 5, 2741–2785.

[40] D. Cao, S. Peng, S. Yan, Planar vortex patch problem in incompressible steady flow. *Adv. Math.* 270 (2015), 263–301.

[41] D. Cao, S. Yan, Infinitely many solutions for an elliptic problem involving critical Sobolev growth and Hardy potential. *Calc. Var. Partial Differential Equations* 38 (2010), no. 3–4, 471–501.

[42] D. Cao, X. Zhu, On the existence and nodal character of solutions of semilinear elliptic equation. *Acta Math. Sci.* 8(1988), 285–300.

[43] A. Capozzi, D. Fortunato, G, Palmieri, An existence result for nonlinear elliptic problems involving critical Sobolev exponent. *Ann. Inst. H. Poincaré Anal. Nonlináire* 2 (1985), 463–470.

[44] G. Cerami, G. Devillanova, S. Solimini, Infinitely many bound states for some nonlinear scalar field equations. *Calc. Var. Partial Differential Equations* 23 (2005), 139–168.

[45] G. Cerami, D. Fortunato, M. Struwe, Bifurcation and multiplicity results for nonlinear elliptic problems involving critical Sobolev exponents. *Ann. Inst. H. Poincaré Anal. Non Linéaire* 1 (1984), no. 5, 341–350.

[46] G. Cerami, S. Solimini, M. Struwe, Some existence results for superlinear elliptic boundary value problems involving critical exponents. *J. Funct. Anal.* 69 (1986), 289–306.

[47] W. Chen, J. Wei, S. Yan, Infinitely many solutions for the Schrödinger equations in $\mathbb{R}^N$ with critical growth. *J. Differential Equations* 252 (2012), no. 3, 2425–2447.

[48] J. M. Coron, Topologie et cas limite des injections de Sobolev. *C. R. Acad. Sci. Paris* 299 (1984), 209–212.

[49] E. N. Dancer, S. Yan, Multipeak solutions for a singularly perturbed Neumann problem. *Pacific J. Math.* 189 (1999), no. 2, 241–262.

[50] E. N. Dancer, S. Yan, A singularly perturbed elliptic problem in bounded domains with nontrivial topology. *Adv. Differential Equations* 4 (1999), no. 3, 347–368.

[51] E. N. Dancer, S. Yan, Effect of the domain geometry on the existence of multipeak solutions for an elliptic problem. *Topol. Methods Nonlinear Anal.* 14 (1999), no. 1, 1–38.

[52] E. N. Dancer, S. Yan, Interior and boundary peak solutions for a mixed boundary value problem. *Indiana Univ. Math. J.* 48 (1999), no. 4, 1177–1212.

[53] E. N. Dancer, S. Yan, On the existence of multipeak solutions for nonlinear field equations on $\mathbb{R}^N$. *Discrete Contin. Dynam. Systems* 6 (2000), 39–50.

[54] E. N. Dancer, S. Yan, On the superlinear Lazer-McKenna conjecture. *J. Differential Equations* 210 (2005), no. 2, 317–351.

[55] E. N. Dancer, S. Yan, On the superlinear Lazer-McKenna conjecture. II. *Comm. Partial Differential Equations* 30 (2005), no. 7–9, 1331–1358.

[56] E. N. Dancer, S. Yan, A new type of concentration solutions for a singularly perturbed elliptic problem. *Trans. Amer. Math. Soc.* 359 (2007), no. 4, 1765–1790.

[57] E. N. Dancer, S. Yan, The Lazer-McKenna conjecture and a free boundary problem in two dimensions. *J. Lond. Math. Soc.* (2) 78 (2008), no. 3, 639–662.

[58] T. D'Aprile, A. Pistoia, Nodal solutions for some singularly perturbed Dirichlet problems. *Trans. Amer. Math. Soc.* 363 (2011), no. 7, 3601–3620.

[59] M. del Pino, P. Felmer, Local mountain passes for semilinear elliptic problems in unbounded domains. *Calc. Var. Partial Differential Equations* 4 (1996), 121–137.

[60] M. del Pino, P. Felmer, Semi-classcal states for nonlinear Schrödinger equations. *J. Funct. Anal.* 149 (1997), 245–265.

[61] M. del Pino, P. Felmer, Multi-peak bound states of nonlinear Schróinger equations. *Ann. Inst. H. Poincaré Anal. Non Linéaire* 15 (1998), 127–149.

[62] M. del Pino, P. Felmer, Spike-layered solutions of singularly perturbed elliptic problems in a degenerate setting. *Indiana Univ. Math. J.* 48 (1999), no. 3, 883–898.

[63] M. del Pino, P. Felmer, Semi-classical states of nonlinear Schröinger equations: a variational reduction method. *Math. Ann.* 324 (2002), 1–32.

[64] M. del Pino, P. Felmer, M. Musso, Two-bubble solutions in the super-critical Bahri-Coron's problem. *Calc. Var. Partial Differential Equations* 16 (2003), no. 2, 113–145.

[65] M. del Pino, M. Kowalczyk, J. Wei, Concentration on curves for nonlinear Schrödinger equations. *Comm. Pure Appl. Math.* 60 (2007), no. 1, 113–146.

[66] M. del Pino, M. Musso, F. Pacard, A. Pistoia, Large energy entire solutions for the Yamabe equation. *J. Differential Equations* 251 (2011), 2568–2597.

[67] M. del Pino, J. Wei, An introduction to the finite and infinite dimensional reduction methods. Geometric analysis around scalar curvatures, 35118, Lect. Notes Ser. Inst. Math. Sci. Natl. Univ. Singap., 31. World Sci. Publ., Hackensack, NJ (2016).

[68] Y. Deng, C.-S. Lin, S. Yan, On the prescribed scalar curvature problem in $\mathbb{R}^N$, local uniqueness and periodicity. *J. Math. Pures Appl.* (9) 104 (2015), no. 6, 1013–1044.

[69] Y. Deng, S. Peng, H. Pi, Bound states with clustered peaks for nonlinear Schrödinger equations with compactly supported potentials. *Adv. Nonlinear Stud.* 14 (2014), no. 2, 463–481.

[70] Y. Deng, S. Peng, L. Wang, Infinitely many radial solutions to elliptic systems involving critical exponents. *Discrete Contin. Dyn. Syst.* 34 (2014), no. 2, 461–475.

[71] G. Devillannova, S. Solimini, Concentration estimates and multiple solutions to elliptic problems at critical growth. *Adv. Differential Equations* 7 (2002), 1257-1280.

[72] W. Ding, W.-M. Ni, On the existence of positive entire solutions of a semilinear elliptic equation. *Arch. Ration. Mech. Anal.* 91 (1986), 283–308.

[73] D. E. Edmunds, D. Fortunato, E. Jannelli, Critical exponents, critical dimensions and the biharmonic operator. *Arch. Ration. Mech. Anal.* 112 (1990), no. 3, 269–289.

[74] L. C. Evans, Partial differential equations. Graduate Studies in Mathematics, 19. American Mathematical Society, Providence, RI (1998).

[75] A. Floer, A. Weinstein, Nonspreading wave packets for the cubic Schröinger equation with a bounded potential. *J. Funct.* Anal. 69 (1986), 397–408.

[76] B. Gidas, W.-M. Ni, L. Nirenberg, Symmetry and related properties via the maximum principle. *Comm. Math. Phys.* 68 (1979), no. 3, 209–243.

[77] D. Gilbarg, N. S. Trudinger, *Elliptic partial differential equations of second order,* 2nd edition. Grundlehren der Mathematischen Wissenschaften, 224. Springer-Verlag, Berlin, (1983).

[78] L. Glangetas, Uniqueness of positive solutions of a nonlinear elliptic equation involving the critical exponent. *Nonlinear Anal.* 20 (1993), no. 5, 571–603.

[79] M. Grossi, On the number of single-peak solutions of the nonlinear Schrödinger equation. *Ann. Inst. H. Poincaré Anal. Non Linéaire* 19 (2002), no. 3, 261–280.

[80] M. Grossi, D. Passaseo, Nonlinear elliptic Dirichlet problems in exterior domains, the role of geometry and topology of the domain. *Comm. Appl. Nonlinear Anal.* 2 (1995), 1–31.

[81] M. Grossi, F. Takahashi, Nonexistence of multi-bubble solutions to some elliptic equations on convex domains. *J. Funct. Anal.* 259 (2010), no. 4, 904–917.

[82] C. Gui, C.-S. Lin, Estimates for boundary-bubbling solutions to an elliptic Neumann problem. *J. Reine Angew. Math.* 546 (2002), 201–235.

[83] C. Gui, J. Wei, Multiple interior peak solutions for some singularly perturbed Neumann problems. *J. Differential Equations* 158 (1999), no. 1, 1–27.

[84] C. Gui, J. Wei, On multiple mixed interior and boundary peak solutions for some singularly perturbed Neumann problems. *Canada J. Math.* 52 (2000), no. 3, 522–538.

[85] C. Gui, J. Wei, M. Winter, Multiple boundary peak solutions for some singularly perturbed Neumann problems. *Ann. Inst. H. Poincaré Anal. Non Linéaire* 17 (2000), no. 1, 47–82.

[86] Y. Guo, B. Li, Infinitely many solutions for the prescribed curvature problem of polyharmonic operator. *Calc. Var. Partial Differential Equations* 46 (2013), no. 3–4, 809–836.

[87] Y. Guo, J. Liu, Z.-Q. Wang, On a Brezis-Nirenberg type quasilinear problem. *J. Fixed Point Theory Appl.* 19 (2017), no. 1, 719–753.

[88] Y. Guo, J. Nie, M. Niu, Z.Tang, Local uniqueness and periodicity for the prescribed scalar curvature problem of fractional operator in $\mathbb{R}^N$. *Calc. Var. Partial Differential Equations* 56 (2017), no. 4, 56–118.

[89] Y. Guo, S. Peng, S. Yan, Local uniqueness and periodicity induced by concentration. *Proc. Lond. Math. Soc.* (3) 114 (2017), no. 6, 1005–1043.

[90] X. Kang, J. Wei, On interacting bumps of semi-classical states of nonlinear Schrödinger equations. *Adv. Differential Equations* 5 (2000), 899–928.

[91] T. Kilpeläinen, J. Malý, The Wiener test and potential estimates for quasilinear elliptic equations. *Acta Math.* 172 (1994), 137–161.

[92] M. K. Kwong, Uniqueness of positive solutions of $\Delta u - u + u^p = 0$ in R^n. *Arch. Ration. Mech. Anal.* 105 (1989), 243–266.

[93] A. C. Lazer, P. J. McKenna, On the number of solutions of a nonlinear Dirichlet problem. *J. Math. Anal. Appl.* 84 (1981), no. 1, 282–294.

[94] A. C. Lazer, P. J. McKenna, Large-amplitude periodic oscillations in suspension bridges: some new connections with nonlinear analysis. *SIAM Rev.* 32 (1990), no. 4, 537–578.

[95] G. Li, S. Peng, S. Yan, A new type of solutions for a singularly perturbed elliptic Neumann problem. *Rev. Mat. Iberoam.* 23 (2007), no. 3, 1039–1066.

[96] G. Li, S. Yan, J. Yang, An elliptic problem with critical growth in domains with shrinking holes. *J. Differential Equations* 198 (2004), 275–300.

[97] G. Li, S. Yan, J. Yang, The Lazer-McKenna conjecture for an elliptic problem with critical growth. *Calc. Var. Partial Differential Equations* 28 (2007), no. 4, 471–508.

[98] Y. Y. Li, Prescribing scalar curvature on $\mathbb{S}^n$ and related problems, Part I. *J. Differential Equations* 120 (1995), 319–410.

[99] Y. Y. Li, On a singularly perturbed elliptic equation. *Adv. Differential Equations* 2 (1997), 955–980.

[100] Y. Y. Li, J. Wei and H. Xu, Multi-bump solutions of $-\Delta u = K(x)u^{\frac{n+2}{n-2}}$ on lattices in $\mathbb{R}^n$. *J. Reine Angew. Math.* 743 (2018), 163–211.

[101] F. Lin, W.-M. Ni, J. Wei, On the number of interior peak solutions for a singularly perturbed Neumann problem. *Comm. Pure Appl. Math.* 60 (2007), no. 2, 252–281.

[102] C.-S. Lin, S. Peng, Segregated vector solutions for linearly coupled nonlinear Schrödinger systems. *Indiana Univ. Math.* J. 63 (2014), no. 4, 939–967.

[103] P. L. Lions, The concentration-compactness principle in the calculus of variations. The locally compact case. I. *Ann. Inst. H. Poincaré Anal. Non Linéaire* 1 (1984), no. 2, 109–145.

[104] P. L. Lions, The concentration-compactness principle in the calculus of variations. The locally compact case. II. *Ann. Inst. H. Poincaré Anal. Non Linéaire* 1 (1984), no. 4, 223–283.

[105] P. L. Lions, The concentration-compactness principle in the calculus of variations. The limit case. I. *Rev. Mat. Iberoamericana* 1 (1985), no. 1, 145–201.

[106] P. L. Lions, The concentration-compactness principle in the calculus of variations. The limit case. II. *Rev. Mat. Iberoamericana* 1 (1985), no. 2, 45–121.

[107] F. Mahmoudi, A. Malchiodi, Concentration on minimal submanifolds for a singularly perturbed Neumann problem. *Adv. Math.* 209 (2007), no. 2, 460–525.

[108] F. Mahmoudi, A. Malchiodi, M. Montenegro, Solutions to the nonlinear Schrödinger equation carrying momentum along a curve. *Comm. Pure Appl. Math.* 62 (2009), no. 9, 1155–1264.

[109] J. Moser, A new proof of De Giorgi's theorem concerning the regularity problem for elliptic differential equations. *Comm. Pure Appl. Math.* 13 (1960), 457–468.

[110] M. Musso, J. Wei, S. Yan, Infinitely many positive solutions for a nonlinear field equation with super-critical growth. *Proc. Lond. Math. Soc.* (3) 112 (2016), no. 1, 1–26.

[111] W.-M. Ni, Diffusion, cross-diffusion, and their spike-layer steady states. *Notices Amer. Math. Soc.* 45 (1998), no. 1, 9–18.

[112] W.-M. Ni, X, Pan, I. Takagi, Singular behavior of least-energy solutions of a semilinear Neumann problem involving critical Sobolev exponents. *Duke Math. J.* 67 (1992), no. 1, 1–20.

[113] W.-M. Ni, I. Takagi, On the shape of least-energy solutions to a semilinear Neumann problem. *Comm. Pure Appl. Math.* 44 (1991), no. 7, 819–851.

[114] W.-M. Ni, J. Wei, On the location and profile of spike-layer solutions to singularly perturbed semilinear Dirichlet problems. *Comm. Pure Appl. Math.* 48 (1995), no. 7, 731–768.

[115] E. S. Noussair, S. Yan, The effect of the domain geometry in singular perturbation problems. *Proc. London Math. Soc.* (3) 76 (1998), no. 2, 427–452.

[116] E. S. Noussair, S. Yan, On positive multipeak solutions of a nonlinear elliptic problem. *J. London Math. Soc.* (2) 62 (2000), no. 1, 213–227.

[117] Y. J. Oh, On positive multi-lump bound states nonlinear Schröinger equations under multiple well potential. *Commun. Math. Phys.* 131 (1990), 223–253.

[118] S. Peng, Q. Wang, Z.-Q. Wang, On coupled nonlinear Schrodinger systems with mixed couplings. *Trans. Amer. Math. Soc.* 371 (2019), 7559–7583.

[119] S. Peng, Z.-Q. Wang, Segregated and synchronized vector solutions for nonlinear Schrödinger systems. *Arch. Ration. Mech. Anal.* 208 (2013), no. 1, 305–339.

[120] S. Peng, C. Wang, Infinitely many solutions for a Hardy-Sobolev equation involving critical growth. *Math. Methods Appl. Sci.* 38 (2015), no. 2, 197–220.

[121] S. Peng, C. Wang, S. Yan, Construction of solutions via local Pohozaev identities. *J. Funct. Anal.* 274 (2018), 2606–2633.

[122] A. Pistoia, E. Serra, Multi-peak solutions for the Hénon equation with slightly subcritical growth. *Math. Z.* 256 (2007), no. 1, 75–97.

[123] S. Pohozaev, On the eigenfunctions of the equation $\Delta u+\lambda f(u)=0$. *Dokl. Akad. Nauk.* 5 (1965), 1408–1411.

[124] P. Rabinowitz, *Minimax methods in critical points theory with applications to differential equations*, CBMS Series, no. 65. Providence, RI (1986).

[125] O. Rey, The role of the Green's function in a nonlinear elliptic equation involving the critical Sobolev exponent. *J. Funct. Anal.* 89 (1990), no. 1, 1–52.

[126] O. Rey, Boundary effect for an elliptic Neumann problem with critical nonlinearity. *Comm. Partial Differential Equations* 22 (1997), no. 7–8, 1055–1139.

[127] O. Rey, An elliptic Neumann problem with critical nonlinearity in three dimensional domains. *Comm. Contemp. Math.* 1 (1999), 405–449.

[128] O. Rey, J. Wei, Blowing up solutions for an elliptic Neumann problem with sub- or supercritical nonlinearity. I. N=3. *J. Funct. Anal.* 212 (2004), no. 2, 472–499.

[129] O. Rey, J. Wei, Blowing up solutions for an elliptic Neumann problem with sub- or supercritical nonlinearity. II. $N \geq 4$. *Ann. Inst. H. Poincaré Anal. Non Linairé* 22 (2005), no. 4, 459–484.

[130] R. Schoen, Variational theory for the total scalar curvature functional for Riemannian metrics and related topics, in 'Topics in Calculus of Variations'. *Lecture Notes in Mathematics,* vol. 1365 (M. Giaquintta, ed.), pp. 120–154, Springer-Verlag, Berlin, New York (1989).

[131] R. Schoen, On the number of constant scalar curvature metrics in a confirmal class, in *Defferential geometry: a symposium in honor of Manfredo Do Carmo* (H. B. Lawson and K. Tenenblat, eds.), pp. 311–320, Wiley, New York (1991).

[132] R. Schoen, The existence of weak solutions with prescribed singular behavior for a confirmally invariant scalar equation. *Comm. Pure Appl. Math.* 41 (1988), 317–392.

[133] S. Solimini, On the existence of infinitely many radial solutions for some elliptic problems. *Revista Mat. Applicadas* 9 (1987), 75–86.

[134] M. Struwe, *Variational methods: Applications to nonlinear partial differential equations and Hamiltonian systems.* Springer-Verlag, Berlin (1990).

[135] M. Struwe, A global compactness result for elliptic boundary value problems involving limiting nonlinearities. *Math. Z.* 187 (1984), no. 4, 511–517.

[136] G. Talenti, Best constant in Sobolev inequality. *Ann. Mat. Pura Appl.* (4) 110 (1976), 353–372.

[137] C. Tintarev, *Concentration analysis and compactness: Concentration analysis and applications to PDE,117–141,* Trends Math., Birkhäuser/Springer, Basel (2013).

[138] C. Tintarev, K. H. Fineseler, Concentration compactness, functional analytic grounds and applications. Imperial College Press, London (2007).

[139] L. Wang, J. Wei, S. Yan, A Neumann problem with critical exponent in nonconvex domains and Lin-Ni's conjecture. *Trans. Amer. Math. Soc.* 362 (2010), no. 9, 4581–4615.

[140] L. Wang, J. Wei, S. Yan, On Lin-Ni's conjecture in convex domains. *Proc. Lond. Math. Soc.* (3) 102 (2011), no. 6, 1099–1126.

[141] Z.-Q. Wang, On the existence of positive solutions for semilinear Neumann problems in exterior domains. *Comm. Partial Differential Equations* 17 (1992), no. 7–8, 1309–1325.

[142] Z.-Q. Wang, Remarks on nonlinear Neumann problem with critical exponent. *Houston. J. Math.* 20 (1994), 671–684.

[143] Z. Q. Wang, The effect of domain geometry on the number of positive solutions of Neumann problems with critical exponents. *Diff. Integ. Equ.* 8 (1995), 1533–1554.

[144] J. Wei, On the construction of single-peaked solutions to a singularly perturbed semilinear Dirichlet problem. *J. Differential Equations* 129 (1996), no. 2, 315–333.

[145] J. Wei, On the boundary spike layer solutions to a singularly perturbed Neumann problem. *J. Differential Equations* 134 (1997), no. 1, 104–133.

[146] J. Wei, S. Yan, Arbitrary many boundary peak solutions for an elliptic Neumann problem with critical growth. *J. Math. Pures Appl.* (9) 88 (2007), no. 4, 350–378.

[147] J. Wei, S. Yan, Lazer-McKenna conjecture: the critical case. *J. Funct. Anal.* 244 (2007), no. 2, 639–667.

[148] J. Wei, S. Yan, Infinitely many positive solutions for the nonlinear Schrödinger equations in $\mathbb{R}^N$. *Calc. Var. Partial Differential Equations* 37 (2010), no. 3–4, 423–439.

[149] J. Wei, S. Yan, Infinitely many solutions for the prescribed scalar curvature problem on $\mathbb{S}^N$. *J. Funct. Anal.* 258 (2010), no. 9, 3048–3081.

[150] J. Wei, S. Yan, Infinitely many positive solutions for an elliptic problem with critical or supercritical growth. *J. Math. Pures Appl.* (9) 96 (2011), no. 4, 307–333.

[151] M. Willem, *Minimax theorems: Progress in nonlinear differential equations and their applications,* 24. Birkhäuser Boston, Boston (1996).

[152] S. Yan, The number of interior multipeak solutions for singularly perturbed Neumann problems. *Topol. Methods Nonlinear Anal.* 12 (1998), no. 1, 61–78.

[153] S. Yan, Concentration of solutions for the scalar curvature equation on $\mathbb{R}^N$. *J. Differential Equations* 163 (2000), no. 2, 239–264.

[154] S. Yan, J. Yang, Infinitely many solutions for an elliptic problem involving critical Sobolev and Hardy-Sobolev exponents. *Calc. Var. Partial Differential Equations* 48 (2013), no. 3–4, 587–610.

[155] S. Yan, J. Yang, X. Yu, Equations involving fractional Laplacian operator: compactness and application. *J. Funct. Anal.* 269 (2015), no. 1, 47–79.

[156] J. Zhao, X. Liu, J. Liu, p-Laplacian equations in $\mathbb{R}^N$ with finite potential via truncation method, the critical case. *J. Math. Anal. Appl.* 455 (2017), 58–88.

[157] X. Zhu, D. Cao, The concentration-compactness principle in nonlinear elliptic equations. *Acta Math. Sci.* (English Ed.) 9 (1989), no. 3, 307–328.

Index